AF760111

Biobased Aerogels
Polysaccharide and Protein-based Materials

Green Chemistry Series

Editor-in-chief:
James H. Clark, Department of Chemistry, University of York, UK

Series editors:
George A. Kraus, *Iowa State University, USA*
Andrzej Stankiewicz, *Delft University of Technology, The Netherlands*
Peter Siedl, *Federal University of Rio de Janeiro, Brazil*

Titles in the series:
1: The Future of Glycerol: New Uses of a Versatile Raw Material
2: Alternative Solvents for Green Chemistry
3: Eco-Friendly Synthesis of Fine Chemicals
4: Sustainable Solutions for Modern Economies
5: Chemical Reactions and Processes under Flow Conditions
6: Radical Reactions in Aqueous Media
7: Aqueous Microwave Chemistry
8: The Future of Glycerol: 2nd Edition
9: Transportation Biofuels: Novel Pathways for the Production of Ethanol, Biogas and Biodiesel
10: Alternatives to Conventional Food Processing
11: Green Trends in Insect Control
12: A Handbook of Applied Biopolymer Technology: Synthesis, Degradation and Applications
13: Challenges in Green Analytical Chemistry
14: Advanced Oil Crop Biorefineries
15: Enantioselective Homogeneous Supported Catalysis
16: Natural Polymers Volume 1: Composites
17: Natural Polymers Volume 2: Nanocomposites
18: Integrated Forest Biorefineries
19: Sustainable Preparation of Metal Nanoparticles: Methods and Applications
20: Alternative Solvents for Green Chemistry: 2nd Edition
21: Natural Product Extraction: Principles and Applications
22: Element Recovery and Sustainability
23: Green Materials for Sustainable Water Remediation and Treatment
24: The Economic Utilisation of Food Co-Products
25: Biomass for Sustainable Applications: Pollution Remediation and Energy
26: From C-H to C-C Bonds: Cross-Dehydrogenative-Coupling
27: Renewable Resources for Biorefineries
28: Transition Metal Catalysis in Aerobic Alcohol Oxidation
29: Green Materials from Plant Oils
30: Polyhydroxyalkanoates (PHAs) Based Blends, Composites and Nanocomposites
31: Ball Milling Towards Green Synthesis: Applications, Projects, Challenges

32: Porous Carbon Materials from Sustainable Precursors
33: Heterogeneous Catalysis for Today's Challenges: Synthesis, Characterization and Applications
34: Chemical Biotechnology and Bioengineering
35: Microwave-Assisted Polymerization
36: Ionic Liquids in the Biorefinery Concept: Challenges and Perspectives
37: Starch-based Blends, Composites and Nanocomposites
38: Sustainable Catalysis: With Non-endangered Metals, Part 1
39: Sustainable Catalysis: With Non-endangered Metals, Part 2
40: Sustainable Catalysis: Without Metals or Other Endangered Elements, Part 1
41: Sustainable Catalysis: Without Metals or Other Endangered Elements, Part 2
42: Green Photo-active Nanomaterials
43: Commercializing Biobased Products: Opportunities, Challenges, Benefits, and Risks
44: Biomass Sugars for Non-Fuel Applications
45: White Biotechnology for Sustainable Chemistry
46: Green and Sustainable Medicinal Chemistry: Methods, Tools and Strategies for the 21st Century Pharmaceutical Industry
47: Alternative Energy Sources for Green Chemistry
48: High Pressure Technologies in Biomass Conversion
49: Sustainable Solvents: Perspectives from Research, Business and International Policy
50: Fast Pyrolysis of Biomass: Advances in Science and Technology
51: Catalyst-free Organic Synthesis
52: Hazardous Reagent Substitution: A Pharmaceutical Perspective
53: Alternatives to Conventional Food Processing: 2nd Edition
54: Sustainable Synthesis of Pharmaceuticals: Using Transition Metal Complexes as Catalysts
55: Intensification of Biobased Processes
56: Sustainable Catalysis for Biorefineries
57: Supercritical and Other High-pressure Solvent Systems: For Extraction, Reaction and Material Processing
58: Biobased Aerogels: Polysaccharide and Protein-based Materials

How to obtain future titles on publication:
A standing order plan is available for this series. A standing order will bring delivery of each new volume immediately on publication.

For further information please contact:
Book Sales Department, Royal Society of Chemistry, Thomas Graham House, Science Park, Milton Road, Cambridge, CB4 0WF, UK
Telephone: +44 (0)1223 420066, Fax: +44 (0)1223 420247
Email: booksales@rsc.org
Visit our website at www.rsc.org/books

Biobased Aerogels
Polysaccharide and Protein-based Materials

Edited by

Sabu Thomas
Mahatma Gandhi University, Kottayam, India
Email: sabuthomas@mgu.ac.in

Laly A. Pothan
Bishop Moore College, Mavelikara, India
Email: lapothan@gmail.com

and

Rubie Mavelil-Sam
Bishop Moore College, Mavelikara, India
Email: mavelilsam.rubie@gmail.com

ROYAL SOCIETY OF CHEMISTRY

Green Chemistry Series No. 58

Print ISBN: 978-1-78262-765-4
PDF ISBN: 978-1-78262-997-9
EPUB ISBN: 978-1-78801-519-6
Print ISSN: 1757-7039
Electronic ISSN: 1757-7047

A catalogue record for this book is available from the British Library

© The Royal Society of Chemistry 2018

All rights reserved

Apart from fair dealing for the purposes of research for non-commercial purposes or for private study, criticism or review, as permitted under the Copyright, Designs and Patents Act 1988 and the Copyright and Related Rights Regulations 2003, this publication may not be reproduced, stored or transmitted, in any form or by any means, without the prior permission in writing of The Royal Society of Chemistry or the copyright owner, or in the case of reproduction in accordance with the terms of licences issued by the Copyright Licensing Agency in the UK, or in accordance with the terms of the licences issued by the appropriate Reproduction Rights Organization outside the UK. Enquiries concerning reproduction outside the terms stated here should be sent to The Royal Society of Chemistry at the address printed on this page.

Whilst this material has been produced with all due care, The Royal Society of Chemistry cannot be held responsible or liable for its accuracy and completeness, nor for any consequences arising from any errors or the use of the information contained in this publication. The publication of advertisements does not constitute any endorsement by The Royal Society of Chemistry or Authors of any products advertised. The views and opinions advanced by contributors do not necessarily reflect those of The Royal Society of Chemistry which shall not be liable for any resulting loss or damage arising as a result of reliance upon this material.

The Royal Society of Chemistry is a charity, registered in England and Wales, Number 207890, and a company incorporated in England by Royal Charter (Registered No. RC000524), registered office: Burlington House, Piccadilly, London W1J 0BA, UK, Telephone: +44 (0) 207 4378 6556.

For further information see our web site at www.rsc.org

Printed in the United Kingdom by CPI Group (UK) Ltd, Croydon, CR0 4YY, UK

Foreword

Never before in human history have so many people populated our planet and the need for sharing and sensibly using natural life resources, water and food as provided by rivers, lakes, oceans, pasture and farm land, been so urgent. Never before was the pace at which fossil and mineral resources, developed and formed far before human beings populated the earth, consumed so fast and the risk of irreversibly unhinging life-sustaining ecological equilibria been so high as it is today.

It is, therefore, particularly important to rediscover the great opportunities, gifts and miracles our biosphere holds for us. Recent advances in powerful instrumental-analytical techniques and an exponentially increasing understanding of biochemical and physical processes, natural building concepts and functional principles have created an unique fundament that just needs to be used with greater determination to make significant progress towards a more renewable-based life and bio-economy that includes bio-based materials, too.

Lightweight, highly open-porous materials, optimized throughout evolution with regard to weight-to-stiffness ratio, mechanical and chemical properties and adapted to humid environments, are ubiquitous in nature and are thoroughly designed for a wide range of functions in different environmental conditions.

This book gives an overview of the current state of research in the field of lightweight, highly open-porous polysaccharide and protein aerogels, conveying an impression of how diverse the field of bio-based aerogels has become in the last decade and addressing some of the major challenges of future research. The particular value of this book, however, is that the 16 chapters describe novel approaches related to modeling, simulation, tailoring, tuning and ageing of aerogels from some of the most abundant

Green Chemistry Series No. 58
Biobased Aerogels: Polysaccharide and Protein-based Materials
Edited by Sabu Thomas, Laly A. Pothan and Rubie Mavelil-Sam
© The Royal Society of Chemistry 2018
Published by the Royal Society of Chemistry, www.rsc.org

terrestrial and marine biopolymers, next to aspects of the impact of source type and processing conditions on morphological properties. This is complemented by a comprehensive discussion of the opportunities of polysaccharide and protein aerogels for a wide range of applications in thermal insulation, water treatment, food engineering, packaging, aeronautics or biomedicine.

Even though most of the presented aerogels are still not much beyond the research state, the book is gripping as it demonstrates that bio-based aerogels can compete with similar materials made from synthetic polymers and open new opportunities for novel applications, as their specific chemical and physical properties can greatly impact the biocompatibility, biodegradability or self-assembly to ordered supramolecular phases.

Bearing in mind that a more bio-based economy will become an ineluctable necessity in near future, I wish this book a broad readership across all communities including science, industry and decision makers.

Falk W. Liebner
Vienna

Contents

Green Chemistry Series No. 58
Biobased Aerogels: Polysaccharide and Protein-based Materials
Edited by Sabu Thomas, Laly A. Pothan and Rubie Mavelil-Sam
© The Royal Society of Chemistry 2018
Published by the Royal Society of Chemistry, www.rsc.org

CHAPTER 1

Polysaccharide and Protein Based Aerogels: An Introductory Outlook

RUBIE MAVELIL-SAM,*[a] LALY A. POTHAN[a] AND SABU THOMAS[b,c]

[a] Department of Chemistry, Bishop Moore College, University of Kerala, Mavelikara, Alappuzha, Kerala, India; [b] International and Inter University Centre for Nanoscience and Nanotechnology (IIUCNN), Mahatma Gandhi University, Kottayam, Kerala, India; [c] School of Chemical Sciences, Mahatma Gandhi University, Kottayam, Kerala, India
*Email: mavelilsam.rubie@gmail.com

1.1 Introduction

Aerogels have been in use for almost nine decades now. Synthetic aerogels are being used as super capacitors, electrodes, insulators, and so forth. Increased aerospace applications and rising energy costs have increased the prominence of insulating nanoporous materials (Figure 1.1). Nanocrystals based on a wide range of biomaterials have found their way into the production of nanoporous materials by replacing synthetic counterparts. In addition, awareness of using non benign, environmental friendly materials is driving researchers to develop novel green materials. Aerogels based on

Green Chemistry Series No. 58
Biobased Aerogels: Polysaccharide and Protein-based Materials
Edited by Sabu Thomas, Laly A. Pothan and Rubie Mavelil-Sam
© The Royal Society of Chemistry 2018
Published by the Royal Society of Chemistry, www.rsc.org

Figure 1.1 The crayons on top of the aerogel are protected from the flame underneath. Reproduced from ref. 1.

several polysaccharides and proteins have been developed recently, involving comprehensive research such as theory modelling and life cycle analysis of such materials.

The aim of this book is to bring together the results of research studies in the area of polysaccharide and protein based aerogels. Aerogels based on these biomaterials, their preparation from various sources, characterisation methods and applications in various domains of science and day to day life will be discussed in detail.

It is indeed commendable that all the contributors in this book have done their best to profoundly review recent advances in this growing area of strong international interest, providing a comprehensive idea on the preparation, properties and applications of polysaccharide and protein based aerogels. In-depth studies and reviews on the technological developments concerning processing strategies and structural analyses have also been included, providing a general idea of the individual types. The chapters are designed in such a way as to edify readers from various realms: academics, students, researchers and industrialists in particular.

1.2 Aerogels: A General Overview

Aerogels are porous ultralight materials derived from gels, in which the liquid component of the gel has been replaced with a gas.[2] These advanced materials are highly porous solids that hold gas (usually air) within the pores or networks of solid substances. Due to their light weight, low density, large surface area and high mechanical strength, aerogels are useful for many applications, such as heat insulators, particle filters, particle trappers and catalyst supports.[3] Aerogels are composed of a network of clustered nanoparticles. The materials usually have unique properties including high strength to density, and high surface area to volume ratios. They are manufactured by subjecting a wet gel precursor to critical point drying in order to remove the liquid through supercritical drying, without disturbing the network.

Well known aerogels are those of silica and metal oxides such as TiO_2 and Fe_2O_3. However, these inorganic aerogels usually lack mechanical strength and tend to collapse easily when subjected to small stresses. In contrast, aerogels made of organic polymers are stronger and can also be used as carbon precursors for pyrolysis, the resorcinol–formaldehyde aerogel being a prominent example. Aerogels fabricated from synthetic polymers are fragile. As a result, such aerogels have limitations to be used in situations that need robustness. As aerogels combine the properties of highly divided solids and metastable characteristics, they can develop very attractive physical and chemical properties that are not achievable by other means of low temperature soft chemical synthesis. In other words, they form a new class of solids showing a sophisticated potential for a range of applications.[4]

Aerogels can be classified according to their appearance (as monoliths, powders and films), their different microstructural characteristics (as microporous, mesoporous and mixed porous) or by defining their composition (Figure 1.2).

Polysaccharide-based aerogels are commonly obtained in the form of cylindrical monoliths, although many other shapes (Figure 1.3) can be found in the literature (such as beads, microspheres, *etc.*).[6,7]

The size and morphology of aerogels can be customized by means of shaping the gel by moulding, extrusion or any other suitable physical techniques. In general, gels take the shape of the mould in which gelation takes place and this shape is preserved in the monolithic aerogels after supercritical drying (Figure 1.4).

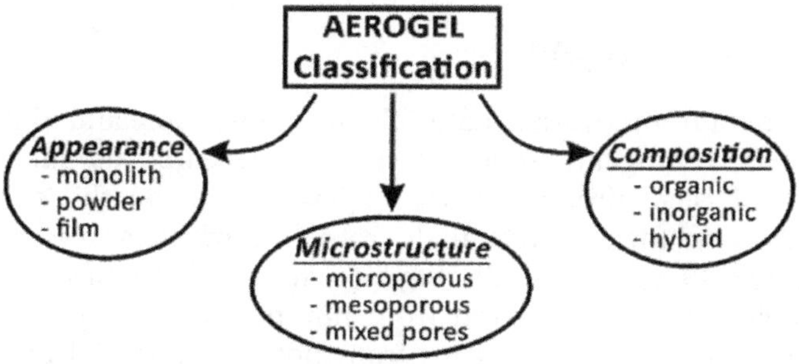

Figure 1.2 Different classification possibilities of aerogels.
Adapted from ref. 5 with permission from Springer Nature, © Springer Science + Business Media New York 2016.

Figure 1.3 Calcium–alginate aerogel obtained in different shapes: (a) monoliths, (b) beads and (c) microparticles.
Adapted from ref. 6 with permission from Elsevier, Copyright 2011.

Figure 1.4 Starch aerogel obtained by supercritical drying from a hydrogel prepared in a mould for Christmas cookies.
Adapted from ref. 6 with permission from Elsevier, Copyright 2011.

1.3 Why Bio-based?

The development of innovative materials from renewable and abundant bio-resources is becoming an important area of research as such materials exhibit high physical properties with a low impact on the environment. Increasing demand for products made from sustainable and non-petroleum based resources is also a major driving force for the development of new bio-based products. Polymers derived from non-petrochemical feedstocks are gaining a great deal of momentum from both commercial and scientific points of view. Biopolymers can be derived from natural sources such as plants, exoskeletons of arthropods, skin, silkworm cocoons, spider webbing, hair, and so forth. These materials are carbon neutral, sustainable, renewable, recyclable, nontoxic and environmental friendly, and can replace petroleum based products.[8,9]

Fundamental research in the production, modification, property enhancement and new applications of these materials is important. The new materials, concepts and utilizations that result from these efforts will shape the future of polymers from renewable resources.[10,11] Various biopolymers, predominantly polysaccharides and proteins, are used in aerogel production because they are ecological materials that can transform numerous industrial processes from being petroleum-dependent into biomaterial-dependent. In particular, they have numerous applications in food and non-food industries.

Biopolymers from various sources such as alginate, cellulose, lignin, pectin, chitosan, proteins and others have been tested as precursors. The resulting aerogels exhibit both the specific inheritable functions of the starting polymer and the distinctive features of aerogels (open porous structure with high specific surface and pore volume). This synergy of properties has prompted researchers to view biopolymer aerogels as promising candidates for a wide range of applications.[12] More recent reports on biopolymer aerogels, as chronicled in this book, describe their use for thermal insulation, tissue engineering, regenerative medicine, drug delivery systems, functional foods, and as catalysts and sensors.

1.4 Applications of Bio-based Aerogels: Highlights

Some of the most studied uses of aerogels include different aerospace-related applications, thermal super insulations, acoustic devices, and so forth. Recently, a growing interest in aerogels has been observed in the fields of pharmaceutical science, food related technology, biosensors, and diagnostics and biotechnology, to name a few (Figure 1.5).[5]

Polysaccharide based aerogels have proved to be useful in scientific domains where biocompatibility and biodegradability are needed, such as for medicinal, cosmetic and pharmaceutical applications, opening up the field of aerogel research even further.[13,14] Polysaccharides that could be used to prepare aerogels include cellulose, marine polysaccharides and starch, all of which have in common the ability to form gels either by themselves in the presence of water or with di-cations, other cross-linking agents, and/or other blended or mixed polysaccharides.[15] Ultraporous nanocellulose aerogels have been used as a separation medium for liquid mixtures of oil/water. Polysaccharide-based aerogels are highly porous ($\varepsilon = 90$–99%), lightweight ($\rho = 0.07$–0.46 g cm^{-3}) drug carriers with a high surface area ($S_a = 70$–680 m^2 g^{-1}), enabling them to provide enhanced drug bioavailability and drug loading capacity.[16]

Several outstanding properties of polysaccharide aerogels make them promising materials for catalysis, especially their high surface area and high density of functional groups (up to 5.8 mmol g^{-1} amino groups for chitosan, 5.6 mmol g^{-1} carboxylic groups for alginic acid, and 2.8 mmol g^{-1} sulphate

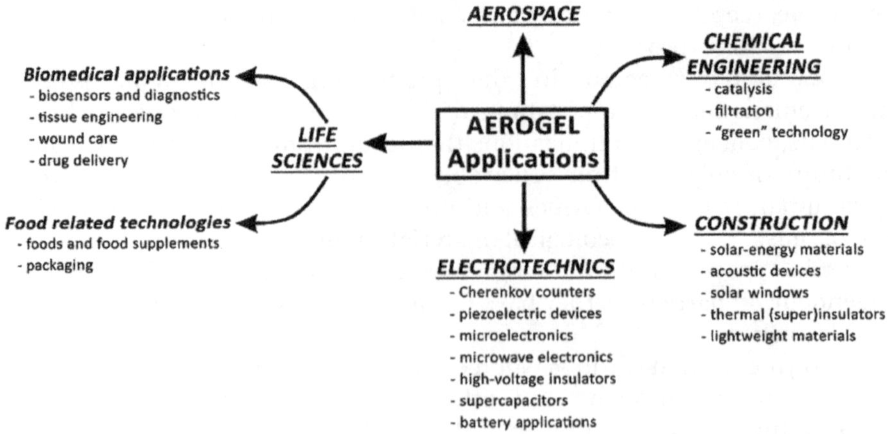

Figure 1.5 Scheme showing potential aerogel applications. Adapted from ref. 5 with permission from Springer Nature, © Springer Science + Business Media New York 2016.

groups for k-carrageenan). With surface areas as high as 500 $m^2 g^{-1}$, polysaccharides can compete with inorganic solids as supports for organometallic or metal catalysts.[13]

Protein based aerogels, though not as advanced as their polysaccharide counterparts, have gone through numerous stages of research in recent years. The foremost aerogels include those created from proteins of plant and animal origin, such as those derived from soy, wheat, silk fibroin, egg white (ovalbumin), milk (caseins and whey proteins), and so forth. They have proved to exhibit superior gel forming properties and ease of availability as they can be readily obtained from the food processing industry as by-products.

Other than the single-component aerogels, those composed of one of the precursors with specific additives have often conferred additional functionalities such as mechanical strength, hydrophobicity and catalytic features to the pristine materials, and this has improved the usefulness for some high-performance applications.[17] The prospect of hybrid aerogels widens the horizon for development and innovations in this field. A broad realm of such aerogels has been established with diverse combinations of organic, inorganic, synthetic and bio-based raw materials.[12,18–22] Aerogels thus developed have found an extensive range of applications in the fields of drug delivery, nutraceuticals and oil absorption, thanks to their superior mechanical properties, thermo-responsiveness, biocompatibility, non-cytotoxicity, porous structure and reduced brittleness.[12,23,24]

1.5 About This Book

Continuing on this flow of ongoing studies, this book documents the history, chemistry and recent developments in the field of bio-based aerogels, which will undoubtedly be an inevitable reference text for both academic and research communities alike. Even though one can find several research publications in the domain of aerogels, to the best of our knowledge, no systematic scientific reference book has been exclusively written on bio-based aerogels.

This multi-author book provides a useful summary of current knowledge in the realm of polysaccharide and protein based aerogels. The book commences with an introduction to different types of bio-based aerogels, their processing techniques and morphological analyses (Chapters 2–7). A detailed description is given in Chapter 8 on theory modelling and simulations of such aerogels. Chapters 9–12 deal with various properties possessed by these aerogels, while highlighting the procedures used for the tuning and tailoring of certain parameters that have influential effects on their properties. Chapters 13–16 shed light on the significance and applicability of bio-based aerogels in various arenas, with special reference to life science, food engineering, insulation, aerospace, packaging and biomedical applications.

The entire book has been brought together with the anticipation of providing a better understanding, and assisting further development, in the field of bio-based aerogels.

References

1. NASA, https://stardust.jpl.nasa.gov/images/gallery/aerogelcrayons.jpg.
2. A. McNaught and A. Wilkinson, *IUPAC Compendium of Chemical Terminology (the 'Gold Book')*, International Union of Pure and Applied Chemistry, Oxford, 2nd edn, 1997.
3. N. Hüsing and U. Schubert, *Angew. Chem.*, 1998, **110**, 22–47.
4. A. C. Pierre and M. Pajonk, *Chem. Rev.*, 2002, **102**, 4243–4265.
5. J. Stergar and U. Maver, *J. Sol-Gel Sci. Technol.*, 2016, **77**, 738–752.
6. C. A. García-González, M. Alnaief and I. Smirnova, *Carbohydr. Polym.*, 2011, **86**, 1425–1438.
7. F. Quignard, F. Di Renzo and E. Guibal, in *Carbohydrates in Sustainable Development 1*, Springer, Berlin, Heidelberg, 2010, pp. 165–197.
8. R. Mohammadinejad, S. Karimi, S. Iravani and R. S. Varma, *Green Chem.*, 2016, **18**, 20–52.
9. M. Mariano and A. Dufresne, in *Nanocelluloses: Their Preparation, Properties, and Applications*, American Chemical Society, 2017, pp. 203–225.
10. C. Williams and M. Hillmyer, *Polym. Rev.*, 2008, **48**, 1–10.
11. R. Xiong, A. M. Grant, R. Ma, S. Zhang and V. V. Tsukruk, *Mater. Sci. Eng., R*, 2018, **125**, 1–41.
12. S. P. Raman, P. Gurikov and I. Smirnova, *J. Supercrit. Fluids*, 2015, **106**, 23–33.
13. F. Quignard, R. Valentin and F. Di Renzo, *New J. Chem.*, 2008, **32**, 1300.
14. L. Heath and W. Thielemans, *Green Chem.*, 2010, **12**, 1448.

15. K. S. Mikkonen, K. Parikka, A. Ghafar and M. Tenkanen, *Trends Food Sci. Technol.*, 2013, **34**, 124–136.
16. C. A. Garcia-Gonzalez, M. Alnaief and I. Smirnova, *Carbohydr. Polym.*, 2011, **86**, 1425–1438.
17. H. Maleki and N. Husing, in *New Polymer Nanocomposites for Environmental Remediation*, Elsevier, 2018, pp. 389–436.
18. L. Zuo, Y. Zhang, L. Zhang, Y.-E. Miao, W. Fan and T. Liu, *Materials*, 2015, **8**, 6806–6848.
19. L. Wang, Universitat Politècnica de Catalunya, 2015.
20. P. Veres, A. M. López-Periago, I. Lázár, J. Saurina and C. Domingo, *Int. J. Pharm.*, 2015, **496**, 360–370.
21. A. El Kadib and M. Bousmina, *Chem. - Eur. J.*, 2012, **18**, 8264–8277.
22. A. A. Alhwaige, T. Agag, H. Ishida and S. Qutubuddin, *RSC Adv.*, 2013, **3**, 16011.
23. M. Ahmadi, A. Madadlou and A. A. Saboury, *Food Chem.*, 2016, **196**, 1016–1022.
24. N. Pircher, S. Veigel, N. Aigner, J. M. Nedelec, T. Rosenau and F. Liebner, *Carbohydr. Polym.*, 2014, **111**, 505–513.

CHAPTER 2

Chitin/Chitosan Based Aerogels: Processing and Morphology

ANA RITA DUARTE,*[a,b] SIMONE S. SILVA[a,b] AND RUI L. REIS[a,b]

[a] 3B's Research Group – Biomaterials, Biodegradables, and Biomimetics, University of Minho, Headquarters of the European Institute of Excellence on Tissue Engineering and Regenerative Medicine, AvePark, 4805-017 Barco, Guimarães, Portugal; [b] ICVS/3B's – PT Government Associate Laboratory, Braga/Guimarães, Portugal
*Email: aduarte@dep.uminho.pt

2.1 Introduction

Chitin and chitosan have been reviewed in a number of different publications in the literature. The importance of these polymers is highlighted in the high number of publications listed on Web of Knowledge, and the applications found for both chitin and chitosan are constantly increasing as the processing methodologies evolve. An extensive review of the properties and application of both polymers was reported by Rinaudo in 2006, and it is still the most cited work.[1] However, other authors have reported the potential of chitin and chitosan in soft-based applications.[2] In this chapter, we will review the most important and particular features of chitin and chitosan, and will focus on the processing methodologies for the preparation of chitin and chitosan aerogels and discuss the different morphologies obtained. We also highlight the possibility of preparing functional aerogels with the bioactive properties of these two polymers.

Green Chemistry Series No. 58
Biobased Aerogels: Polysaccharide and Protein-based Materials
Edited by Sabu Thomas, Laly A. Pothan and Rubie Mavelil-Sam
© The Royal Society of Chemistry 2018
Published by the Royal Society of Chemistry, www.rsc.org

2.2 Chitin and Chitosan: Structure and Properties

After cellulose, chitin is the second most abundant natural polymer. It is present in many marine origin organisms, among the most common are the shells of crustaceans such as crabs and shrimps, but it is also present in cephalopods.[3,4] Marine origin biopolymers have recently gained more attention as an alternative source to obtain biopolymers with different characteristics.[4,5] The structure of chitin is formed by *N*-acetyl-D-glucosamine linked by strong intermolecular β-(1,4)-glycosidic bonds (Figure 2.1). The strength of these bonds limit its solubility and hinders the use of chitin due to the difficulties encountered in processing. Moreover, depending on its source, chitin can be characterized by α and β-forms, distinguished by strong and weak inter- and intramolecular bonds, respectively.[5,6] Chitin is insoluble in most solvents and the most commonly used methods to solubilize chitin is through complexation with lithium chloride in dimethylacetamide or *N*-methyl-2-pyrrolidone. However, the use of volatile organic solvents, their emissions and disposal remain a huge problem, which hampers the widespread use of this natural polymer. In 2002, Swatloski *et al.* reported, for the first time, the ability of ionic liquids (ILs) to dissolve naturally occurring polymers such as cellulose, chitin, starch and lignin.[7,8] The use of this class of solvents, so called green solvents due to their chemical and thermal stability, low vapour pressure, high ionic conductivity and easy recyclability, may hereafter boost the

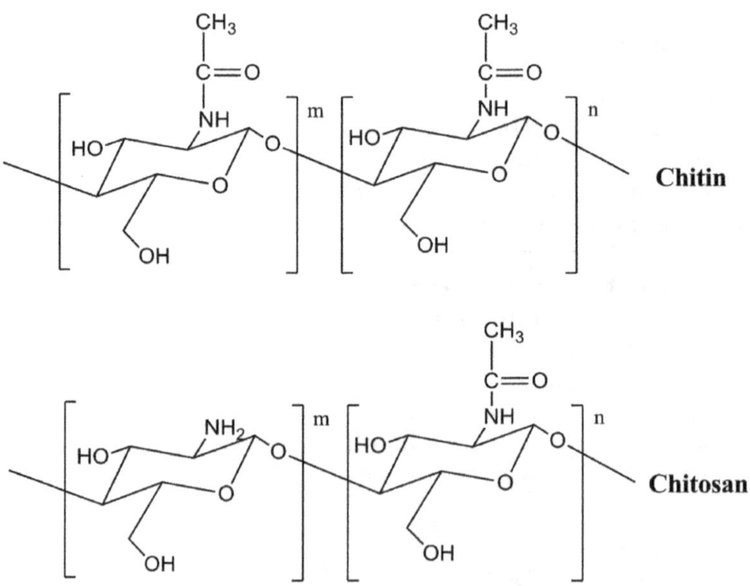

Figure 2.1 Chemical structure of chitin and chitosan.

ability of these materials to be used as renewable, biodegradable polymers in large scale applications.

Chitosan is obtained from the partial deacetylation of chitin. It is a linear, cationic polymer formed by β-(1,4)-glycosidic bonds between D-glucosamine and *N*-acetyl-D-glucosamine units (Figure 2.1).[1,3] The properties of chitosan can therefore be modulated through modifications to the deacetylation degree, molecular weight, and also the sequence of repeating units. Chitosan is insoluble in water, in basic solutions and most organic solvents. However, chitosan, in contrast to chitin, can be solubilized in acidic conditions, as the amino groups of chitosan can be protonated, making it soluble in dilute acid solutions and is therefore easier to process.[1,9] For this reason, the applications of chitosan are much broader than chitin. Chitosan may find application in the fields of agriculture, water and waste treatment, food and beverage technology, cosmetics, pharmaceuticals and biomedical applications.[1,6,10,11] The inherent characteristics of chitosan make it suitable for all the applications listed, and although some applications have been explored since the late 1970's, the potential of chitosan as a promising raw material for biomedical applications was only suggested in the early 1990's. The physico-chemical and biological characteristics of chitosan make it an attractive biopolymer for the development of new biomaterials with applications in tissue engineering and regenerative medicine in many fields, from skin to bone or cartilage.[9,11,12] Chitosan has been processed in several forms to be used in tissue engineering, from particles, fibres, membranes, 3D structures and fibre meshes. Pharmaceutical applications and drug delivery methods using chitosan have also been proposed in a great number of publications.[9,13,14]

2.3 Chitin and Chitosan Aerogels: Processing and Morphology

Organic aerogels made from polysaccharides such as chitin and chitosan are of particular importance as they are renewable feedstocks. Therefore, diverse strategies have been applied to chitin and chitosan to create tailor made aerogels (see Figure 2.2). Most of them are based on the dissolution of these polysaccharides in adequate solvents, followed by drying. Drying is one of the most critical steps in the creation of an aerogel. Therefore, conventional techniques, namely supercritical drying (the most common technique for producing aerogels), freeze-drying and sol–gel methodologies have been associated with the drying process. In general, these techniques are inexpensive and flexible enough to produce the desired structures by adjusting the parameters of the processing, and also by allowing the possibility of incorporating pharmaceutical agents. The aerogels obtained have many applications in several areas from drug delivery to thermal insulators (see Figure 2.2).[15–23]

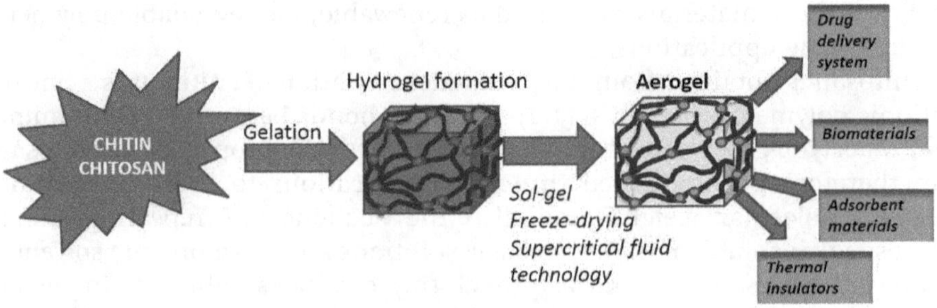

Figure 2.2 Schematic representation of the preparation and potential applications of chitin/chitosan-based aerogels.

2.3.1 Supercritical Fluid Technology

Supercritical fluid technology (SCF) is an interesting green technology which used for the processing of aerogels as it does not require the use of large amounts of organic solvents, prevents shrinking of the structures and it adds the possibility of impregnating active compounds during the drying step, in a one-step process approach.[24] Supercritical drying consists of exposing the wet gels at conditions above the supercritical point of the liquid solvent, promoting its removal.[12] The most commonly used gas in supercritical fluid processing is carbon dioxide (CO_2), which has the advantages of mild critical conditions (31.1 °C and 73.8 bars), availability and low cost. The preparation of chitin aerogels can be achieved through the dissolution in mixtures of solvents including *N*,*N*-dimethylacetamide/lithium chloride (DMA/LiCl)[25] and sodium hydroxide-urea solutions,[19] followed by the drying step under CO_2 supercritical conditions. By using supercritical CO_2 drying, the network structure in the hydrogel is well preserved in the aerogel. These chitin aerogels exhibit a low density (0.23–0.27 $g\,cm^{-3}$), large surface area (up to 366 $m^2\,g^{-1}$), moderate thermal stability and high physical integrity up to 270 °C, and significantly high mechanical properties.[14]

Recent progress in overcoming the solubility limitations of chitin have been achieved by the use of ILs.[21,26,27] ILs are organic salts that are liquid at room temperature, and they are known as green solvents as they have a negligible vapour pressure, thermal stability, and high polarity, among other properties.[28] Therefore, the combination of ILs as a solvent for chitin with SCF has demonstrated the possibility of developing chitin aerogels. Silva *et al.* dissolved chitin in 1-butyl-3-imidazolium acetate (BMIMAc) at high temperature, followed by regeneration of the polymer in ethanol in specific moulds.[21] The ionic liquid was removed using Soxhlet extraction and successive steps of extraction with SCF using carbon dioxide/ethanol. The developed chitin aerogels have a very low density (0.039–0.063 $g\,l^{-1}$) and high porosity (84.1–90.2%). Moreover, the morphology of the samples displayed a heterogeneous porous formation with pores around 1 μm

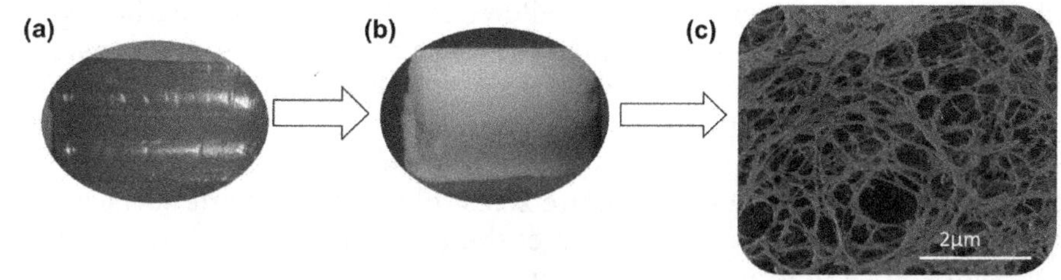

Figure 2.3 (a) Chitin gel as prepared, (b) after soxhlet extraction + SCF drying and (c) SEM image of dried aerogel.
Adapted from *Acta Biomaterialia*, 7, S. S. Silva, A. R. C. Duarte, A. P. Carvalho, J. F. Mano and R. L. Reis, Green processing of porous chitin structures for biomedical applications combining ionic liquids and supercritical fluid technology, 1166–1172, Copyright 2011, with permission from Elsevier.[21]

(Figure 2.3). Based on these findings, several potential biomedical applications of the obtained chitin aerogels as biomaterials have been suggested.

Chitin nanowhiskers are structured into mesoporous aerogels by the sonication-assisted assembly of chitin nanowhiskers in water, followed by solvent exchange with ethanol and drying with supercritical CO_2.[15] The rod-like crystalline nature of the nanowhisker was retained during the aerogel production process, truly making the aerogel an assembled structure of chitin nanocrystals. These aerogels showed highly porous, low densities and moderate surface areas, and could find potential application as thermal insulators, catalyst supports, and biomedical materials. In another work, chitin nanofibers/nanowhishers hydrogels were prepared using the "gas phase coagulation" gelation method.[29] These hydrogels were freeze-dried to form aerogels with fibre-like structures.

Recent trends have demonstrated the relevance and efficient use of renewable biopolymers as catalyst supports. The fabrication and direct use of derived chitin nanofibrils (ChNF) aerogels as base catalysts for the effective production of useful chemicals has been described by Tsutsumi and coworkers.[23] They used freeze-drying or a supercritical drying process to obtain the chitin-based aerogels, and then examined their performance as catalysts in a typical Knoevenagel condensation reaction at room temperature. Both the mesoporous aerogel structure and the surface-exposed C_2-amines on the crystalline ChNF surface were effective for use in continuous flow catalysis.

The chitosan aerogel is dissolved in aqueous acetic acid to form a homogeneous solution at the molecular level, followed by hydrogel formation through physical and chemical interaction (*e.g.* crosslinking[16]), and then dried under CO_2 supercritical drying conditions.[18] Translucent chitosan aerogels have been proposed as thermal insulators by Takeshita and coworkers.[16] In this approach, formaldehyde was used as a mild crosslinker. Formaldehyde reacts with a NH_2 group to form a Schiff base, $-NCH_2$, which reacts with another NH_2 group to form a N–C–N crosslinking bond.

Subsequently, the solvent exchange allows the formation of methanogels, which are extracted with supercritical CO_2. The aerogels obtained consisted of entangled nanofibers of 5–10 nm in diameter with mesopores of 10–50 nm in size and a high porosity up to ~97% (see Figure 2.4).[16] The area of pores observed in the SEM images increases as the apparent density decreases. For instance, the samples shown in Figure 2.4e and f have a specific surface area of 545 $m^2 g^{-1}$, which is 4–30 times larger than those of chitosan cryogels,[30,31] and comparative to those of nanocellulose aerogels.[32]

Crosslinked chitosan aerogels have been prepared using different cross-linkers including glutaradehyde (GTA), glyoxal and formaldehyde, followed by the supercritical CO_2 drying process. Chang and co-workers used glutaraldehyde, glyoxal, and formaldehyde to create aerogels with a solid fibrillar network, high surface areas, and mesopores.[33] The complete removal of water in the pores by displacement with absolute ethanol can be a key factor for producing the aerogels. The obtained aerogels have a superior adsorption performance for anionic surfactant sodium dodecylbenzene-sulfonate (SDBS) in acidic solution, as compared with other reported

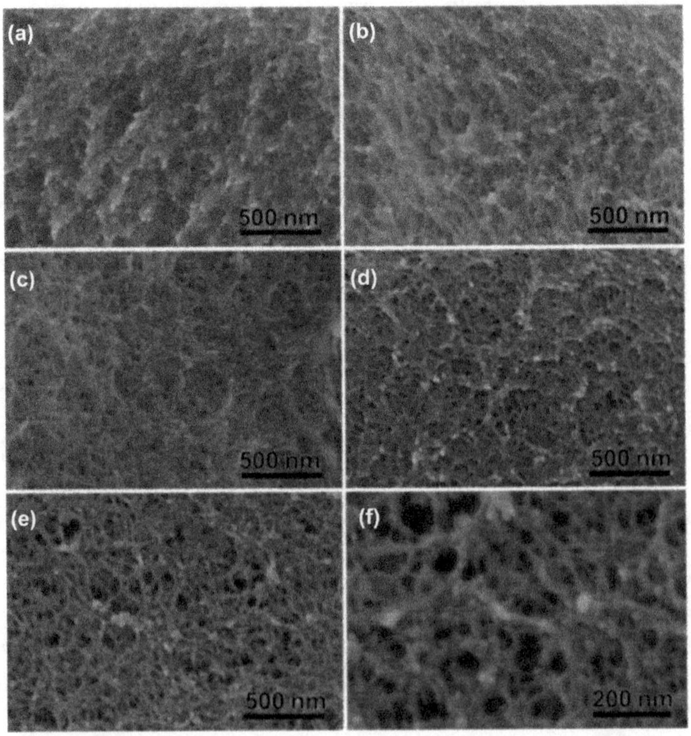

Figure 2.4 SEM images of chitosan aerogels: (a) C16F7, (b) C16F2, (c) C8F7, (d) C8F2, and (e and f) C4F7, prepared using the crosslinking gelation method.
Reprinted with permission from S. Takeshita and S. Yoda, *Chem Mater*, 2015, **27**, 7569–7572, Copyright 2016 American Chemical Society.[16]

materials. In another work, a supercritical carbon dioxide gel drying process was used to crosslink chitosan and to simultaneously promote the elimination of unreacted GTA, due to its solubility in supercritical mixtures (supercritical CO_2 + ethanol).[34] GTA is the most commonly used crosslinking agent for chitosan.[13] However, it is a highly cytotoxic crosslinking agent that, if not removed from the polymeric structure, impairs the proliferation of cells. In this work, a study of the release of GTA residues showed the presence of negligible residues of GTA at the end of the release experiments. These findings suggest that this approach can be suitable to obtain improved chitosan aerogels with minimal toxicity for tissue engineering applications. Moreover, the crosslinked chitosan aerogels showed a nanofibrous structure characterized by an average diameter of about 100 nm.

SCF can also be used to produce chitosan aerogel microparticles and allow the formation of aerogel microparticles as drug delivery carriers.[35] Chitosan aerogel microparticles loaded with salbutamol have been prepared using SCF technology and could be suitable for use as a pulmonary drug delivery system.[35] Additionally, composite natural aerogels composed of chitosan–gelatin (CSG) have been produced by supercritical gel drying. The CSG mixtures formed uniform gels during the preparation step, while the supercritical gel drying preserved the nanostructured morphology due to the near zero surface tension of the supercritical mixture (CO_2 + organic solvent) formed during the drying process. Aerogel-based scaffolds, such as the obtained CSG, can be used for tissue engineering applications, as their nanofibrous structure is suitable for cell adhesion, proliferation, and migration.

2.3.2 Sol–Gel Technique

Chitin and chitosan-based aerogels can also be synthetized *via* the sol–gel technique. This technique involves two steps, namely (i) formation of the wet gel through the hydrolysis of silicon alkoxide precursors, suitable solvents, catalysts and water, which are stirred into a homogeneous solution; and (ii) drying of the wet gel.[36] This last step consists of the removal of the liquid in a gel above its critical temperature and pressure, ensuring that the gel matrix (*i.e.*, the networked structure) remains intact without large shrinkage upon drying. Particularly for chitin and chitosan aerogels, the most commonly used precursors in the sol–gel technique are silicon alkoxides, including tetramethoxysilane (TMOS)[37] and tetraethoxysilane (TEOS).[22,26,38] Chitosan-silica hybrid aerogels were described for the first time by Ayers and Hunt.[38] In their publication, they prepared different hybrid materials with various ratios of chitosan to silica following the sol–gel methodology. Functional chitosan-silica hybrid aerogels can be applied in various fields namely, in pharmaceutical and biomedical areas and in waste water treatment. More details about functional aerogels obtained using the sol–gel technique will be described throughout this section 2.4 of this chapter.

2.3.3 Freeze-drying

Freeze-drying is also a useful technique for obtaining aerogel. This technique is a drying method largely applied to reducing the water activity and the susceptibility of the materials to bacterial attack. It is based on freezing the polymeric solution at temperatures varying between −20 and −196 °C to allow the growth of ice crystals, followed by the removal of solvent, which is generally water, through lyophilization.[39] Different variables affect the morphology of the structures obtained. For example, fast-freezing could allow the better preservation of the fine structure of the gels. Some authors suggest that the textural properties of the chitosan aerogels are largely controlled by the viscosity of the polymer solution, which in turns depends on the molecular weight, the deacetylation degree and the natural source of the chitin. For instance, aerogels with a higher surface area can be obtained from hydrogels of chitosan obtained by the deacetylation of α-chitin from crab shells, compared to those obtained by the deacetylation of β-chitin from squid pen.[17] In another study, crosslinked chitosan aerogels were fabricated using Debus–Radziszewski imidazole synthesis, and glyoxal and propinaldehyde to connect the amine groups of chitosan *via* imidazolium crosslinking.[40] The imidazolium-crosslinked chitosan (ICC) aerogels were produced after freeze-drying. The aerogel exhibited irregularly shaped macrostructures with very smooth surfaces, and was highly hydrophilic and exhibited an excellent adsorption ability (481–351 mg g^{-1}), even under alkaline conditions.[40]

Recent approaches involving the development of chitosan-based aerogels by dispersion of graphene[20] or graphene oxide (GO)[41] into a chitosan matrix using the freeze-drying technique, suggested that these approaches could increase the surface area, and could also promote an improvement in the mechanical and physical properties of the aerogel.

Chitosan/cellulose composite aerogels can be prepared using the sol–gel method and freeze-drying.[42] This approach has many advantages including low-cost and the possibility of scale-up. Moreover, these composites have antifouling and self-cleaning properties due to the rough surface and hydrophilic groups. Furthermore, they can separate water from oil/water mixtures selectively and efficiently.

Table 2.1 summarizes the textural characterization and drying methods used in the preparation of chitin and chitosan aerogels.

2.4 Functional Aerogels

Functional chitin and chitosan aerogels may broaden the application of these type of materials.[43] Different strategies have been followed and are reviewed here to provide a perspective of the different methodological approaches. The effect on the functionality of aerogels is also discussed by the differentiation of the preparation of aerogel monoliths and aerogel microspheres, which also provides various features and properties, and hence different applications can be envisaged.

Table 2.1 Textural properties and drying methods used in the preparation of chitin and chitosan aerogels.

Sample	Conditions/Drying method	Surface area ($m^2 g^{-1}$)	Porosity (%)	Ref.
Chitin aerogel	Supercritical carbon dioxide	220	88	25
	Supercritical carbon dioxide	366	85	19
	Supercritical carbon dioxide	108–145	84.1–90.2	21
Chitin nanowhiskers aerogel	Supercritical carbon dioxide	up to 261	92	15
Chitin nanofibrils aerogel	Freeze-drying or Supercritical carbon dioxide	up to 289	—	23
Chitosan aerogel	Supercritical carbon dioxide	545	up to 97	16
	Chemical crosslinking/ supercritical carbon dioxide	up to 845	—	33
Chitosan-silica hybrid aerogel	Sol–gel process/supercritical carbon dioxide	470–750	—	38

Few studies report the functionalization of chitin aerogels. Garcia and coworkers have presented an interesting methodology to modify aerogels that have been loaded with carbon nanotubes.[44] These have been prepared after a chemical modification of chitin into chitin nanowhiskers and sonicated assisted assembly, and the mesoporous material obtained could have applications as thermal insulators, catalyst supports or biomedical materials.

Chitosan has been much further explored than chitin by far. In terms of monoliths, silica–chitosan aerogels are the most obvious hybrid materials to be prepared. Silica aerogels have long been known for their specific properties. However, they lack mechanical integrity. In this sense, blending the polymers with biopolymers may be an interesting and easy way to overcome this drawback. Furthermore, they can help balance the hydrophobicity–hydrophilicity of the structures.

Chitosan-based materials have been proposed as adsorbents in different areas, particularly due to their low cost, high availability, ease of manufacture, environmentally friendly properties, and versatility. The previous literature reports various strategies to enhance the potential of absorbency of chitosan aerogels and the ability to target different molecules.

Chitosan–silica hybrid aerogels prepared using sol–gel process have been described since 2001,[38] as mentioned previously.[36] Diverse applications have been suggested, from drug delivery to waste water treatment. Since then, the application development of this type of material has also been extended to other fields.[45–47]

Hydrophobic chitosan-silica aerogels have been described by Ma *et al.*, in this work, the authors prepared the aerogels by a typical sol–gel process and a two-step hydrophobic treatment.[22] The gels were formed by dissolving chitosan in an acidic solution and adding the TEOS solution, under vigorous stirring, after 12 hours. Hydrophobicity was further increased by the addition of hexamethyl disilazane as a modifier agent, the samples were then freeze dried. Low density, high porosity matrices were obtained. These systems were evaluated for oil absorbency and were shown to have a superior performance as a recyclable oil absorber.

Wang and coworkers prepared chitosan-silica aerogels for dye absorption. In this work, different ratios of chitosan and TMOS were tested.[37] The solutions of chitosan and TMOS were left to gelate for 48 h, under stirring. Increasing the silica content in the matrices led to a decrease in the bulk density and an increase in the pore size and surface area. At the same time, the structure is more homogeneous with increasing silica content. However, the differences observed do not significantly affect the performance of the aerogels as an adsorbent. Another strategy for the preparation of dye adsorbents based on chitosan was described by another group.[47] In this case, the methodology for the development of the hybrid aerogels involved a two-step approach, firstly the graphene oxide was dispersed in an acidic solution and chitosan was added. In the second step, methyl aldehyde was added dropwise to the solution at 50 °C to promote gelation. The gelled matrices were freeze-dried to obtain the aerogels. The aerogels obtained had a high porosity and low density and were proven to be effective for the adsorption of dyes, presenting a significantly higher adsorption capacity when compared with other materials, for the same pigments.

Graphene loaded chitosan aerogels were described by Ji and co-workers for other purposes.[48] The objective of this work was to prepare supercapacitors blending graphene with different carbohydrates. The aerogels were prepared following a hydrothermal route. Chitosan dissolved in acetic acid solution was mixed with a suspension of graphene. The solution was placed in an autoclave at 180 °C for 18 h, after which the hydrogels were formed. The materials were then dried by freeze-drying. The results suggested that a high charge density of protonated chitosan in acidic solution may be beneficial for the modification of both the surface and bulk properties of graphene, and this therefore represents a simple method for the preparation of hybrid graphene-based aerogels.

Hao *et al.*, on the other hand, described the preparation of graphene-based nitrogen doped aerogels from an initial chitosan solution.[20] Chitosan aerogels were prepared from an acidic solution by freeze-drying. Afterwards the samples were carbonized in a furnace at 800 °C for 3 h under a nitrogen atmosphere and activated using potassium hydroxide. In this work, samples with a high degree of graphitization, high specific surface area, and pore volume, and homogeneous nitrogen doping were obtained, demonstrating the potential scalability of the process towards industrial and large scale production.

Other authors suggest that doping chitosan monoliths with graphene oxide could be used for carbon dioxide capture with greater efficiency that chitosan *per se*.[41] In this case, the bio-based hybrid aerogels were prepared by freeze-dying.

Table 2.2 Summary of functionalized chitosan monoliths and their applications.

Functional material	Application	Ref.
Chitosan–silica hybrid aerogels	Oil absorber	22
	Dye adsorbent	37,47
Graphene loaded chitosan aerogels	Supercapacitors	41,48
	CO_2 capture	

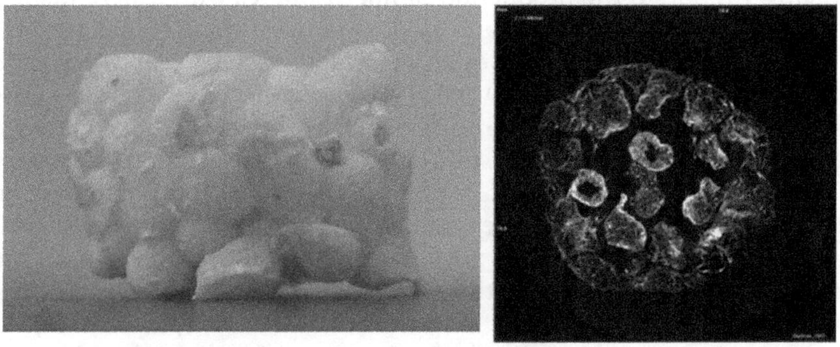

Figure 2.5 3D chitin scaffold prepared by particle agglomeration.

Table 2.2 summarizes the type of functionalization described in the literature and the main applications envisaged.

Another approach which has been followed and reported in the literature has to do with the preparation of aerogel particles. Chitin progress has been highly hindered by its low solubility in almost any solvent. In our group, we have processed chitin microspheres and functionalized them by dissolution in an ionic liquid.[26] Chitin hybrid beads were prepared with TEOS to confer the particles with bioactivite behaviour. The objective was to provide osteoinductive properties to the microspheres, which are then able to induce the formation of an apatite layer. The systems could also be functionalized with dexamethasone; an anti-inflammatory agent used to direct stem cell differentiation into the osteogenic lineage. The main breakthrough of this work was the development of a new particle agglomeration technique which allows the design of three dimensional structures (see Figure 2.5). This technique consists of confining the microparticles in a mould with or without the addition of a biological glue and drying the construct using supercritical fluid carbon dioxide. This system, with both functional and structural characteristics, may potentiate the use of microparticles, thereby taking advantage of their high surface area and of the 3D structure which confers the desired mechanical properties.

El Kadib *et al.* have reviewed the preparation and functionalization of chitosan-based microspheres, both organic and inorganic in nature.[49] Polymerization by sol–gel chemistry seems to be the most effective way to produce chitosan-inorganic hybrid materials. The main advantages of this type of approach are the fact that it operates under mild conditions and provides a good mixing between the organic and inorganic phase, leading to very homogenous structures. Different metal alkoxides have been described in the literature, and the common objective is the development of functional aerogels for heterogeneous catalysis. The ability to functionalise microspheres with an acid–base system provides the particles with a dual function, which may act synergistically in catalysis, and this is the major advantage of these type of systems.

Molvinger reported the production of microspheres with different surface areas depending on the procedure used to prepare them.[46] In their work, they focused on the production of chitosan–SiO_2 particles and the surface area varied between 56 and 149 $m^2 g^{-1}$, depending on the route chosen. These systems can be prepared from previously prepared chitosan aerogels, which are functionalized in a subsequent step, or a simpler procedure with fewer steps can be followed. In any case, it is possible to prepare both bulk chitosan–SiO_2 microparticles and core–shell microparticles. However, the systems prepared present poor stability, and in this sense, other approaches may involve the substitution of silica by titanium, which is easier in terms of production as the modification can take place in aqueous solutions, rather than the acid solutions required for the processing of the silica-based microspheres.[50] This may expand the applications from mere catalytic supports, as chitosan–TiO_2 aerogels also find relevancy as pigments and bioactive devices in other fields of study. Likewise, the possibility to perform the sol–gel polymerization with other metal precursors could lead to the production of new types of hybrid systems that are more thermal and chemically stable, namely the preparation of aluminium, zirconium and tin derivative chitosan-based microspheres such as those reported in the literature.[51]

Chitosan-clay hybrid systems in particulate form have been described and can be applied in different areas from food packaging to adsorbents. Frindy *et al.* report the development of chitosan microspheres loaded with three different types of clay, montmorillonite (MMT), nanotubular halloysite (HNT) and micro-fibrillar sepiolite (SP).[52] In this work, the authors found that the structure and properties of the final materials is greatly dependent on the type of clay used in their preparation, as it depends on the hydrophilicity of the clay, which will ultimately influence the interfacial tension between the two materials present in the composite and the mechanical properties of the final structure.

2.5 Conclusions and Future Perspectives

In this chapter, we reviewed the methodologies adopted by researchers in different fields to process chitin and chitosan aerogels. Aerogels based on chitin and chitosan have interesting features useful for a broad range of applications from oil absorbers to biomaterials. From the literature, it can be concluded that most of the research reported is related to the processing of chitosan alone, rather than chitin, which is due to the difficulties encountered in chitin solubilization. The polymers, alone or in combination with other polymers, or even incorporation of graphene, are prepared mostly through the application of different green technologies, such as supercritical fluid technology, freeze-drying and the sol–gel process. The methodologies described give rise to different architectures, in particular those with surface areas ranging from 100 up to 800 $m^2 g^{-1}$ and with porosities over 80% in most cases. Most research has been centered on the development of adsorbents or catalytic supports, hence the functionalization of chitin and chitosan based aerogels has been performed with silica or other inorganic compounds. In this regard, for certain biomedical applications, despite the promising findings, there has been little reported regarding the *in vitro* and *in vivo* biocompatibility of chitin and chitosan-based aerogels, which could compromise their applicability as biomaterials. The versatility of the new methodologies adopted may help boost the application of these type of aerogels, namely by assisting their implementation at a larger scale.

Abbreviations

1-Butyl-3-imidazolium acetate	BMIMAc
Carbon dioxide	CO_2
Chitosan–gelatin	CSG
Chitin nanofibrils	ChNF
Imidazolium-crosslinked chitosan	ICC
Ionic liquids	ILs
Graphene oxide	GO
Glutaradehyde	GTA
Micro-fibrillar sepiolite	SP
Montmorillonite	MMT
Nanotubular halloysite	HNT
N,N-dimethylacetamide/lithium chloride	DMA/LiCl
Supercritical fluid technology	SCF
Sodium dodecylbenzene-sulfonate	SDBS
Tetraethoxysilane	TEOS
Tetramethoxysilane	TMOS

Acknowledgements

The research leading to these results has received funding from the European Union Seventh Framework Programme (FP7/2007–2013) under grant agreement number REGPOT-CT2012-316331-POLARIS and from the project "Novel smart and biomimetic materials for innovative regenerative medicine approaches" RL1 – ABMR – NORTE-01-0124-FEDER-000016) co-financed by the North Portugal Regional Operational Programme (ON.2 – O Novo Norte), under the National Strategic Reference Framework (NSRF), through the European Regional Development Fund (ERDF). The authors would also like to acknowledge the financial support of the Associate Laboratory, Life and Health Sciences Research Institute/Biomaterials LA ICVS-3Bs (2015–2017). The authors would like also to thank the Portuguese Foundation for Science and Technology (FCT) for the financial support for the fellowship grant of Simone S Silva (SFRH/BPD/112140/2015), "Fundo Social Europeu" - FSE and "Programa Diferencial de Potencial Humano POPH".

References

1. M. Rinaudo, *Prog. Polym. Sci.*, 2006, **31**, 603–632.
2. A. El Kadib, M. Bousmina and D. Brunel, *J. Nanosci. Nanotechnol.*, 2014, **14**, 308–331.
3. F. Croisier and C. Jérôme, *Eur. Polym. J.*, 2013, **49**, 780–792.
4. Y. Lu, Q. Sun, X. She, Y. Xia, Y. Liu, J. Li and D. Yang, *Carbohydr. Polym.*, 2013, **98**, 1497–1504.
5. L. Reys, S. Silva, J. Oliveira, S. Caridade, J. Mano, T. Silva and R. Reis, *Biomed. Mater.*, 2013, **8**, 045002.
6. T. H. Silva, A. Alves, B. M. Ferreira, J. M. Oliveira, L. L. Reys, R. J. F. Ferreira, R. A. Sousa, S. S. Silva, J. F. Mano and R. L. Reis, *Int. Mater. Rev.*, 2012, **57**, 276–306.
7. R. P. Swatloski, S. K. Spear, J. D. Holbrey and R. D. Rogers, *J. Am. Chem. Soc.*, 2002, **124**, 4974–4975.
8. O. A. El Seoud, A. Koschella, L. C. Fidale, S. Dorn and T. Heinze, *Biomacromolecules*, 2007, **8**, 2629–2647.
9. K. Pal, B. Behera, S. Roy, S. Sekhar Ray and G. Thakur, *Soft Mater.*, 2013, **11**, 125–142.
10. S. S. Silva, J. F. Mano and R. L. Reis, *Crit. Rev. Biotechnol.*, 2010, **30**, 200–221.
11. M. Dash, F. Chiellini, R. M. Ottenbrite and E. Chiellini, *Prog. Polym. Sci.*, 2011, **36**, 981–1014.
12. S. K. Shukla, A. K. Mishra, O. A. Arotiba and B. B. Mamba, *Int. J. Biol. Macromol.*, 2013, **59**, 46–58.
13. E. Mirzaei, B. A. Ramazani, S. A. M. Shafiee and M. Danaei, *Int. J. Polym. Mater. Polym. Biomater.*, 2013, **62**, 605–611.
14. J. Zhang, W. Xia, P. Liu, Q. Cheng, T. Tahirou, W. Gu and B. Li, *Mar. Drugs*, 2010, **8**, 1962–1987.
15. L. Heath, L. Zhu and W. Thielemans, *ChemSusChem*, 2013, **6**, 537–544.
16. S. Takeshita and S. Yoda, *Chem. Mater.*, 2015, **27**, 7569–7572.
17. F. Quignard, R. Valentin and F. Di Renzo, *New J. Chem.*, 2008, **32**, 1300–1310.
18. K. Rinki, P. K. Dutta, A. J. Hunt, D. J. Macquarrie and J. H. Clark, *Int. J. Polym. Mater. Polym. Biomater.*, 2011, **60**, 988–999.
19. B. B. Ding, J. Cai, J. C. Huang, L. N. Zhang, Y. Chen, X. W. Shi, Y. M. Du and S. Kuga, *J. Mater. Chem.*, 2012, **22**, 5801–5809.

20. P. Hao, Z. Zhao, Y. Leng, J. Tian, Y. Sang, R. I. Boughton, C. P. Wong, H. Liu and B. Yang, *Nano Energy*, 2015, **15**, 9–23.
21. S. S. Silva, A. R. C. Duarte, A. P. Carvalho, J. F. Mano and R. L. Reis, *Acta Biomater.*, 2011, 7, 1166–1172.
22. Q. Ma, Y. F. Liu, Z. Dong, J. L. Wang and X. Hou, *J. Appl. Polym. Sci.*, 2015, 132.
23. Y. Tsutsumi, H. Koga, Z. D. Qi, T. Saito and A. Isogai, *Biomacromolecules*, 2014, **15**, 4314–4319.
24. A. R. C. Duarte, V. E. Santo, A. Alves, S. S. Silva, J. Moreira-Silva, T. H. Silva, A. P. Marques, R. A. Sousa, M. E. Gomes, J. F. Mano and R. L. Reis, *J. Supercrit. Fluids*, 2013, **79**, 177–185.
25. C. Tsioptsias, C. Michailof, G. Stauropoulos and C. Panayiotou, *Carbohydr. Polym.*, 2009, **76**, 535–540.
26. S. S. Silva, A. R. C. Duarte, J. F. Mano and R. L. Reis, *Green Chem.*, 2013, **15**, 3252–3258.
27. S. S. Silva, A. R. C. Duarte, J. M. Oliveira, J. F. Mano and R. L. Reis, *J. Bioact. Compat. Polym.*, 2013, **28**, 481–491.
28. J. Anthony, J. Brennecke, J. Holbrey, E. Maginn, P. Mantz, A. V. Trulove and T. Welton, in *Ionic Liquids in Synthesis*, ed. P. Wasserscheid and T. Welton, Wiley-VCH, 2002, pp. 41–126.
29. L. Liu, R. Wang, J. Yu, J. Jiang, K. Zheng, L. Hu, Z. Wang and Y. Fan, *Biomacromolecules*, 2016, **17**, 3773–3781.
30. G. L. Rorrer, T. Y. Hsien and J. D. Way, *Ind. Eng. Chem. Res.*, 1993, **32**, 2170–2178.
31. Y.-C. Kuo and C.-Y. Lin, *Biotechnol. Bioeng.*, 2006, **95**, 132–144.
32. Y. Kobayashi, T. Saito and A. Isogai, *Angew. Chem., Int. Ed.*, 2014, **53**, 10394–10397.
33. X. Chang, D. Chen and X. Jiao, *J. Phys. Chem. B*, 2008, **112**, 7721–7725.
34. L. Baldino, S. Concilio, S. Cardea, I. De Marco and E. Reverchon, *J. Supercrit. Fluid*, 2015, **103**, 70–76.
35. R. M. Obaidat, B. M. Tashtoush, M. F. Bayan, R. T. Al Bustami and M. Alnaief, *AAPS PharmSciTech*, 2015, **16**, 1235–1244.
36. M. Sachithanadam and S. C. Joshi, in *Silica Aerogel Composites: Novel Fabrication Methods*, Springer Singapore, 2016, DOI: 10.1007/978-981-10-0440-7_2, pp. 5–14.
37. Y. Wang, G. M. Xia, C. Wu, J. Sun, R. Song and W. Huang, *Carbohydr. Polym.*, 2015, **115**, 686–693.
38. M. R. Ayers and A. J. Hunt, *J. Non-Cryst. Solids*, 2001, **285**, 123–127.
39. Q. Hou, D. W. Grijpma and J. Feijen, *J. Biomed. Mater. Res., Part B*, 2003, **67B**, 732–740.
40. J. A. Sirvio, M. Visanko and H. Liimatainen, *RSC Adv.*, 2016, **6**, 56544–56548.
41. A. A. Alhwaige, T. Agag, H. Ishida and S. Qutubuddin, *RSC Adv.*, 2013, **3**, 16011–16020.
42. H. Peng, J. Wu, Y. Wang, H. Wang, Z. Liu, Y. Shi and X. Guo, *Appl. Phys. A: Mater. Sci. Process.*, 2016, **122**, 516.

43. L. Carro, E. Hablot and T. Coradin, *J. Sol-Gel Sci. Technol.*, 2014, **70**, 263–271.
44. I. Garcia, I. Azcune, P. Casuso, P. M. Carrasco, H. J. Grande, G. Cabanero, D. Katsigiannopoulos, E. Grana, K. Dimos, M. A. Karakassides, I. Odriozola and A. Avgeropoulos, *J. Appl. Polym. Sci.*, 2015, 132.
45. C. J. Yao, X. Liu and W. M. Risen, in *Aerogels Handbook*, ed. M. A. Aegerter, N. Leventis and M. M. Koebel, Springer New York, NY, 2011, DOI: 10.1007/978-1-4419-7589-8_18, pp. 385–401.
46. K. Molvinger, F. Quignard, D. Brunel, M. Boissière and J.-M. Devoisselle, *Chem. Mater.*, 2004, **16**, 3367–3372.
47. J. Wang, Q. Zhou, D. Song, B. Qi, Y. Zhang, Y. Shao and Z. Shao, *J. Sol-Gel Sci. Technol.*, 2015, **76**, 501–509.
48. C.-C. Ji, M.-W. Xu, S.-J. Bao, C.-J. Cai, Z.-J. Lu, H. Chai, F. Yang and H. Wei, *J. Colloid Interface Sci.*, 2013, **407**, 416–424.
49. A. El Kadib and M. Bousmina, *Chem. – A Eur. J.*, 2012, **18**, 8264–8277.
50. A. E. Kadib, K. Molvinger, T. Cacciaguerra, M. Bousmina and D. Brunel, *Microporous Mesoporous Mater.*, 2011, **142**, 301–307.
51. A. E. Kadib, K. Molvinger, M. Bousmina and D. Brunel, *J. Catal.*, 2010, **273**, 147–155.
52. S. Frindy, A. Primo, A. e. k. Qaiss, R. Bouhfid, M. Lahcini, H. Garcia, M. Bousmina and A. El Kadib, *Carbohydr. Polym.*, 2016, **146**, 353–361.

CHAPTER 3

Cellulose Based Aerogels: Processing and Morphology

YUN LU,*[a] RU-NAN GAO,[b] SHAOLIANG XIAO,[b] YAFANG YIN,[a] QI LIU[c,d] AND JIAN LI[b]

[a] Research Institute of Wood Industry, Chinese Academy of Forestry, Beijing 100091, China; [b] Material Science and Engineering College; Key Laboratory of Bio-based Material Science and Technology, Ministry of Education, Northeast Forestry University, Harbin 150040, China; [c] National Engineering Laboratory for Crop Efficient Water Use and Disaster Mitigation and Key Laboratory of Dryland Agriculture, Ministry of Agriculture, Institute of Environment and Sustainable Development in Agriculture, Chinese Academy of Agricultural Sciences, Beijing 100081, China; [d] TERRA Research Center and Laboratory of Gastronomical Science, Gembloux Agro-Bio Tech-University of Liège, Passage des Déportés 2B-5030 Gembloux, Belgium
*Email: y.lu@criwi.org.cn

3.1 Introduction of Cellulose-based Aerogels

Cellulose, the most abundant biopolymer obtained from terrestrial plants and many aquatic species, has the features of being low cost, degradable and being biocompatible. Cellulose is the major component of plant cell walls and is usually combined with hemicellulose, pectin and lignin. Native cellulose-based nanomaterials, such as nanofibrillated cellulose (NFC), nanocrystalline cellulose (NCC), 2,2,6,6-tetramethypiperidine-1-oxy radical (TEMPO)-mediated oxidised cellulose (TO-NFC) and regenerated nanocellulose (RNC), have attracted increasing attention in nano-manufacturing owing to their great

Green Chemistry Series No. 58
Biobased Aerogels: Polysaccharide and Protein-based Materials
Edited by Sabu Thomas, Laly A. Pothan and Rubie Mavelil-Sam
© The Royal Society of Chemistry 2018
Published by the Royal Society of Chemistry, www.rsc.org

performance as 'green' nano-building blocks. Cellulose hydrogels/aerogels are flexible and highly porous materials with many potential applications, such as for tissue engineering,[1] controllable delivery,[2] blood purification,[3] sensing,[4] agriculture,[5] water purification[6] and thermal superinsulation.[7]

Research published on cellulose hydrogels/aerogels has focused on assembly strategies, chemical modification, cellulose-derived structures and applications. Studies show that the fine structure of cellulose-based aerogels can be controlled by modifying the building blocks, density, surface charge, coagulation bath composition, crosslinking agents and drying process. In particular, cellulose-derived carbon aerogels can inherit the physical morphology of cellulose precursors after being carbonised. This feature allows cellulose hydrogels/aerogels to serve as excellent templates for creating functional hierarchical 3D nanostructures with large porosity, and as promising resources of macroscopic carbon nanomaterials. Furthermore, various composite systems in which the cellulose matrix is functionalised with an inorganic nano-sized architecture have been developed to create specific properties. Previous studies have developed various cellulose aerogel-like materials, including soft flexible foams based on entangled cellulose I nanofibers obtained from plant cellulose,[8] highly porous hard aerogels prepared by cellulose dissolution/regeneration in various solutions followed by regeneration and controlled drying,[9,10] bacterial cellulose gels produced directly by *Gluconacetobacter xylinum*[11] and monolithic gels synthesised by crosslinking cellulose derivatives *via* the sol–gel route.[7] This chapter reviews the recent findings and applications of native, modified and recycled cellulose-based functional hydrogels and aerogels (Figure 3.1).

Figure 3.1 Summary of cellulose-based aerogels. (A) SEM of cellulose aerogel, (B) inorganic–cellulose hybrid illustration, (C) ultralight cellulose aerogels, (D) TEM of 1D nanocellulose building blocks, (E) cellulose-derived carbon aerogel, and (F) regenerated cellulose hydrogels.

3.2 Cellulose Hydrogels/Aerogels Assembled from 1D Nanocellulose Building Blocks from Native Cellulose

3.2.1 Chemical Pre-treatment of Native Cellulose

Chemical processes have been implemented to obtain purified precursors of cellulose-based nanomaterials. Methods depend on the different components of raw materials. For cellulose from terrestrial plants, lignin, hemicellulose and pectin must be eliminated. First, pectic substances are removed using a mixture of benzene and ethyl alcohol. Then, delignification is achieved by oxidation with acidified sodium chlorite. Finally, hemicellulose is removed by alkali treatment (KOH/NaOH). Different from the cellulose obtained from terrestrial plants, the cellulose from algae lacks lignin and hemicellulose, but contains proteins and inorganic salt. Lipid can be extracted from algae by using a mixture of acetate and ethanol. Polysaccharides are removed by sodium carbonate under heating conditions. The remaining proteins and inorganic salts are eliminated by potassium and hydrochloric acid, respectively.[12]

3.2.2 One-dimensional Nanocellulose Building Blocks from Native Cellulose and Recycled Cellulose

Given their high aspect ratio, good mechanical properties and modifiability, nanocellulose building blocks are ideal for creating 3D network aerogels. Therefore, researchers have developed different strategies to isolate nanocellulose from purified raw materials and to fabricate nanocellulose building blocks. NFC and NCC can be obtained by mechanical treatments, such as homogenisation and ultrasonication. Nanocellulose obtained using mechanical treatments has a wide size distribution and large energy consumption. To reduce the energy consumption, TEMPO-mediated oxidation is combined with mild mechanical treatment. Nanocellulose obtained using this method is designated as TO-NFC. Compared with NFC and NCC, TO-NFC possesses a uniform width range of 3–5 nm and can disperse as individual nanofibrils in water.[13] RNC, which is the most common building block for recycled aerogels, can be obtained by directly dissolving waste paper into ionic liquids (ILs) followed by adding an anti-solvent.[14]

3.2.3 Assembled Cellulose Hydrogels from Nano-building Blocks

Assembly is crucial in the construction of cellulose aerogels. This process refers to the arrangement of individual cellulose building elements into a hierarchical structure (bottom-up approach). Cellulose nanofibers with abundant hydroxyls are excellent 1D materials for preparing 3D aerogels and their derivative functional products due to their high elastic modulus, large specific area and high aspect ratio.

Self-assembly without manual intervention spontaneously organises small building blocks into the nanostructure through interactions such as hydrogen bonding, Van der Waals force and weak ionic bonding. This method is a common and effective means of constructing the original interconnected network of cellulose hydrogel and aerogels. During dissolution–regeneration, the cellulose monomolecular chain serves as the elementary building block. When the cellulose regenerates from the anti-solvent, cellulose monomolecular chains with a diameter of 5 Å interconnect with the fibril network. When the solvent is fully exchanged with the anti-solvent, cellulose is completely precipitated into the hydrogel with a dense fibril network structure. The entire process is controlled by the mutual diffusion between the solvent and the anti-solvent. A nanofibril network is formed through the hydrogen bonding between the cellulose molecular chains. During aerogel preparation, drying affects the porosity and specific area of the final cellulose aerogel because of the ice crystal growth during freezing, or because of the force caused by capillary pressure. Nevertheless, self-assembly is the process responsible for the construction of a 3D pristine network.

Aside from self-assembly, interface assembly has also been developed to fabricate novel cellulose based hydrogels.[15] He *et al.* prepared tubular multi-layered hydrogels and onion-like hydrogels through cellulose assembly on the solid–liquid interface.[16] In this process, a template core is alternately immersed in cellulose solution (dissolved using a NaOH-urea system) and acetic acid. The solid–liquid interface assembly between the template core and the cellulose solution is caused by the acid-induced disruption of the inclusion complex around the cellulose. This phenomenon exposes the hydroxyl groups of the cellulose and induces the self-aggregation of cellulose to form multi-layered hydrogels. The layer thickness and inter-layer space could be controlled by adjusting the cellulose concentration and the contact time of the solid–liquid interface (Figure 3.2).

A polyelectrolyte multi-layer incorporated with cellulose has been synthesised through electrostatic layer-by-layer (LBL) self-assembly; this technique facilitates the self-assembly of oppositely charged polyelectrolytes and cellulose *via* electrostatic interactions.[17] Composite multi-layer materials have been prepared using this method. Cranston *et al.* successfully fabricated multi-layered films of cationic poly(allylamine hydrochloride) and anionic cellulose nanocrystals (sulphuric acid hydrolysed from cotton) by spin-coating and solution dipping.[18] Cellulose sulphate and multi-layered polyelectrolyte have also been incorporated in other composites. Aulin *et al.* fabricated cationic/anionic microfibrillated cellulose (MFC) multi-layers and polyethyleneimine/anionic MFC by using the LBL technique.[19] Wågberg *et al.* realised the build-up of multi-layers on a silica substrate by using MFC as an anionic colloid.[20] Different assembly methods provide versatile ways to prepare cellulose-based aerogels for different applications. However, the

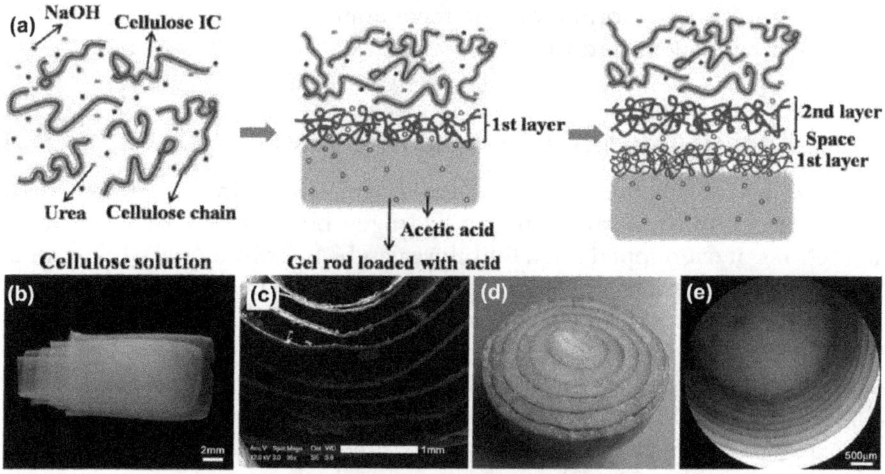

Figure 3.2 (a) Schematic model describing the formation of the multi-layered cellulose hydrogel *via* fast solution–solid interfacial interaction. (b) Multi-layered tubular cellulose hydrogel. (c) Cross-section SEM images of freeze-dried gels. (d) Anion and (e) multi-layered cellulose hydrogels. Reprinted with permission from M. He, Y. Zhao, J. Duan, Z. Wang, Y. Chen, L. Zhang, Fast contact of solid–liquid interface created high strength multi-layered cellulose hydrogels with controllable size, *ACS Applied Materials & Interfaces*, 2014, **6**(3), 1872–1878, Copyright 2014 American Chemical Society.

precise mechanisms underlying the assembly processes need further investigation. Moreover, new strategies should be developed to design novel functional cellulose-based hydrogels and aerogels for a wide range of applications.

3.3 Cellulose Aerogels Regenerated from Dissolving Cellulose

Researchers have explored the intractable dissolution of cellulose as a novel method to solve insolubility and have achieved great progress. Non-derivating solvent systems, such as amine oxide, ILs, aqueous alkali solutions and organic solvents in the presence of an inorganic salt, have been developed to dissolve cellulose without chemical modification pre-treatments.[21,22] The development of solvents for cellulose dissolution has propelled the extensive investigation on the fabrication of regenerated cellulose hydrogels and aerogels from corresponding solvent systems. Cellulose-based hydrogels can appear in different forms, such as slabs, membranes, beads and microspheres, by modifying the regeneration process.[23] Regeneration after cellulose dissolution causes molecular-level reorganisation, which is induced by adding anti-solvents or directly placing the cellulose solution into an anti-solvent bath. The effect of various anti-solvents on the morphology of regenerated cellulose during regeneration must be investigated to design structured materials precisely.

3.3.1 *N*-methylmorpholine-*N*-oxide (NMMO)-H_2O

The interaction between cellulose and *N*-methylmorpholine-*N*-oxide (NMMO) leads to the formation of a hydrogen bond complex that contains interactions of overlapped ions. Cellulose can be dissolved only when a small amount of water exists in the composite systems. The regeneration of cellulose fibre from NMMO solution can be achieved by adding water, the most commonly used anti-solvent for the industrial production of Lyocell fibres. Meanwhile, different anti-solvents, such as alcohols with different molecular weights, improve the properties of Lyocell fibres and the derived materials.[24] Liebner *et al.* prepared cellulose aerogels from NMMO solution with ethanol, dimethyl sulfoxide (DMSO), or mixtures of both as the anti-solvents to compare aerogels regenerated with H_2O.[25] The ESEM image in Figure 3.3 shows that ethanol-regenerated aerogels possess a more intact pore structure than water-regenerated ones. Moreover, ethanol-regenerated aerogels have a smaller specific density than water-regenerated ones, indicating the higher porosity of the former compared with the latter type of aerogels. However, the morphology of aerogels is affected not only by the type of anti-solvent used, but also by the drying technique applied during fabrication. This topic will be further discussed in the latter part of this chapter. DMSO, which presents high miscibility with a supercritical drying (SCD) medium and high recoverability, is another suitable anti-solvent for aerogels prepared from NMMO solution systems. DMSO can also completely remove the by-products and stabilisers during drying.[25]

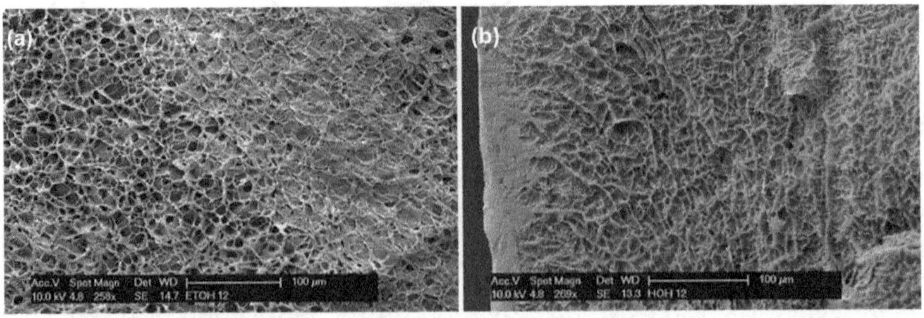

Figure 3.3 (a) ESEM images of cellulose aerogel, intact pore structure after ethanol regeneration and drying with super CO_2. (b) Partially collapsed pore structure after water regeneration, solvent exchange (ethanol) and $scCO_2$ drying.
Reproduced from ref. 25, http://dx.doi.org/10.1155/2007/7372, under the terms of the CC BY 3.0 licence, https://creativecommons.org/licenses/by/3.0/.

3.3.2 Ionic Liquids

Ionic liquids are salts comprised of large dissymmetrical organic cations and relatively small inorganic/organic anions, which melt below 100 °C.[26,27] Anions of ILs can form hydrogen bonds with the hydroxyl groups of cellulose, which destroy the hydrogen bonding networks and thus dissolve the cellulose macromolecule.[28,29] Cellulose can be regenerated from an IL solution by adding water or other anti-solvents, including ethanol and acetone. After regeneration, hydrogels composed of cellulose II are obtained. Polymerisation and polydispersity are almost the same with pristine cellulose. However, the morphology of cellulose significantly changes after regeneration, and regenerated nanofibrils are fused into a relatively homogeneous macrostructure.[30]

Jin *et al.* prepared cellulose-based aerogels from waste paper. The waste paper was dissolved in AMImCl and then regenerated with water.[14] After freeze drying, recycled aerogels were obtained. Subsequently, trimethylchlorosilane was deposited onto the aerogels. The aerogel exhibited a good absorption performance for oils and some organic solvents. Lu *et al.* prepared highly porous lignocellulose aerogels by direct dissolving wood in AMImCl, subsequent freezing–thawing treatment, regeneration and critical CO_2 drying.[31,32] Their results indicated that AMImCl is an effective solvent for the direct dissolution of wood materials.

3.3.3 NaOH-related Systems

Kamida reported that a weak layer of bound water that is surrounded by free water exists around OH^- and Na^+ in a NaOH aqueous solution.[33] During immersion in NaOH aqueous solution, the water in the outer layer of hydrate associates tightly with the hydroxyl groups in the amorphous region of cellulose; in addition, this water is loose due to the attraction of cellulose. Unimolecular water is free to permeate cellulose, leading to the destruction of the intramolecular hydrogen bond in the amorphous region. Zhang conducted a series of studies on cellulose dissolution in an alkali/urea (thiourea) aqueous solution and developed various anti-solvents for gel regeneration, including H_2SO_4, HOAc, H_2SO_4/Na_2SO_4, Na_2SO_4, $(NH_4)_2SO_4$, H_2O, C_2H_5OH and $(CH_3)_2CO$.[34] Structural changes in the presence of different anti-solvents were investigated. A homogeneous porous structure can be constructed in both the free surface and the fracture surface (cross section). In the free surface, membranes regenerated from acids exhibit a relatively small pore size and narrow distribution compared with those regenerated from water and salts. Meanwhile, membranes regenerated from organic anti-solvents possess a relatively large pore size and wide distribution. Compared with the free surface, the fracture surface obtains a denser structure, slightly smaller pore size and narrower distribution because of the phase separation from the surface to the back in the gel body. Moreover, membranes coagulated with acidic and salt aqueous solutions exhibit a greater tensile strength and elongation at break, than those with water and organic coagulants.[35] During coagulation, cellulose molecules form a physical cross-linkage. Recently, Zhao *et al.* have constructed double-cross-linked cellulose hydrogels by adding epichlorohydrin as the cross-linker into a cellulose/LiOH/urea solution with sequential chemical and physical cross-linking.[36] The novel cellulose hydrogel presents superior mechanical properties, with tensile strength, elongation at break and fracture energy as high as 2.7 MPa, 81% and 0.85 MJ m^{-3}, respectively.

Li *et al.* analysed cellulose hydrogels and aerogels regenerated from the powerful and environmentally-friendly NaOH/PEG solvent system.[37–40] In this system, hydrochloric acid is commonly used as the anti-solvent for the regeneration of cellulosic aerogels.[41] Macro images of regenerated cellulose hydrogel from powerful NaOH/PEG solution and the freeze-dried aerogels are presented in Figure 3.4, and the microstructure is presented in Figure 3.4e and f.[40]

3.3.4 LiCl/DMAc

The dissolution mechanism between cellulose and LiCl/DMAc is due to chloride ions (Cl^-) being freed from the macroactions formed by the complexation between lithium cations (Li^+) and the carbonyl oxygen of DMAc. The Cl^- ions are liable to form hydrogen bonds with cellulose hydroxyl protons and lead to the formation of competitive hydrogen bonds and disruption of the existing intermolecular hydrogen bonded structure in cellulose, which may cause the dissolution of cellulose.

Oliveira and Glasser prepared cellulosic hydrogels in bead form by dissolving cellulose in a LiCl/DMAc solution and adding to a regeneration bath of azetotropic methanol and isopropanol.[23] Water-miscible organic solutions, such as acetone, methanol, ethanol, 1-propanol, acetonitrile, tetrahydrofuran and DMSO, were chosen for coagulation. The physical properties of the obtained aerogels are affected by different types of anti-solvents. For instance, the transmittance of hydrogels reaches a maximum of 97% when the content of acetone is about 60% and then drops gradually. When other anti-solvents are used, a high hydrogel transmittance of over 90% is obtained. The density and tensile strength of the hydrogels increases with increasing acetone concentration.

3.4 Drying Strategies of Cellulose Aerogels from Hydrogels

(Ligno-) cellulose aerogels are converted from their corresponding hydrogel precursors. Drying techniques are performed to substitute gas for solvent media to obtain the final products. However, (ligno-) cellulose hydrogels featuring abundant nano-sized pores are filled with a liquid solvent that forms liquid-vapour menisci in the gel network. As the liquid evaporates, the

Figure 3.4 Preparation of cellulose aerogel. (a) Purified cellulose from waste wheat straw; (b) cellulose solution; (c) cellulose hydrogel; and (d) cellulose aerogel. (e) Low-magnification and (f) high-magnification SEM images of cellulose aerogels.
Reproduced from *Journal of Sol-Gel Science and Technology*, Preparation of mechanically strong and lightweight cellulose aerogels from cellulose-NaOH/PEG solution, **74**, 2015, C. Wan, Y. Lu, J. Cao, Q. Sun and J. Li, 256, © Springer Science + Business Media New York 2015, with permission of Springer.

meniscus caused by the intensive surface tension builds a capillary pressure gradient that can compel the extrusion and aggregation of cellulose building blocks, resulting in damage to the pore structure and reduction of the specific area.[42] These data indicate that the drying process is important in keeping the aerogel structure intact from its starting hydrogels. The various drying methods and aerogels obtained from different drying techniques are compared below.

3.4.1 Supercritical Drying

Supercritical drying is the mostly preferred technique for liquid removal from hydrogels.[43–46] When the drying medium is in the critical condition, the gas–liquid interface disappears, the surface tension no longer exists and the deformation and collapse of the network can be avoided. Critical point parameters of common fluids are listed in Table 3.1. Liquid carbon dioxide is extensively used to dry (ligno-) cellulose hydrogels because of its conditional advantages, such as low critical temperature and moderate critical pressure over other media.

3.4.2 Direct Freeze-drying

Direct freeze-drying is a low-temperature dehydration process that eliminates water by sublimation after freezing treatment. Freezing, primary drying and secondary drying are implemented successively during this process. Ice crystals formed during freezing serve as templates for pore walls of porous materials.[47,48] Amorphous ice crystals are sublimated during primary drying, and the temperature during this process is below the melting point of ice; temperatures below the glass-transition condition avoid the collapse of the pore structure.[49] Secondary drying begins after the ice crystals are completely removed, and the bound water absorbed on the surface of materials is eliminated.

3.4.3 Organic Solvent-mediated Freeze-drying

In this technique, the solvent in the hydrogels is replaced by *t*-BuOH, followed by freezing and sublimation. The freeze-drying of *t*-BuOH avoids the liquid–gas boundaries and fabricates cellulose aerogels with a comparable surface area. Thus, this technique is the most promising alternative to SCD.

3.4.4 Atmospheric Drying

Atmospheric drying of (ligno-) cellulose aerogels is still in the initial stages. Considerable shrinkage of the network caused by the liquid meniscus and pressure gradient is the main problem that impedes the development of atmospheric drying for aerogels. Under the same regeneration conditions, aerogels obtained through vacuum drying demonstrate severe shrinkage and

Table 3.1 Critical constants for some common fluids.

Fluid	Formula	T_c (°C)	P_c (MPa)
Methanol	CH_3OH	239.4	8.09
Ethanol	C_2H_5OH	243.0	6.3
Acetone	$(CH_3)_2O$	235.0	4.66
2-Propanol	C_3H_8O	235.0	4.7
Water	H_2O	374.1	22.04
Carbon dioxide	CO_2	31.0	7.37
Nitrous oxide	N_2O	36.4	7.24

collapse in comparison with aerogels obtained from supercritical CO_2 drying. ESEM images show that the capillary force during vacuum drying destroys the porous structure.

Drying techniques adjust and control the structure of cellulose aerogels. In this section, the material structures constructed by the above four drying techniques are compared. As shown in Figure 3.5a, mesopore aerogels with high porosities and large specific areas can be obtained by $ScCO_2$ drying. The *t*-BuOH drying can obtain similar results, but the porous structure produced is less uniform than that generated by $ScCO_2$ drying. Both of these drying techniques reduce the aggregation of NFCs and maintain the intactness of the inner structure of the aerogels obtained from corresponding hydrogels. The products of direct freeze-drying are foams with more macropores than micropores, but they are not technically aerogel materials. With direct drying, the bulky ice crystals formed during freezing serve as the macropore template for the cellular structure with open pores. When the NFC concentration is high, the cavum wall formed during the generation of NFC creates a fibril structure that is similar to the NFC membrane, as shown in Figure 3.5d. In addition, a low concentration of NFC causes relatively large spacing, where almost no entanglement structure forms and relative displacement is prone to occur, leading to a poor mechanical

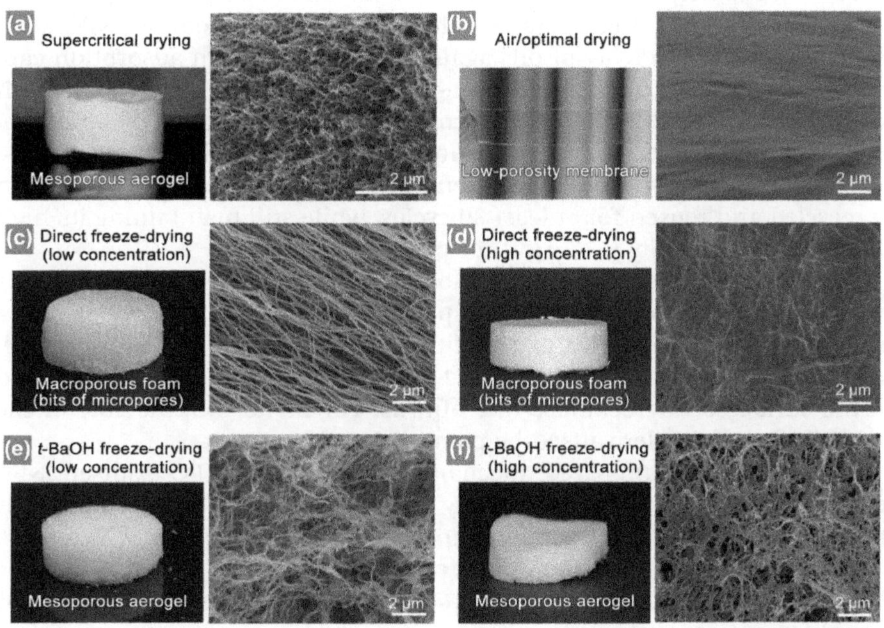

Figure 3.5 Images and hierarchical structures of various NFC porous materials prepared by different drying processes from NFC aqueous dispersion. (a) Mesoporous aerogel obtained by $ScCO_2$ drying; (b) low-porosity membrane obtained from air/optimal drying; (c) macroporous foam obtained by direct freeze-drying of low NFC concentration and (d) high NFC concentration; (e) mesoporous aerogel obtained by *t*-BaOH freeze-drying of low NFC concentration and (f) high NFC concentration.

performance on the hole. Figure 3.5b shows that a high-transmittance NFC membrane with a dense structure can be obtained by applying ambient drying. Thus, the morphology and 3D framework of NFC aerogels assembled from 1D nanoscale building blocks can be fabricated by various drying techniques, thereby extending the applications of these aerogels.

3.5 Cellulose-based Aerogels with a 3D Superstructure

3.5.1 Chemically Modified Aerogels

Various chemical modifications of cellulose aerogels have been developed to extend their intractable intrinsic properties. Sulphonation, esterification, etherification, oxidation, silylanisation and polymer grafting have been applied for the chemical modification of hydroxyl groups.

Chemically modified aerogels can be obtained by assembling cellulose fibre precursors from the above-mentioned methods. Tingaut *et al.* designed and prepared silylated nanocellulose sponges for selective oil–water separation.[50,51] Methyltrimethoxysilane solution (MTMS) serves as a silylating agent, and the cellulose nanofilaments are covered with polysiloxanes. The adsorption capacity of various organic solvents and oils can reach 102 $g\,g^{-1}$. Wang *et al.* also prepared hierarchically porous MTMS-modified NFC aerogels with a high selectivity of oils against water and a high adsorption capacity of 88–228 $g\,g^{-1}$ due to their ultra-low density (0.0024 $g\,cm^{-3}$) and high porosity (up to 99.84%).[52] The adsorbed oil can be readily and rapidly recovered by simple mechanical squeezing, while the superabsorbent could be reused immediately without the need for treatment. The superabsorbent can be recycled and reused for at least 30 cycles, while still maintaining its high oil adsorption capacity (137 $g\,g^{-1}$ for pump oil).

Fumagalli *et al.* fabricated cellulose tripalmitate microfibril aerogels through gas-phase esterification with palmitoyl chloride.[53,54] Samples with a gradually increasing degree of substitution (DS) from 0 to 2.36 were prepared. At a relatively low DS range (0.1–0.4), a hydrophobic skin is endowed on the surface of the cellulose microfibrils. For higher DS, the core of the microfibrils is fully derivatised by palmitoylation.

Feng and Hsieh prepared mechanically strong TEMPO-oxidised NFC aerogels with a high water absorbency of 104 $g\,g^{-1}$.[55] The obtained aerogels show a superior water adsorption capacity and shape recovery. Water 100 times heavier than that of the weight of the aerogels was adsorbed, and the shape of the aerogels can recover in 4 s. In addition, the aerogels can be reused for five repeated cycles.

3.5.2 Cellulose-based Inorganic Aerogels

With the increased emphasis on 'green' chemistry, researchers interested in sustainable processes have aimed to minimise the use of toxic chemicals, solvents and energy. Accordingly, inorganic replication of biotemplates with polysaccharides (cellulose, chitosan and chitin, agarose, starch and carrageenan) have been actively explored in metal oxide synthesis.[56,57] Using native cellulose as a template can effectively obtain inorganic materials in different sizes, dimensions and with controllable morphology. The template method is highly effective for the synthesis of multi–micro–nanostructure materials. This method involves three steps. The first is the preparation of template materials, the second is the synthesis of a multi-level structure, and the third is the template removal through chemical etching or calcination. This method allows the inorganic precursor to permeate the biomass materials. After precipitation, amorphous, polycrystalline or rare monocrystalline ceramics are formed. As shown in Figure 3.6, the template as a mould provides a closed space for mineralisation. Atomic layer deposition (ALD) is a sequential, self-limiting chemical vapour deposition method which enables the control of the deposition thickness by varying the number of cycles instead of the deposition time.[58,59] ALD is a facile method of preparing hollow inorganic nanotubes by coating nanocellulose aerogels with different oxide materials, followed by the temperature-induced decomposition of cellulose through calcination.[60]

Inorganic cellulose aerogels can be calcinated or chemically etched to obtain purely inorganic hollow nanotube aerogels. This type of inorganic aerogel has many advantages conferred by their 3D nanometric structure. The obtained inorganic aerogels have been considered for use in medical and life science devices.[50,61] In addition, the cellulose aerogel structure can be modified to produce inorganic functional aerogels with novel structures.

3.5.3 Cellulose-based Carbon Aerogels

Cellulose-based carbon aerogels inherit the physical morphology of cellulose aerogels after carbonisation. Carbon aerogels are composed of interconnected 3D network structures with desirable properties, including good thermal conductivities, high specific surface area, chemical inertness and outstanding mechanical properties. Carbon aerogels exhibit tremendous potential applications in supercapacitors and catalyst carriers. In fact, their low density, high porosity and inherent hydrophobicity render them a good degreasing material for oil–water separation. As-synthesised organic aerogels (or cryogels) may be converted into the carbon aerogel equivalent through controlled thermal annealing/carbonisation under an inert atmosphere (*e.g.* flowing Ar).[62,63] However, organic aerogels have many disadvantages, such as high density, brittleness, toxicity and environmental pollution. Cellulosic aerogels are excellent precursors for producing carbon aerogels because of their low cost, environmental friendliness and the process is scalable, allowing industrial applications. The obtained hierarchically porous carbon aerogels (with micro-, meso- and macro-porosity in one structure) have also been widely used for energy storage in supercapacitors and lithium–ion batteries.

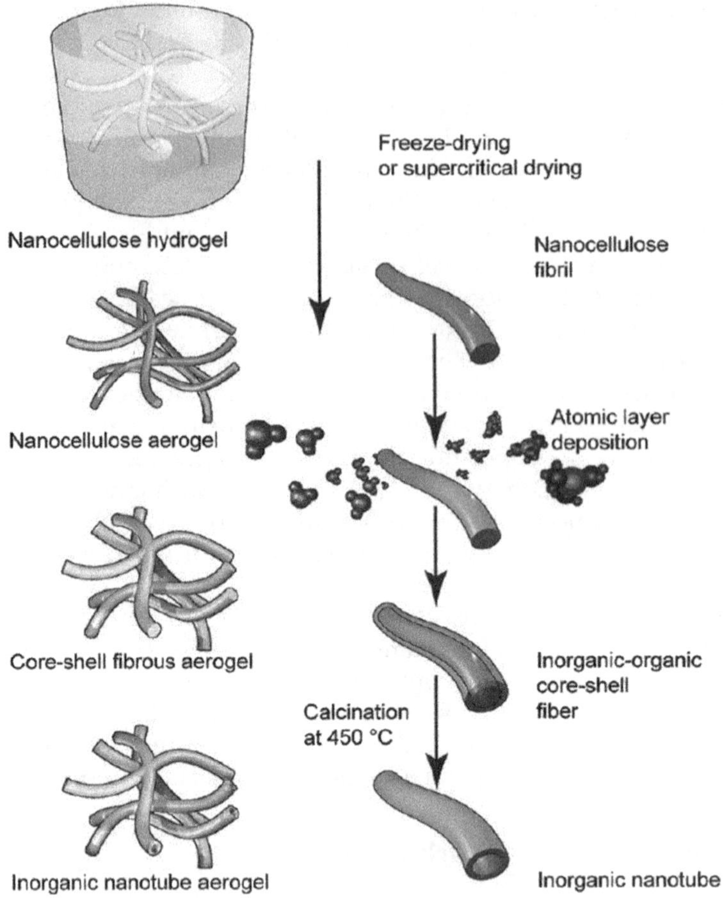

Figure 3.6 Schematic of the preparation of cellulose-based inorganic aerogels. Nanocellulose hydrogel is dried to an aerogel, which is then coated with inorganic oxides using ALD to form composite organic/inorganic nanofibers, and finally calcinated to inorganic hollow nanotubes. Reprinted with permission from J. T. Korhonen, P. Hiekkataipale, J. Malm, M. Karppinen, O. Ikkala and R. H. A. Ras, Inorganic Hollow Nanotube Aerogels by Atomic Layer Deposition onto Native Nanocellulose Templates, *ACS Nano,* 2011, 5(3), 1967–1974, Copyright 2011 American Chemical Society.

3.6 Concluding Remarks and Prospects

Cellulose-based aerogels from terrestrial plants, many aquatic species and recycled resources have emerged as an attractive 3D network for a range of porous structures. These materials, cellulose and lignocellulose, are not only biocompatible and earth abundant, but also have natural intrinsic structures for the assembly of functional structures. Several purification and assembly techniques have been demonstrated, but grand challenges hinder the acceleration of native and recycled cellulose resources toward tuneable 3D networks, especially for specific aerogel morphologies, such as uniform porous structures and hierarchically porous structures. These challenges include: (1) the highly efficient extraction of cellulose from various resources; (2) the assembly of cellulose molecules and 1D nano-building blocks to develop unique properties; and (3) the further modification of 3D cellulose scaffolds for emerging applications. With continued worldwide effort, new developments of native and recycled cellulose aerogels will provide the chemical, biological, physical and engineering communities with a virtual cornucopia of opportunities for new advancements and discoveries.

Acknowledgements

This work was financially supported by the National Key Research and Development Program of China [2017YFD0600204] and the National Natural Science Foundation of China [No. 31500468].

References

1. C. Vinatier, O. Gauthier, A. Fatimi, C. Merceron, M. Masson, A. Moreau, F. Moreau, B. Fellah, P. Weiss and J. Guicheux, *An Injectable Cellulose-based Hydrogel for the Transfer of Autologous Nasal Chondrocytes in Articular Cartilage Defects*, 2009.
2. C. Chang, B. Duan, J. Cai and L. Zhang, *Eur. Polym. J.*, 2010, **46**, 92.
3. S. H. Ye, J. Watanabe, Y. Iwasaki and K. Ishihara, *Biomaterials*, 2003, **24**, 4143.
4. A. Sannino, S. Pappadà, L. Giotta, L. Valli and A. Maffezzoli, *J. Appl. Polym. Sci.*, 2007, **106**, 3040.
5. S. M. Ibrahim, K. M. E. Salmawi and A. H. Zahran, *J. Appl. Polym. Sci.*, 2007, **104**, 2003.
6. N. T. Cervin, C. Aulin, P. T. Larsson and L. Wågberg, *Cellulose*, 2012, **19**, 401.
7. F. Fischer, A. Rigacci, R. Pirard, S. Berthon-Fabry and P. Achard, *Polymer*, 2006, **47**, 7636.
8. M. Paakko, J. Vapaavuori, R. Silvennoinen, H. Kosonen, M. Ankerfors, T. Lindstrom, L. A. Berglund and O. Ikkala, *Soft Matter*, 2008, **4**, 2492.
9. H. Jin, Y. Nishiyama, M. Wada and S. Kuga, *Colloids Surf., A*, 2004, **240**, 63.
10. J. Cai, S. Kimura, M. Wada, S. Kuga and L. Zhang, *ChemSusChem*, 2008, **1**, 149.
11. H. Sehaqui, M. Salajková, Q. Zhou and L. A. Berglund, *Soft Matter*, 2010, **6**, 1824.
12. W. Zhao, Y. Lu, H. Ouyang, X. Yan, R. Pan, D. Yang, X. She, J. Li and Y. Xia, *Sci. Technol. Rev.*, 2014, **32**, 34.
13. S. Fujisawa, Y. Okita and H. Fukuzumi, *Carbohydr. Polym.*, 2011, **84**, 579.
14. C. Jin, S. Han, J. Li and Q. Sun, *Carbohydr. Polym.*, 2015, **123**, 150.

15. J. Duan, R. Hou, X. Xiong, Y. Wang, Y. Wang, J. Fu and Z. Yu, *J. Mater. Chem. B*, 2012, **1**, 485.
16. M. He, Y. Zhao, J. Duan, Z. Wang, Y. Chen and L. Zhang, *ACS Appl. Mater. Interfaces*, 2014, **6**, 1872.
17. L. Sui, L. Huang, P. Podsiadlo, N. A. Kotov and J. Kieffer, *Macromolecules*, 2010, **43**, 9541.
18. E. D. Cranston and D. G. Gray, *Biomacromolecules*, 2006, **7**, 2522.
19. C. Aulin, E. Johansson, L. Wågberg and T. Lindström, *Biomacromolecules*, 2010, **11**, 872.
20. L. Wågberg, G. Decher, M. Norgren, T. Lindström, A. Mikael Ankerfors and K. Axnäs, *Langmuir*, 2008, **24**, 784.
21. S. Sen, J. D. Martin and D. S. Argyropoulos, *ACS Sustainable Chem. Eng.*, 2013, **1**, 858.
22. T. Rosenau, A. Potthast, I. Adorjan, A. Hofinger, H. Sixta, H. Firgo and P. Kosma, *Cellulose*, 2002, **9**, 283.
23. W. D. Oliveira and W. G. Glasser, *J. Appl. Polym. Sci.*, 1996, **60**, 63.
24. A. J. Uddin, A. Yamamoto, Y. Gotoh, M. Nagura and M. Iwata, *Text. Res. J.*, 2010, **80**, 1846.
25. F. Liebner, A. Potthast, T. Rosenau, E. Haimer and M. Wendland, *Res. Lett. Mater. Sci.*, 2007, **2007**, 4.
26. M. J. Earle and K. R. Seddon, *Pure Appl. Chem.*, 2000, **72**, 10.
27. N. V. Plechkova and K. R. Seddon, *Chem. Soc. Rev.*, 2008, **37**, 123.
28. H. Wang, G. Gurau and R. D. Rogers, *Chem. Soc. Rev.*, 2012, **41**, 1519.
29. J. Zhang, H. Zhang, J. Wu, J. Zhang, J. He and J. Xiang, *Phys. Chem. Chem. Phys.*, 2010, **12**, 1941.
30. A. Farrán, C. Chao, M. Sandoval, Y. Xu, L. Jian, M. J. Hernáiz and R. J. Linhardt, *Chem. Rev.*, 2015, 115.
31. J. Li, Y. Lu, D. Yang, Q. Sun, Y. Liu and H. Zhao, *Biomacromolecules*, 2011, **12**, 1860.
32. Y. Lu, Q. Sun, D. Yang, X. She, X. Yao, G. Zhu, Y. Liu, H. Zhao and J. Li, *J. Mater. Chem.*, 2012, **22**, 13548.
33. K. Kamida, K. Okajima, T. Matsui and K. Kowsaka, *Polym. J.*, 1984, **16**, 857.
34. C. Chang, L. Zhang, J. Zhou, L. Zhang and J. F. Kennedy, *Carbohydr. Polym.*, 2010, **82**, 122.
35. Y. Mao, J. Zhou, J. Cai and L. Zhang, *J. Membr. Sci.*, 2006, **279**, 246.
36. D. Zhao, J. Huang, Y. Zhong, K. Li, L. Zhang and J. Cai, *Adv. Funct. Mater.*, 2016, **26**, 6279.
37. C. Wan and J. Li, *Mater. Des.*, 2015, **83**, 620.
38. C. C. Wan and J. Li, *Carbohydr. Polym.*, 2016, **146**, 362.
39. C. Wan, Y. Lu, C. Jin, Q. Sun and J. Li, *Appl. Phys. A: Mater. Sci. Process.*, 2015, **119**, 45.
40. C. Wan, Y. Lu, Y. Jiao, J. Cao, Q. Sun and J. Li, *J. Sol-Gel Sci. Technol.*, 2015, **74**, 256.
41. J. Li, C. Wan, Y. Lu and Q. Sun, *Front. Agric. Sci. Eng.*, 2014, **1**, 46.
42. A. Pfenning, *Chem. Ing. Tech.*, 1995, **67**, 595.

43. E. J. Beckman, *J. Supercrit. Fluids*, 2004, **28**, 121.
44. Y. Kobayashi, T. Saito and A. Isogai, *Angew. Chem.*, 2014, **126**, 10394.
45. C. Rudaz, R. Courson, L. Bonnet, S. Calasetienne, H. Sallée and T. Budtova, *Biomacromolecules*, 2014, **15**, 2188.
46. J. D. Rusmirović, J. Z. Ivanović, V. B. Pavlović, V. M. Rakić, M. P. Rančić, V. Djokić and A. D. Marinković, *Carbohydr. Polym.*, 2017, **164**, 64.
47. C. Jiménez-Saelices, B. Seantier, B. Cathala and Y. Grohens, *Carbohydr. Polym.*, 2017, **157**, 105.
48. S. Flauder, T. Heinze and F. A. Müller, *Cellulose*, 2014, **21**, 97.
49. T. A. Jennings, *Lyophilization: Introduction and Basic Principles*, CRC Press, 1999.
50. P. Tingaut, T. Zimmermann and G. Sèbe, *J. Mater. Chem.*, 2012, **22**, 20105.
51. Z. Zheng, G. Sèbe, D. Rentsch, T. Zimmermann and P. Tingaut, *Geogr. Res.*, 2009, **26**, 2659.
52. A. M. Pourrahimi, D. Liu, V. Ström, M. S. Hedenqvist, R. T. Olsson and U. W. Gedde, *J. Mater. Chem. A*, 2015, **3**, 17190.
53. M. Fumagalli, D. Ouhab, S. M. Boisseau and L. Heux, *Biomacromolecules*, 2013, **14**, 3246.
54. M. Granström, M. K. N. Pääkkö, H. Jin, E. Kolehmainen, I. Kilpeläinen and O. Ikkala, *Polym. Chem.*, 2011, **2**, 1789.
55. J. Feng and Y. L. Hsieh, *J. Mater. Chem. A*, 2013, **2**, 350.
56. R. A. Caruso and M. Antonietti, *Chem. Mater.*, 2001, **13**, 3272.
57. J. Huang and Y. Gu, *Curr. Opin. Colloid Interface Sci.*, 2011, **16**, 470.
58. M. Leskelä and M. Ritala, *Angew. Chem.*, 2004, **35**, 5548.
59. R. L. Puurunen, *J. Appl. Phys.*, 2005, **97**, 121301.
60. J. T. Korhonen, P. Hiekkataipale, J. Malm, M. Karppinen, O. Ikkala and R. H. Ras, *ACS Nano*, 2011, **5**, 1967.
61. D. Klemm, F. Kramer, S. Moritz, T. Lindström, M. Ankerfors, D. Gray and A. Dorris, *Angew. Chem., Int. Ed.*, 2011, **50**, 5438.
62. J. Lee, J. Kim and T. Hyeon, *Adv. Mater.*, 2006, **18**, 2073.
63. C. Moreno-Castilla and F. J. Maldonado-Hódar, *Carbon*, 2005, **43**, 455.

CHAPTER 4

Starch Based Aerogels: Processing and Morphology

SIJI K. MARY*[a,b] AND LALY A. POTHAN[a]

[a] Department of Chemistry, Bishop Moore College, Mavelikara, Kerala, India; [b] Department of Chemistry, CMS College, Kottayam, Kerala, India
*Email: sijikmary2010@gmail.com

4.1 Introduction

The last decade has seen the development of green materials, intended to reduce the impact from humans on the environment. Starch, as an agro-sourced polymer has received much attention.[1] The concept of polysaccharides as aerogel-forming materials is relatively novel and has many advantageous aspects in food and non-food areas. Recently, aerogels based on natural or synthetic polymers, called polymer or organic aerogels, have been widely explored due to their porous structures and unique properties, such as high specific surface area, low density, low thermal conductivity and dielectric constant.[2] In this chapter we will discuss the different processing methodologies for the preparation of starch-based aerogels and also the morphology of aerogels.

4.2 Starch

Starch is the most abundant plant polysaccharide after cellulose and the hemicelluloses, with numerous applications in different industries, including as a functional and structural component of various important food products.

Green Chemistry Series No. 58
Biobased Aerogels: Polysaccharide and Protein-based Materials
Edited by Sabu Thomas, Laly A. Pothan and Rubie Mavelil-Sam
© The Royal Society of Chemistry 2018
Published by the Royal Society of Chemistry, www.rsc.org

Starch is a heterogeneous material containing two microstructures—linear (amylose) and branched (amylopectin). Amylopectin is the component of the crystalline parts of starch, with a short linear backbone also comprised of α-(1→4)-linked glucopyranose units, but highly branched by side chains (similar glucosyl chains) connected to the backbone by α-(1→6) bonds (Figure 4.1).

The linear structure of amylose makes its behaviour more closely resemble that of conventional synthetic polymers. Depending on its source and the processing conditions employed during its extraction, the molecular weight of amylose is 10 times higher than conventional synthetic polymers.

Although starch is abundant, it is not as widely studied as an aerogel matrix as alginate and cellulose; however, some studies have been published, which we have reviewed here.

4.3 Processing Techniques for the Preparation of Starch Based Aerogels

The processing steps needed for the production of polysaccharide-based aerogels are summarized in Figure 4.2 and described in the following subsections. Briefly, aerogel processing starts with the formation of a gel from an aqueous solution (the sol–gel process, described in step 1 in Figure 4.2), in example, a hydrogel. The next step is the replacement of the water filling the pores of the gel structure by ethanol (the solvent exchange is described in step 2 in Figure 4.2) to lead to an alcogel. Finally, ethanol is extracted from the gel by super critical drying (SCD) (step 3 in Figure 4.2) and the aerogel end material is obtained.

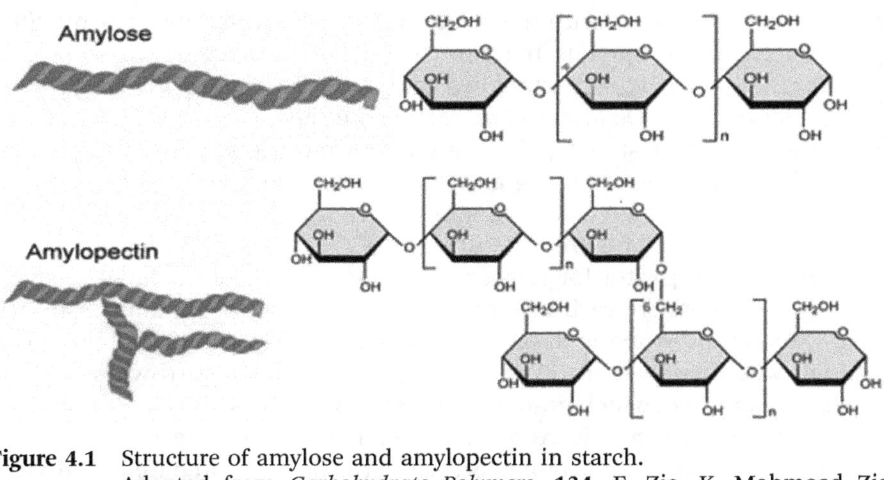

Figure 4.1 Structure of amylose and amylopectin in starch.
Adapted from *Carbohydrate Polymers*, **134**, F. Zia, K. Mahmood Zia, M. Zuber, S. Kamal and N. Aslam, Starch based polyurethanes: A critical review updating recent literature, 784–798, Copyright 2015, with permission from Elsevier.[3]

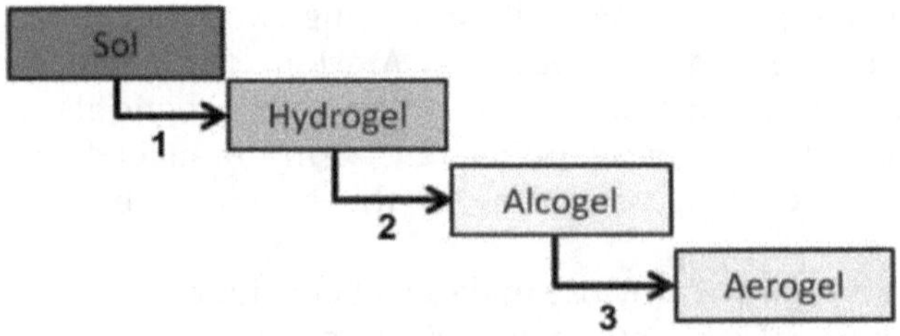

Figure 4.2 Schematic depiction of the supercritical drying process for the preparation of aerogels: 1: Sol–gel transition; 2: Solvent exchange from water to ethanol; 3: Supercritical drying of gels.
Adapted with permission from Garcia-Gonzalez *et al.*, 2011.[8]

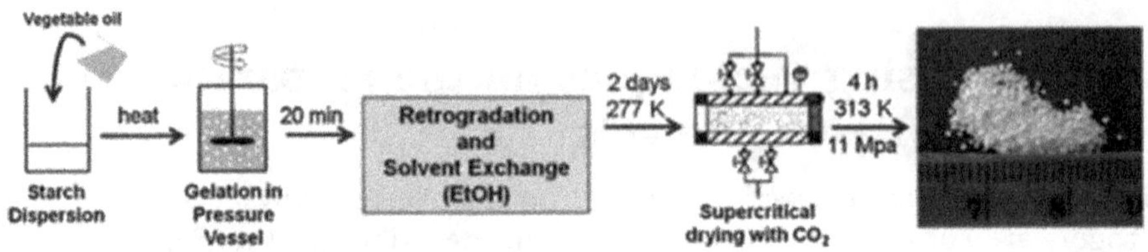

Figure 4.3 Diagram of the processing method used for the preparation of starch aerogel microspheres.
Adapted from *Carbohydrate Polymers*, **88**, C. A. García-González, J. J. Uy, M. Alnaief and I. Smirnova, Preparation of tailor-made starch-based aerogel microspheres by the emulsion-gelation method, 1378–1386, Copyright 2012, with permission from Elsevier.[6]

4.3.1 Sol–Gel Process: Gelation

Aerogel processing starts with the gelation of starch, which involves melting of the starch in an aqueous medium to induce changes in the structure (gelatinization step) and rearranging the structure during a cooling step (retrogradation step). De Marco *et al.* carried out the procedure of gelation by shaking solutions of starch (with starch concentrations equal to 5, 10 or 15% wt) in distilled water; using a magnetic stirrer (about 24 h).[4] Then, the solutions were heated up to 110 °C and poured into cylindrical moulds with an internal diameter of 2 cm and a height of 1 cm. The samples were then placed in the refrigerator for retrogradation at 4 °C for three days.

Chang *et al.* prepared carbon aerogels from starch aerogels using a sol–gel process, followed by drying at ambient pressure. In a typical synthesis, 10 g of starch was dissolved in 50 mL boiling water under stirring to form a translucent solution, which gradually transformed into white wet-gels within 0.5 h. After the wet-gels were aged for 24 h at 25 °C, the solvents were exchanged with absolute acetone. Starch aerogels were prepared by drying the acetone-gels at 50 °C under ambient pressure.[5] After retrogradation, starch particles were transferred to a fresh ethanol solution for solvent exchange as shown in Figure 4.3.

4.3.2 Gel–Aerogel Transition: Drying

For aerogels, the drying step is very important in order to preserve the highly porous structure. The gel, still embedded in the solvent, which consists mainly of alcohol with some water, is called an alcogel or hydrogel. To remove the liquid, several extraction methods can be employed including SCD (using CO_2, acetone, or ethanol), atmospheric drying, freeze-drying (Figure 4.4), microwave drying and vacuum drying.

4.3.3 Super Critical Drying

The SCD process is an alternative drying technique assisted by the use of supercritical fluids, usually super critical CO_2, that overcomes the problems encountered with conventional drying methods to preserve the high open porosity and superior textural properties of the wet gel in a dry form. The SCD process leads to the presence of supercritical fluid mixtures in the gel pores without remnants of any liquid phase. This drying procedure thus avoids the presence of any intermediate vapour liquid transition and surface tensions in the gel pores, preventing the porous gel structure from collapsing during solvent elimination. SCD can be classified into two types, depending on the contact regime between the gel and the supercritical fluid: drying of the gel with a continuous flow of super critical CO_2 throughout the process (called dynamic or continuous SCD), and in batches (called static or batch SCD).[7]

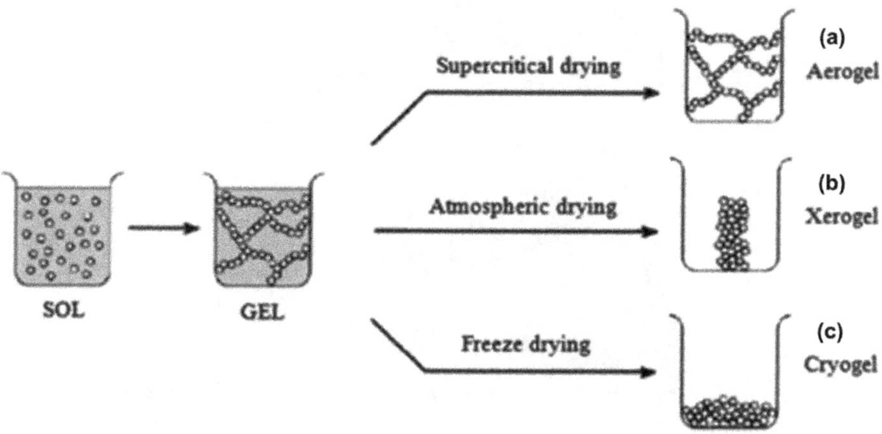

Figure 4.4 Methods of drying a wet gel to give (a) an aerogel; (b) a xerogel and; (c) a cryogel.
Adapted from *The Journal of Supercritical Fluids*, **66**, C. A. García-González, M. C. Camino-Rey, M. Alnaief, C. Zetzl and I. Smirnova, Supercritical drying of aerogels using CO2: Effect of extraction time on the end material textural properties, 297–306, Copyright 2012, with permission from Elsevier.[7]

SCD is a process by which the liquid in a matrix is transformed into a gas in the absence of surface tension and capillary stress, as shown in Figure 4.5 (upper curved arrow). This is achieved by placing the gel in an autoclave in which the temperature and pressure exceed the critical points. The surface tension between the solvent and solute molecules ceases in a supercritical fluid. The liquid phase in the gel is finally converted to a gas phase without destroying the solid delicate networks. The vapours are then slowly released from the autoclave, until the pressure in the autoclave reaches atmospheric pressure.

Carbon dioxide (CO_2) is usually chosen as the gas media due to its low supercritical temperature and nontoxicity. In the process of supercritical CO_2 drying, the liquid in the gel (usually ethanol) is first replaced by liquid CO_2 in a sealed vessel, then CO_2 is converted to a supercritical state by raising the temperature and the pressure, and finally the vessel is isothermally depressurized to release the CO_2 gas.

Traditional drying procedures, for example air drying, are not able to preserve the gel structure which leads to pore-collapse and massive shrinkage. This is because air-drying leads to liquid–vapour menisci in the pores of the gel, which recedes during the emptying of the pores of the wet gels. Upon solvent removal, the surface tension of the liquid contained in

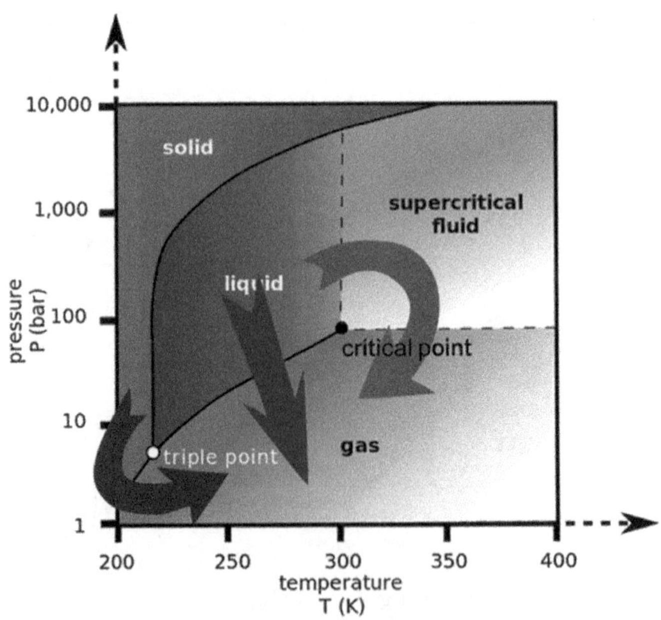

Figure 4.5 Three phase diagram: freeze drying route (lower curved arrow, solid to gas); conventional drying (straight arrow, liquid to gas); supercritical drying (upper curved arrow, liquid to gas *via* critical point).
Adapted from ref. 12 under the terms of the CC BY 3.0 licence, https://creativecommons.org/licenses/by/3.0/.

the gel nanopores will create a capillary pressure gradient in the pore walls that is able to collapse the pores.

The effect of the concentration of polysaccharide in the solution on the resulting aerogel properties was studied for corn starch.[8] The higher mechanical stability of the aerogels processed from hydrogels with a higher starch content highly influenced the volume reduction of the cylinders after the overall process (Figure 4.6). The total shrinkage after corn starch aerogel processing fell from 49 to 25% as the starch content in the original solution increased from 7 to 15%.

De Marco *et al.* produced different starch sources (maize, potato and wheat) aerogels using a supercritical carbon dioxide-based process.[4] First, an aqueous gel (hydrogel) was formed, then water was replaced by ethanol to form an alcogel, finally, a supercritical gel-drying step formed the aerogel. The different starch sources were processed and characterized from a macroscopic and a microscopic point of view. The analyses confirmed that the supercritical gel drying process is suitable to form polymeric nanostructured matrices which can be used for drug delivery systems.

Corn starch aerogel microspheres, a special class of nanoporous materials, were prepared by the combination of an emulsion-gelation method and supercritical drying without the use of chemical cross linkers. The effects of the gelation temperature (368, 393 and 413 K), oil-to-aqueous starch solution ratio (1 : 1, 2 : 1, and 3 : 1) and surfactant content (3, 6 and 10% (w/w)) on the textural and morphological properties of the aerogel material were studied.[6] Santos *et al.* also processed starch aerogels in the form of one-micron particles by a combination of emulsion-gelation and supercritical drying techniques. A white powder composed of starch aerogel particles was obtained with a particle size in between 200 nm to 1.5 μm

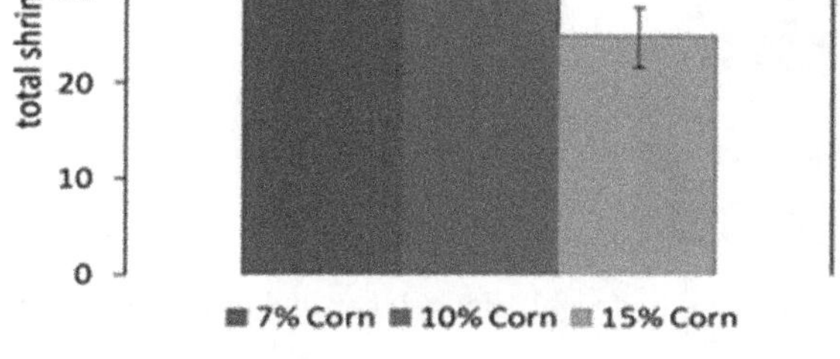

Figure 4.6 Total shrinkage of aerogel monoliths obtained from solutions of corn starch (7, 10 and 15 wt %).
Adapted with permission from Garcia-Gonzalez *et al.*, 2011.[8]

in range. The preparation of aerogels in the form of a fine powder could facilitate the incorporation of the material as an admixture in pharmaceutical and biomedical formulations.[9] Biodegradable nanoporous aerogels were obtained from wheat starch using a simple and green method based on supercritical carbon dioxide drying (Figure 4.7). The effects of processing parameters (temperature, wheat starch concentration and mixing rate during gelatinization; temperature, pressure, and flow rate of CO_2 during SCD) on the aerogel formation were investigated, and optimized for the highest surface area and smallest pore size of the aerogels. At the optimized conditions (optimized for the highest surface area and smallest pore size of the aerogels) the wheat starch aerogels had surface areas between 52.6–59.7 $m^2 g^{-1}$ and densities ranging between 0.05–0.29 $g cm^{-3}$. The average pore size of the starch aerogels was 20 nm. Starch aerogels were stable up to 280 °C. Due to the high surface area and nanoporous structure, the wheat starch aerogels are promising carrier systems for bioactives and drugs in the food and pharmaceutical industries.[10]

4.3.4 Ambient Pressure Drying

Ambient pressure drying is a promising technique that can be applied on the large scale for industrial purposes. This method relies on the passivation of the pore surface inside the gel, so as to impede further formation of new chemical bonds after condensation reactions, when the gel network is compressed under the drying stresses. At the end of the solvent evaporation

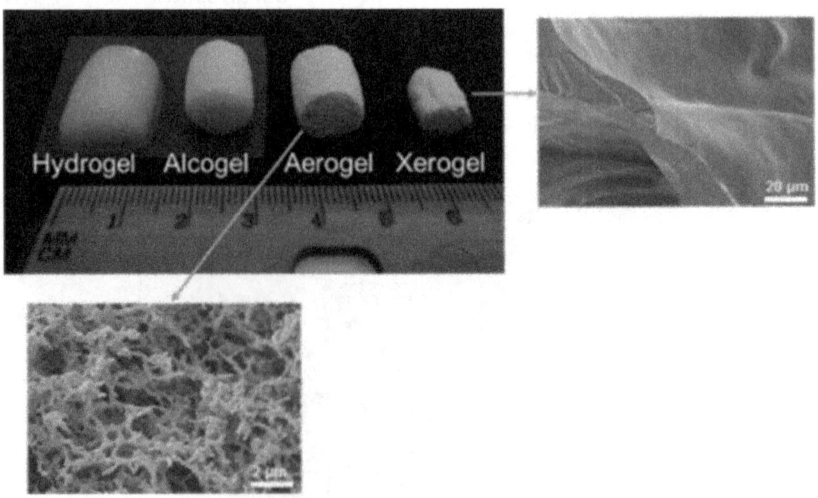

Figure 4.7 Pictures of the hydrogel, alcogel, aerogel and xerogel formed from wheat starch.
Adapted from *Carbohydrate Polymers*, **147**, A. Ubeyitogullari and O. N. Ciftci, Formation of nanoporous aerogels from wheat starch, 125–132, Copyright 2016, with permission from Elsevier.[10]

process, an aerogel monolith is no longer submitted to capillary stresses so that it can resume its wet size by a spring-back effect. Although ambient pressure drying is a simple and energy-saving process to prepare aerogels, simple evaporation of solvent from the hydrogels under ambient conditions may cause significant shrinkage or even form solid films that have no porosity.[2]

4.3.5 Freeze Drying

In general, the freeze-drying technique is a simple, more economical and environmentally friendly process than ambient pressure drying and SCD to obtain aerogels with good porous structures. In this method, liquid in the wet gel is first frozen and then evaporated by sublimation under low pressures. The materials are also termed cryogels. These freeze-dried "cryogels" show a porosity of 80% at the maximum and only half the inner surface of a comparable aerogel. Special precautions for freeze-drying are: long aging periods to stabilize the gel body, a solvent exchange to provide a low expansion coefficient and high sublimation pressure, and the addition of salts to achieve low freezing rates and freezing temperatures. Compared to aerogels prepared with supercritical CO_2 drying, cryogels require more macroporous aging periods to stabilize the gel body and a solvent exchange to provide a low expansion coefficient and structures with larger shrinkage and Brunauer–Emmett–Teller (BET) values.

4.4 Morphological Analysis

4.4.1 Porosity and Pore Distribution

The effect of the concentration of the polysaccharide in the sol (precursor concentration) on the resulting end aerogel properties was studied for two thermotropic gels: corn starch and high methoxyl (HM) pectin. Other processing parameters (*e.g.* temperature) may also strongly influence the aerogel end properties. The higher mechanical stability of the aerogels processed from hydrogels with a higher polysaccharide content highly influenced the volume reduction of the cylinders after the overall process (*i.e.*, total shrinkage: shrinkage after solvent exchange plus supercritical drying). The total shrinkage after corn starch aerogel processing fell from 49 to 25% as the starch content in the original sol increased from 7 to 15%. The extensive volume shrinkage obtained for starch aerogels with a lower precursor concentration partially compensated for the influence in the aerogel density of the initial precursor concentration. Consequently, the density of the starch aerogel cylinders increased with the precursor concentration, but not linearly as it would have initially been expected. The porosity of starch aerogels correlated well with the aerogel density and gave the highest values (90%) for the aerogel with the lowest density (Figure 4.8a and b).

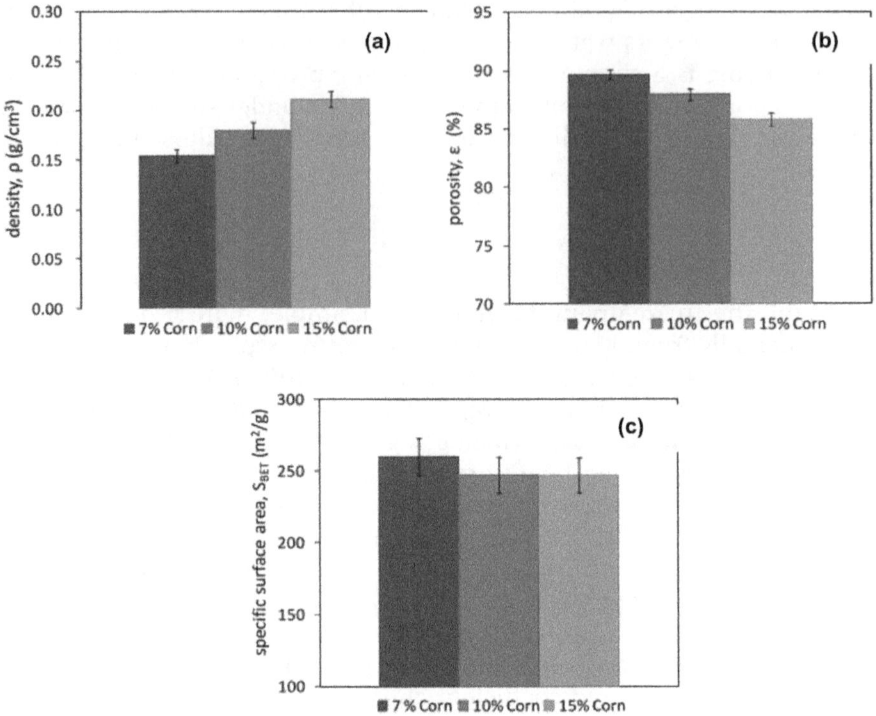

Figure 4.8 Effect of the corn starch concentration on the end aerogel properties: (a) density, (b) porosity and (c) specific surface area.
Adapted with permission from Garcia-Gonzalez *et al.*, 2011.[8]

The specific surface area of the resulting starch aerogels were in the range of 240–260 $m^2 g^{-1}$ (Figure 4.8c).[8]

The textural properties of the dried microspheres (*i.e.*, density, pore size, surface area, *etc.*) determine the adsorption behaviour of chemicals in the matrix, as well as the maximum adsorption capacity of the microspheres. In general, the high surface area and pore volume of the microspheres favour their loading capacity. The drying of the starch microspheres is regarded as the most critical step in obtaining starch micro particles in the dry form while preserving the original nanoporous structure of the wet gel. The air drying of starch gels cannot preserve the gel structure, leading to xerogels with significant pore collapse, shrinkage of the gel structure, formation of cracks and appearance of macropores.[6]

4.4.2 SEM Analysis

Santos *et al.* prepared nanoporous starch microspheres by super critical fluid technology.[9] The SEM images in Figure 4.9 show the presence of spherical microparticles. Primary particles were loosely joined together forming agglomerates, this was likely due to the presence of an emulsifier. The aerogels

Figure 4.9 SEM images of starch aerogel microspheres (top). Higher magnifications show the nanoporous structure of the particles (bottom).
Adapted from ref. 9 under the terms of the CC BY 4.0 licence, https://creativecommons.org/licenses/by/4.0/.

prepared at different starch concentrations were very different from a morphological point of view. At a concentration of 5% w/w, the aerogel consisted of micro particles with a nanoporous structure, as shown in the figure, whereas, at higher concentrations (15 wt%), the wheat starch aerogel showed a lenticular structure with closed pores.[9]

In the case of the maize starch aerogel[4] prepared by SCD, for all the concentrations in the range 5–15% w/w a nanoporous structure was obtained, as shown in Figure 4.10a and b.

Mehling *et al.* prepared starch based aerogels and then loaded them with ibuprofen and paracetamol.[11] Release kinetics were studied *in vitro*. Micrographs for drug-loaded and unloaded starch aerogels are shown in Figure 4.11. SCD of polysaccharide gels results in highly porous biodegradable aerogel matrices with large surface areas. The structural properties of the polysaccharide aerogels depend on the preparation method and

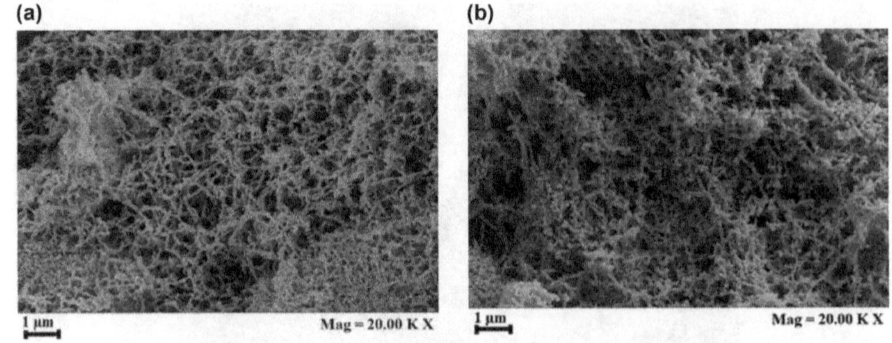

Figure 4.10 FE-SEM images of maize starch aerogels obtained with supercritical gel drying at 200 bar, 45 °C at different starting concentrations: (a) 5% w/w; (b) 15% w/w.
Adapted from ref. 4.

Figure 4.11 SEM pictures of starch aerogels; potato starch aerogel (left) and paracetamol loaded potato starch aerogel (right).
Adapted from *Journal of Non-Crystalline Solids*, **355**, T. Mehling, I. Smirnova, U. Guenther, R. H. H. Neubert, Polysaccharide-based aerogels as drug carriers, 2472–2479, Copyright 2009, with permission form Elsevier.[11]

chemical nature of the gel phase. In this work different polysaccharide precursors (starch, alginate, *etc.*) were used to produce aerogels.

4.5 Conclusion

Natural polymer-based materials are preferable for their renewability and for environmental reasons; among them, starch is the second most abundant after cellulose and it is attractive for its low cost. In this chapter, different steps for the preparation of starch based aerogels were presented. First, a gel is formed by an aqueous solution (hydrogel), then water is replaced by ethanol to form an alcogel; finally, different drying steps can be adopted for the preparation of aerogels. The structural properties of the final starch aerogel are dependent on the preparation method and nature of the material used for the preparation of aerogel.

Abbreviations

Super Critical Drying	SCD
Carbon dioxide	CO_2
Brunauer–Emmett–Teller	BET
High Methoxyl	HM

References

1. P. M. Visakh, A. P. Mathew, K. Oksman and S. Thomas, *Polysaccharide Building Blocks: A Sustainable Approach to the Development of Renewable Biomateriables*, 2012, pp. 287–306.
2. L. Zuo, Y. Zhang, L. Zhang, Y.-E. Miao, W. Fan and T. Liu, *Polymer/Carbon-Based Hybrid Aerogels: Preparation, Properties and Applications*, 2015, vol. 8.
3. F. Zia, K. M. Zia, M. Zuber, S. Kamal and N. Aslam, *Carbohydr. Polym.*, 2016, **134**, 784–798.
4. I. De Marco, L. Baldino, S. Cardea and E. Reverchon, *Chem. Eng. Trans.*, 2015, **43**, 307–312.
5. X. Chang, D. Chen and X. Jiao, *Polymer (Guildf).*, 2010, **51**, 3801–3807.
6. C. A. García-González, J. J. Uy, M. Alnaief and I. Smirnova, *Carbohydr. Polym.*, 2012, **88**, 1378–1386.
7. C. A. García-González, M. C. Camino-rey, M. Alnaief, C. Zetzl and I. Smirnova, *J. Supercrit. Fluids*, 2012, **66**, 297–306.
8. C. A. Garcia Gonzalez, M. Alnaief and I. Smirnova, *Proc. 13th Eur. Meet. Supercrit. Fluids 2011*.
9. J. Santos, P. del Gaudio, M. Landí and C. A. García-González, in *Mol2Net Proceedings*, 2015, vol. 1, pp. 1–5.
10. A. Ubeyitogullari and O. N. Ciftci, *Carbohydr. Polym.*, 2016, **147**, 125–132.
11. T. Mehling, I. Smirnova, U. Guenther and R. H. H. Neubert, *J. Non-Cryst. Solids*, 2009, **355**, 2472–2479.
12. W. He and X. Zhang, *Conducting Polymer Aerogels, Conducting Polymers*, ed. F. Yılmaz, InTech, 2016.

CHAPTER 5

Alginate and Carrageenan Based Aerogels: Processing and Morphology

ASJA PETTIGNANO, ALYSSON DUARTE-RODRIGUES, FRANÇOISE QUIGNARD AND NATHALIE TANCHOUX*

Institut Charles Gerhardt, CNRS-ENSCM-UM, 8 Rue Ecole Normale, 34296 Cedex 5, Montpellier, France
*Email: nathalie.tanchoux@enscm.fr

5.1 Introduction

Alginates and carrageenans are linear bio-sourced polysaccharides extracted from seaweeds and have a large number of academic and industrial applications. "Alginate" is a collective term for a family of linear polysaccharides, widely distributed in the cell walls of brown algae. Alginates can be described as linear non-repeating copolymers of $1 \rightarrow 4$ linked M and G residues, arranged in homopolymeric blocks of consecutive mannuronate residues (M-blocks), consecutive guluronate residues (G-blocks) and in regions of alternating M and G residues (MG-blocks) (Figure 5.1). Although some alginates may exist predominantly as one type block, all three blocks may be present within a single alginate molecule. The most interesting feature of these biopolymers is their ability to form hydrogels. The monocationic salt forms of alginic acid (often Na^+, K^+ or NH_4^+ alginates) are hydrophilic

Green Chemistry Series No. 58
Biobased Aerogels: Polysaccharide and Protein-based Materials
Edited by Sabu Thomas, Laly A. Pothan and Rubie Mavelil-Sam
© The Royal Society of Chemistry 2018
Published by the Royal Society of Chemistry, www.rsc.org

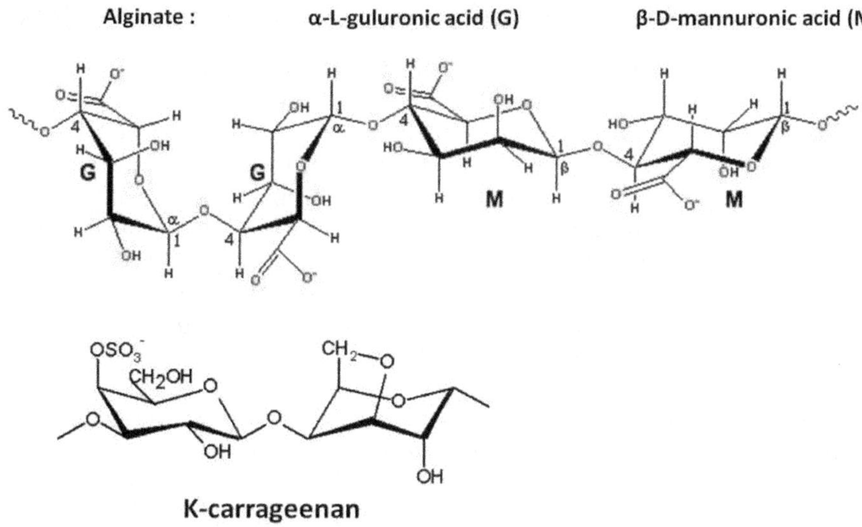

Figure 5.1 Chemical structures of alginate and carrageenan.

colloids and behave in solution as flexible coils.[1] However, in the presence of multivalent cations, such as alkaline-earth (*e.g.* Ca^{2+}, Sr^{2+}, Ba^{2+}), transition metals (*e.g.* Mn^{2+}, Fe^{2+}, Co^{2+}, Ni^{2+}, Cu^{2+}, Zn^{2+}, Cd^{2+}), trivalent cations (*e.g.* Al^{3+}, Fe^{3+}, Ce^{3+}, Eu^{3+} or La^{3+}) or acidic media, alginates form ionotropic gels with a highly-ordered structure.[2]

Carrageenans are a large family of sulphate-bearing polygalactoses extracted from red marine algae (Rhodophyceae). Their three major types, designated by the Greek letters κ, λ and ι differ by the number of sulphate groups and the frequency of anhydrogalactose groups. For instance, the structure of κ-carrageenan is constituted of linear chains of alternating (1→3)-linked α-galactose-4-sulphate and (1→4)-linked 3,6-β-anhydrogalactose (Figure 5.1).[3] κ-carrageenan presents better ionotropic and thermotropic gelation properties than the other types of carrageenans.[4]

Drying of the hydrogels by evaporation of water usually brings about a complete collapse of their secondary structure. Aerogel formulation is a drying method allowing shape and, to a certain extent, size and texture preservation of the samples. This chapter will mainly focus on alginate aerogels, carrageenan aerogels are rarely described in the literature.

The preparation of polysaccharide aerogels being a multi-step procedure starting from a hydrogel, it is worth considering that all the parameters which affect the formation of the hydrogel can affect the formation of the aerogel. The techniques of manufacture and the parameters affecting the formation of alginates and carrageenan aerogels will be discussed. The textural properties, the main interest of aerogel formulation, will also be described.

5.2 Preparation and Main Features of Alginate Gels

Alginate hydrogels can be described as having a highly dispersed structure, able to contain up to 98% of water, in which alginate chains are assembled in a three-dimensional polymer network. This gel-forming ability has also attracted the attention of the scientific community for the development of more high-value applications in different fields, prompted by the possibility to easily tailor their shape and properties for each application. The gel formulation, in fact, allows the easy manufacture of different shapes and several reports describe the preparation of alginate-based hydrogels in the form of beads,[5] microspheres,[6] membranes,[7] fibers[8] and hollow fibers,[9] micro-tubes,[10,11] monoliths,[12] 3-D scaffolds[13,14] and so on.

Two main methods are used to induce gelation of the alginate solution, the diffusion method and the internal setting method.[15]

5.3 Gel Formation by the Diffusion Method

5.3.1 Processing

When the diffusion method is used, the cross-linking ion diffuses into the polysaccharide solution from a large reservoir, usually by direct contact or through a dialysis membrane. The diffusion method can be affected by slight concentration gradients across the thickness, as the high affinity of the cations induces a slight polymer migration to the interface, leading to a higher density closer to the gel surface.[16] Nevertheless, its operative ease renders it the most common procedure for the preparation of alginate hydrogels.

Up to now, bead shaping is the most diffused shape of alginate hydrogels, and thus, it will be discussed as a representative example of alginate gelation by diffusion method.

The preparation of alginate beads consists in the dripping of a viscous sodium alginate solution (*e.g.* 2% w/V) into a mono-, di- or trivalent cation (ionotropic gelation) or acidic (acidic gelation) aqueous solution of appropriate concentration. As soon as the droplets come into contact with the cationic bath, an instantaneous gelation and consequent formation of hydrogel beads is induced. The dropwise addition is often performed by means of a dropping funnel or a syringe and the dropping rate and the distance between the tip of the sodium-alginate feeding container and the surface of the metal-containing solution are adjusted to guarantee the formation of uniform spherical beads. The set-up used for the preparation of alginate beads at room temperature is shown in Figure 5.2.

The properties of the obtained alginate beads are affected by several preparation parameters, such as the nature of the gelling cation and the concentration of the two solutions used.

One study focused on the effect of gelation parameters on the formation of alginate beads and reported that the use of different sodium alginate

Figure 5.2 Preparation of cobalt alginate beads, an example of an experimental setup and hydrogel beads.

concentrations can result in the formation of spherical beads, but also flakes or non-structured gels. In order to keep the droplet integrity when it comes into contact with the cation solution, a sufficient entanglement of the polymer chains is in fact needed. At the same time, the concentration of the cation in the gelation medium was reported as having an effect both on the Young's modulus of the prepared beads and on the gel formation rate, with higher concentrations generally leading to higher moduli and faster kinetics.[17,18] Once the beads are formed, some time is necessary to obtain a stabilized gel network. So-called "maturation" of the beads occurs slowly, to reach a steady-state equilibrium after some hours of ripening in the gelling solution.[17]

5.3.2 Aerogel Formation

An important advantage of alginate gels in their wet state is the high dispersion and accessibility of the hydrophilic polysaccharide chains. On the other hand, considering the improved manageability of a dried material (*e.g.* easier handling, storage and transport), a drying method that allows researchers to maintain this advantageous property in the dry state is highly desirable.

Evaporative drying consists in the removal of the gel solvent by evaporation and, thanks to its low cost and simple procedure, represents a widely employed technique. As a drawback, most secondary structures of the hydrogels are compliant enough to be drawn together by capillary tension, which acts on the polymeric network when a liquid–vapour interface is formed.[19,20] This results in the almost complete shrinkage of the material, leading to a compact solid with an average volume of 2.7% of the average volume of the corresponding hydrogel beads.[21] The structure of the wet gel is, thus, not retained and a virtually non-porous material, "xerogel", is obtained.

Drying by supercritical fluid, in fact, is an effective method to preserve the high open porosity and superior textural properties of the wet gel in a dry form. Thanks to the near zero surface tension of the supercritical solvent, the structure is prevented from collapsing during drying, yielding a fast and complete solvent elimination.[22] Supercritical CO_2, in particular, is the most appropriate fluid for supercritical drying of polysaccharide based aerogels due to its mild critical point conditions (304 K, 7.4 MPa), which avoids any temperature-dependent change in the alginate structure, along with its GRAS (Generally Recognized As Safe) status.[23]

Due to the low miscibility of water and liquid carbon dioxide, direct drying of the hydrogel is not possible and an intermediate solvent exchange step has to be used. A thorough dehydration is necessary as the presence of even small amounts of water can cause a dramatic change in the initially highly porous polysaccharide network upon supercritical drying. The chosen exchange solvent should not be able to dissolve or alter the gel network and should be completely miscible with both water and CO_2. Ethanol has historically been the solvent of choice for supercritical drying, even if other solvents like acetone, acetonitrile, and so forth, have been suggested for this application.[24] In order to limit the solvent induced shrinkage, a step-wise replacement of water, through multiple exchanges with alcoholic solutions of increasing ethanol concentration, has to be performed prior to drying.

The obtained alcogel is then introduced into a pressure vessel and washed with liquid CO_2, in order to replace the ethanol in the gel. The CO_2 impregnated gel is then compressed and heated above the critical point (for CO_2, 31 °C and 73.8 bar) to attain the supercritical state. Release of pressure above the critical temperature then allows the removal of CO_2 without forming any liquid–vapour interface. As a result, a highly porous solid retaining the dispersion of the wet state, "aerogel", is obtained.

5.3.3 Morphology and Texture of Aerogels

Optical or scanning electron microscopy (SEM) is the classical tool used to observe the morphology of the aerogel. Physisorption of nitrogen at 77 K allows the determination of the textural properties such as the specific surface area or mesoporosity of the materials.[19,25]

When the hydrogel beads are subjected to solvent exchange, a partial shrinkage of the polymeric structure, whose extent depends on the type of solvent used and of the rapidity of the exchange, must be taken into account.[24] The sudden increase of organic solvent in the gel interferes with the concomitant release of water from the gel structure, leading to a reduction in the surface tension in the gel pores.[23] This decrease in the capillary pressure of the gel structure is responsible for the reduction in volume of the gel, which follows second-order kinetics of shrinkage for alginate gels.[26] A multi-step and low frequency solvent exchange decreases the diffusion rate of water out of the gel, mitigating the shrinkage, and is consequently often preferred. With respect to the hydrogels, alginate aerogel beads are characterized by a slight shrinkage and corresponding decrease in the gel volume, often this is only negligible when compared with the corresponding evaporative drying procedure. The shrinkage of the alginate aerogels, however, essentially takes place during the water-to-ethanol solvent exchange and can be controlled and minimized by following a proper protocol (*e.g.*, exchange steps and frequency).

The SEM image in Figure 5.3 shows an exemplificative alginate aerogel, which is characterized by an open framework of isolated nanometric fibrils.

The effectiveness of the supercritical drying method depends on the mechanical properties of the polymer. The decrease in gel volume during the preparation of the aerogel is reported in Table 5.1 for several polysaccharide materials.[21] The differences in shrinkage behavior is paralleled by the porosity of the aerogels.

As shown in Table 5.1, the total porosity of the alginate aerogels is close to 38 $cm^3 g^{-1}$, while the pore volume of carrageenan is 40 times smaller. Alginate aerogels present an extremely open structure and are essentially macroporous. Carrageenan hydrogels undergo nearly complete shrinkage and the aerogel of carrageenan presents a completely different texture (Figure 5.4).

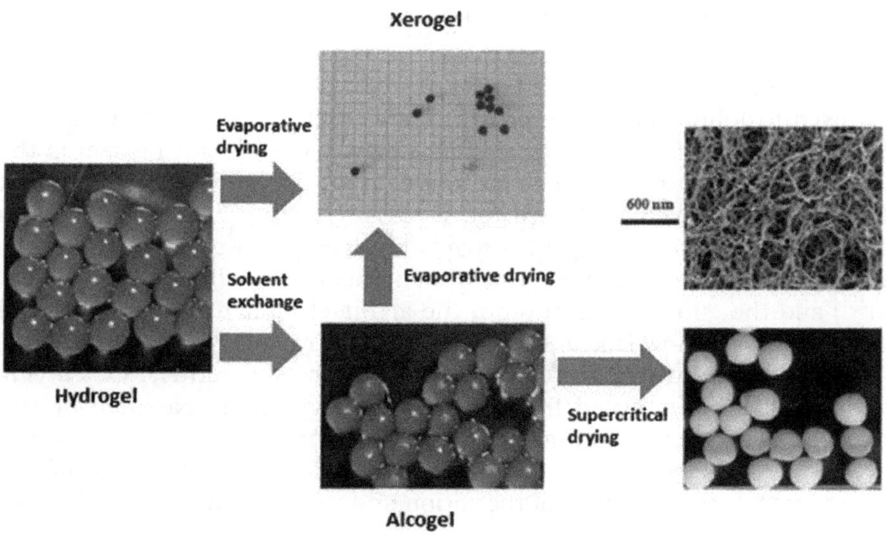

Figure 5.3 Morphology modification during aerogel processing for a cobalt alginate aerogel.

Table 5.1 Example of shrinkage upon supercritical CO_2 drying and textural properties.

Aerogel	Shrinkage upon drying (vol%)	Porosity ($cm^3 g^{-1}$)	Mesopore volume ($cm^3 g^{-1}$)	Average mesopore size (nm)	Surface area ($m^2 g^{-1}$)
Ca-alginate	20	39	1.16	40	570
Alginic acid	22	38	0.89	20	390
κ-Carrageenan	95	1	0.76	18	200

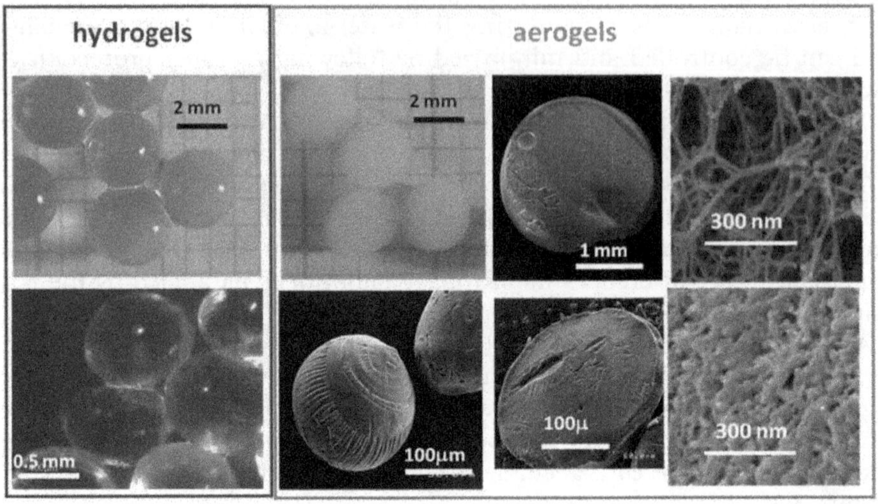

Figure 5.4 Comparison between hydrogels and aerogels of complete beads and cross sections. First row: Copper-alginate; second row: κ-carrageenan.

A small angle X-ray scattering (SAXS) study on several alginate samples was used to follow all of the steps of a hydrogel–alcogel–aerogel transformation, and confirmed that the fibrillar network was already present in the parent hydrogel and that the shape and connectivity of the fibrils were not affected during solvent exchange and successive drying.[27] The morphologies of several ionotropic alginate hydrogels and aerogels were investigated by SAXS according to the nature of the divalent metal cation (Mn^{2+}, Co^{2+}, Zn^{2+}, Cu^{2+}) and the guluronic fraction of the alginate. All alginate hydrogel and aerogel samples show isotropic SAXS. In the size range investigated by SAXS (~10–200 Å), the structure of aerogels obtained by supercritical CO_2 drying was found to be inherited from the morphology of the parent hydrogel, whatever the initial structural regime.[28] As a consequence, the texture of aerogels is affected by the parameters affecting the formation of the hydrogel, such as the nature of the cation and the G/M ratio in the alginate. This is exemplified in Figure 5.5.

These differences in texture are reflected in the measurements of the surface areas, as reported in Table 5.2. Generally, high guluronate

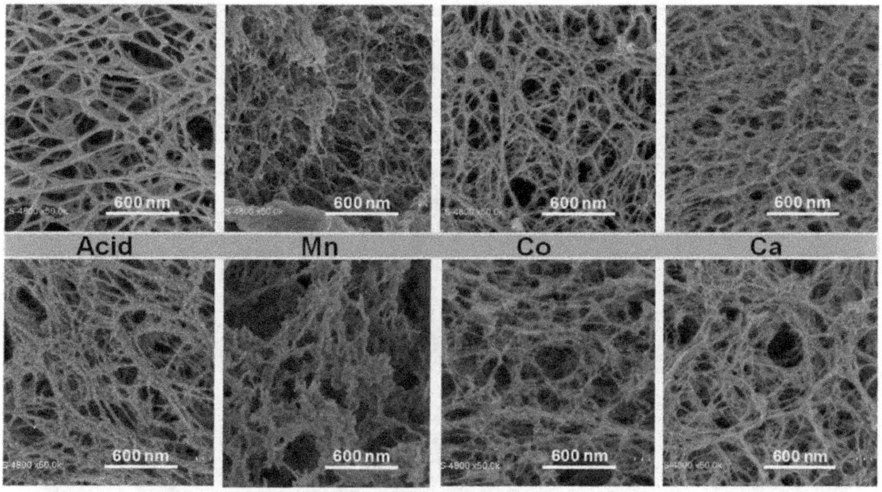

Figure 5.5 Texture of alginate aerogels obtained with different cations, and two alginates with high (first row) or low (second row) guluronic content.

Table 5.2 Surface area ($m^2 g^{-1}$; values ± 5 $m^2 g^{-1}$) of alginate aerogels with high (HG) or low (LG) guluronic content gelled with different cations.

Alginate type	Acid	Mn	Co	Zn	Cu	Ca
HG	240	200	330	340	520	500
LG	260	90	190	240	500	380

containing alginate gives higher surface areas, irrespective of the cations. This level of dispersion of the aerogel can also be related to the structural regime of the alginate hydrogel.[28]

5.3.4 Stability

According to SAXS analysis, aerogels can be directly re-impregnated with an organic solvent (*i.e.* ethanol), without significant changes to the structure.

On the other hand, as reported by SEM (Figure 5.6) and nitrogen adsorption tests, direct re-hydration of alginate beads leads to a dramatic modification of the internal texture, producing the collapse of the fibrils to form hollow spheres.[29,30] However, the hydrogel can be reformed by impregnating the aerogel in ethanol, and then, with a multi-step ethanol-to-water exchange.

5.4 Alginate Hydrogels Formation by the Internal Setting Method

5.4.1 Processing

The internal setting method consists of the introduction of an inert precursor into the polysaccharide solution and is then followed by the

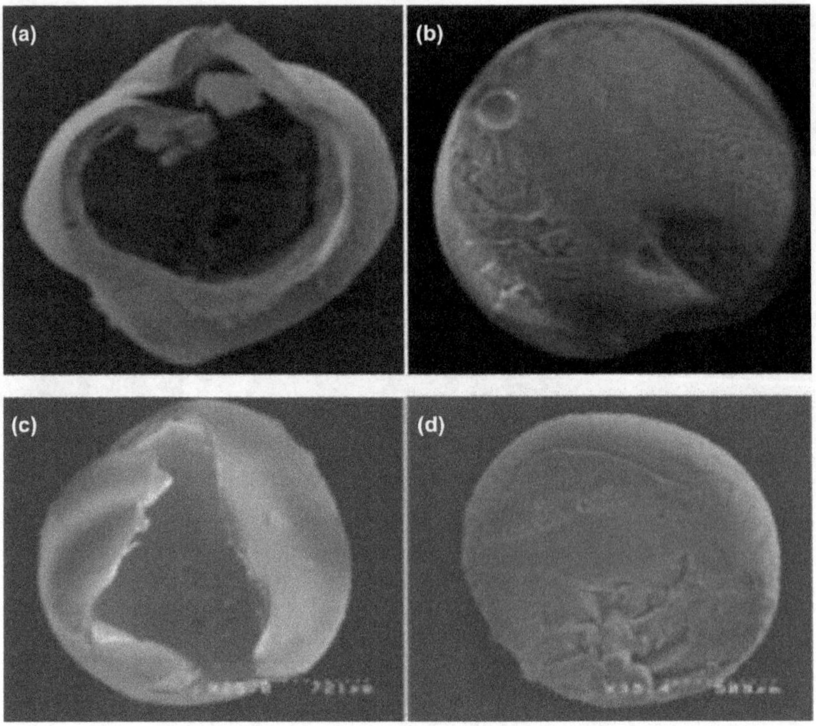

Figure 5.6 Comparison of SEM micrographs of Ca-alginate: (a) xerogel; (b) aerogel; (c) aerogel subjected to direct rehydration; and (d) aerogel rehydrated by a multi-step ethanol-to-water exchange.
Reprinted with permission from R. Valentin, K. Molvinger, C. Viton, A. Domard and F. Quignard, *Biomacromol.*, 2005, **6**(5), 2785, Copyright 2005 American Chemical Society.

controlled release of the crosslinking ions stimulated by pH or solubility triggers. The internal setting method consists in mixing the alginate solution with an insoluble source of divalent cations and a gelation inducer agent.[31,32] In this process, $CaCO_3$ is often used as the precursor, as the low solubility in pure water allows its uniform distribution in alginate solution before gelation occurs. The release of the crosslinking cation into the alginate solution is usually initiated by lowering the pH, leading to the formation of soluble $Ca(HCO_3)_2$. D-glucono-D-lactone (GDL) is a common gelation inducer, which slowly hydrolyses in water reducing the pH of the solution. The consequent time-delayed release of crosslinking calcium ions is particularly advantageous as it allows controlling of the gelation speed, depending on the desired application.[32] Moreover, this technique enables the preparation of mechanically strong and complex-shaped 3-D structures (Figure 5.7), as it allows molding of the solution into the desired shape and size before the gelation occurs.[33]

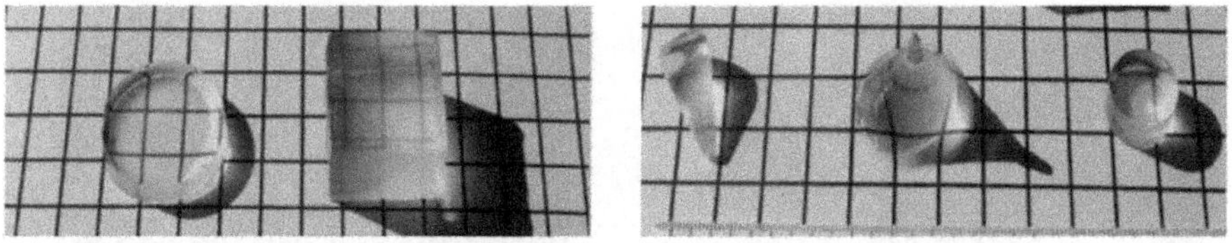

Figure 5.7 Alginate gels of various molded shapes obtained by the internal setting method.
Reprinted from *Biomaterials*, **22**, C. K. Kuo and P. X. Ma, Ionically crosslinked alginate hydrogels as scaffolds for tissue engineering: Part 1. Structure, gelation rate and mechanical properties, 511–521, Copyright 2001, with permission from Elsevier.

The internal setting method has been recently expanded by the use of pressurized CO_2, acting as a weak acid, instead of conventional pH reducers.[34]

5.4.2 Aerogel Formation

The same procedure as previously described can be used, but recently a novel approach for the preparation of alginate aerogels with high pressure CO_2 was developed. All the steps leading to the formation of the alginate aerogel can be performed in the same high pressure autoclave and this is the first example of a one-pot process.[34]

Alginate solution is mixed with the desired amount of $CaCO_3$ and introduced into a high-pressure autoclave. The autoclave is then pressurized with gaseous CO_2 up to 5 MPa at 298 K. After gelation, the autoclave is charged with supercritical CO_2 to reach 12 MPa, and the water-ethanol exchange is realized by successively introducing 30, 60 and 90% ethanol–water solvent for 2.5 h. After this anhydrous ethanol is used. Finally supercritical-drying is carried out with pure supercritical CO_2. This technique allows the formation of gels with a very low alginate concentration down to 0.25 wt%.

5.4.3 Morphology and Texture

Even if the shape of the materials is maintained, an overall linear shrinkage close to 45% is observed (Figure 5.8). Nevertheless, the material is porous and the pore volume lies between 4 and 16 $cm^3 g^{-1}$, depending on the alginate concentration and $CaCO_2$/alginate ratio. In this case, up to 60% of the porosity is attributed to the volume of the mesopores. The material presents dual meso and macro porosity.[34]

5.4.4 Stability and Properties

Interestingly, the aerogels produced by this procedure can be compressed to flexible materials which retain the mesoporous characteristics of the parent

Figure 5.8 (a) Comparison of CO_2 induced Ca-alginate hydrogel and the corresponding aerogel; (b) SEM image of the internal texture of the aerogel.

aerogel. The behavior of the compressed material is then dramatically different from the parent aerogel and does not swell in water, even after several days.[34]

5.5 Conclusion

Aerogels of alginate and carrageenan are obtained by a multi-step procedure, including gel formation by diffusion of the gelling agent or controlled release of the gelling agent mixed in the polysaccharide solution, water–solvent exchange, and drying by supercritical CO_2. The second procedure can be coupled with drying in a ''one-pot'' process, where gelation is induced by CO_2 pressure.

In both procedures, the shape and, to a certain extent, the size of the object are maintained and the obtained aerogels are meso and macroporous materials. When the shrinkage is important, a part of the macroporosity is lost, but the material becomes highly mesoporous and surface areas as high as 600 $m^2 g^{-1}$ can be reached. In terms of texture, the aerogels obtained by the diffusion method can be considered as a correct image of their hydrogel parent, the fibrillar morphology of the gel reflects the self-assembling of the polymer chains during the gelation step.

Polysaccharide aerogels are amazing materials which will open up new fields of application for these biopolymers in non-aqueous media.

References

1. O. Smidsrod, *Carbohydr. Res.*, 1970, **13**, 359.
2. O. Smidsrod and A. Haug, *Acta Chem. Scand.*, 1968, **22**, 1989.
3. J. N. Barbotin and J. E. Nava Saucedo, in *Polysaccharides*, ed. S. Dumitriu, Marcel Dekker, New York, 1998, p. 749.
4. D. Slootmaekers and M. Mandel, *Int. J. Biol. Macromol.*, 1990, **13**, 17.
5. C. Ouwerx, N. Velings, M. M. Mestdagh and M. A. V. Axelos, *Polym. Gels Netw.*, 1998, **6**(5), 393.
6. M. Alnaief, M. A. Alzaitoun, C. A. García-González and I. Smirnova, *Carbohydr. Polym.*, 2011, **84**(3), 1011.

7. H. Dai, X. Li, Y. Long, J. Wu, S. Liang, X. Zhang and J. Xu, *Soft Matter*, 2009, **5**(10), 1987.
8. C. E. Pegg, G. H. Jones, T. J. Athauda, R. R. Ozer and J. M. Chalker, *Chem. Commun.*, 2014, **50**(2), 156.
9. K. H. Lee, S. J. Shin, Y. Park and S. H. Lee, *Small*, 2009, **5**(11), 1264.
10. S. Sugiura, T. Oda, Y. Aoyagi, M. Satake, N. Ohkohchi and M. Nakajima, *Lab Chip*, 2008, **8**(8), 1255.
11. A. Blaeser, D. D. Campos, M. Köpf, M. Weber and H. Fischer, *RSC Adv.*, 2014, **4**(87), 46460.
12. T. Mehling, I. Smirnova, U. Guenther and R. H. H. Neubert, *J. Non-Cryst. Solids*, 2009, **355**(50), 2472.
13. I. T. Ozbolat, H. Chen and Y. Yu, *Robot. Comput.-Integr. Manuf.*, 2014, **30**(3), 295.
14. C. Colosi, M. Costantini, R. Latini, S. Ciccarelli, A. Stampella, A. Barbetta and M. Dentini, *J. Mater. Chem. B*, 2014, **2**(39), 6779.
15. S. N. Pawar and K. J. Edgar, *Biomaterials*, 2012, **33**(11), 3279.
16. G. Skjåk-Bræk, H. Grasdalen and O. Smidsrød, *Carbohydr. Polym.*, 1989, **10**(1), 31.
17. N. M. Velings and M. M. Mestdagh, *Polym. Gels Netw.*, 1995, **3**(3), 311.
18. C. Ouwerx, N. Velings, M. M. Mestdagh and M. A. V. Axelos, *Polym. Gels Netw.*, 1998, **6**(5), 393.
19. R. Valentin, K. Molvinger, F. Quignard and F. Di Renzo, *Macromol. Symp.*, 2005, **222**(1), 93.
20. F. Di Renzo, R. Valentin, M. Boissière, A. Tourrette, G. Sparapano, K. Molvinger and F. Quignard, *Chem. Mater.*, 2005, **17**(18), 4693.
21. F. Quignard, F. Di Renzo and E. Guibal, in *Carbohydrates in Sustainable Development I*, ed. A. P. Rauter, P. Vogel and Y. Queneau, Springer Berlin Heidelberg, 2010, pp. 165–197.
22. G. Della Porta, P. Del Gaudio, F. De Cicco, R. P. Aquino and E. Reverchon, *Ind. Eng. Chem. Res.*, 2013, **52**(34), 12003.
23. C. A. García-González, M. Alnaief and I. Smirnova, *Carbohydr. Polym.*, 2011, **86**(4), 1425–1438.
24. R. Subrahmanyam, P. Gurikov, P. Dieringer, M. Sun and I. Smirnova, *Gels*, 2015, **1**(2), 291.
25. F. Rouquerol, J. Rouquerol and K. S. W. Sing, *Adsorption by Powders and Porous Solids*, Academic Press, San Diego, 1999.
26. B. H. Rehm, Alginate production: precursor biosynthesis, polymerization and secretion, in *Alginates: Biology and Applications*, Springer Berlin Heidelberg, 2009, pp. 55–71.
28. M. Robitzer, L. David, C. Rochas, F. Di Renzo and F. Quignard, *Langmuir*, 2008, **24**(21), 12547.
29. P. Agulhon, M. Robitzer, L. David and F. Quignard, *Biomacromolecules*, 2012, **13**(1), 215.
30. R. Valentin, K. Molvinger, C. Viton, A. Domard and F. Quignard, *Biomacromolecules*, 2005, **6**(5), 2785.

31. K. I. Draget, K. Østgaard and O. Smidsrød, *Microbiol. Biotechnol.*, 1989, **31**(1), 79.
32. T. Mehling, I. Smirnova, U. Guenther and R. H. H. Neubert, *J. Non-Cryst. Solids*, 2009, **355**(50), 2472.
33. C. K. Kuo and P. X. Ma, *Biomaterials*, 2001, **22**, 511.
34. P. Gurikov, S. P. Raman, D. Weinrich, M. Fricke and I. Smirnova, *RSC Adv.*, 2015, **5**(11), 7812.

CHAPTER 6

Protein-based Aerogels: Processing and Morphology

SARAH E. FITZPATRICK,[a] MARK P. STAIGER,*[b] SANTANU DEB-CHOUDHURY[a] AND STEVE RANFORD[a]

[a] AgResearch Limited, Lincoln Research Centre, Private Bag 4749, Christchurch 8140, New Zealand; [b] Department of Mechanical Engineering, University of Canterbury, Private Bag 4800, Christchurch 8140, New Zealand
*Email: mark.staiger@canterbury.ac.nz

6.1 Protein-based Aerogels: An Overview

Kistler introduced the concept of aerogels based on either inorganic compounds or organic materials (including cellulose, gelatine and ovalbumin) in the early 1930s.[1] Since that time, research studies have been largely focussed on exploring the synthesis and properties of aerogels based on inorganic compounds. In contrast, organic-based aerogels, including those based on natural compounds (*aka* bio-aerogels) received very little attention until the late 1980s and early 1990s.[2–4] The precursor materials used in the synthesis of bio-aerogels have been mainly based on various polymers (*e.g.* cellulose) or proteins. Protein-based aerogels are a relatively new trend in the aerogel field, although the idea was also first explored by Kistler, using ovalbumin (egg white) and collagen (gelatine) proteins.[1]

General interest in naturally-sourced materials has increased dramatically in recent years, leading to a retracing of the steps made by Kistler in the earliest studies of bio-based aerogels. The supercritical drying technique has been used by researchers developing specific protein-based gels, as has the

Green Chemistry Series No. 58
Biobased Aerogels: Polysaccharide and Protein-based Materials
Edited by Sabu Thomas, Laly A. Pothan and Rubie Mavelil-Sam
© The Royal Society of Chemistry 2018
Published by the Royal Society of Chemistry, www.rsc.org

more recent technique of freeze-drying. For example, the first protein-based aerogels were reported by Kaplan *et al.* in 2004–2005 using the freeze-drying technique. Kaplan and co-workers produced silk fibroin protein-based aerogels in order to expand the application of biopolymer implant materials based on silk fibroin hydrogels.[5,6]

Proteins of both animal and plant origin have been used for aerogel production including ovalbumin (egg white protein),[7–10] collagen (and gelatine),[1,7,11] silk fibroin (SF),[5,6,12–17] wheat gluten,[18] soy protein isolates (SPI),[19–24] corn zein[25,26] and milk proteins (whey protein isolates (WPI)[27–30] and casein[31]). The above proteins have been selected for their inherent fibrous/gel forming properties (collagen, gluten, silk fibroin, and ovalbumin) and/or availability (gluten, whey and soy protein isolates).

The exploration of protein-based bio-aerogels is largely motivated by the development of aerogels with characteristics of biodegradability and/or biocompatibility. The biodegradability of these materials could be considered obvious, although some researchers are endeavouring to measure the biodegradability of bio-aerogels in practice.[32] Many researchers have targeted naturally-occurring proteins due to their abundance and biodegradability.[8,19] The biocompatibility and low toxicity of many proteins is also a consequence of their natural origins that may be coupled with benign processing routes. Biocompatibility has already been established for many of these proteins in the areas of medical and food research, where the toxicity of the wet-gel precursors has been extensively studied.[11,33]

One of the main applications of protein aerogels is tissue engineering, specifically the development of cell-culture scaffolds for supporting the growth of various tissues (see Table 6.1).[5,6,11,12,15,34] Drug delivery is a further medical application that is receiving great attention in the field of organic aerogels where biocompatibility is also a requirement.[13,16,28,29,35] The biodegradation of the materials is important for medical applications where there is a requirement for the material to be eventually absorbed by the physiological system. The remainder of the protein aerogel field sees applications in the laboratory (*e.g.* biomolecular catalysis and separation platforms),[30] environmental clean-up (*e.g.* waste-water treatment)[22] and industrial applications as bio-friendly versions of their silica and synthetic counterparts, for which the biodegradability, permeability and/or chemical specificity of the aerogel are important properties.[14,18–20,27]

The current literature on protein-based aerogels indicates a similar level of research activity associated with *pure* and *hybrid* protein aerogels. A hybrid gel consists of two or more constituents that make up the solid network.[27] It should be noted that aerogels that include the use of crosslinking molecules for bridging protein molecules in their synthesis are not categorised here as hybrid protein aerogels. Consequently, only those materials that are able to inherently form a gel network (*e.g.* silica, clay, carbohydrates, *etc.*) are categorised as hybrid protein aerogels, where additives are not required for the gel network formation.[36] Hybrid protein aerogels have been produced for each of the protein types already used in aerogel research (ovalbumin,

Table 6.1 Comparison of processing methods and applications across protein based aerogels in the literature from 2004 up to the present day (organised by protein type).

Protein	Other	Protein %	Gelation method	Drying method	Application	Year	Ref.
Whey proteins	—	20	pH/heat	SC-CO_2	Medical – drug delivery	2012	28
	—	20	pH/heat	Freeze-drying			
	—	10–25	Heat/salt	Freeze-drying	Industry – insulation	2013	27
	Alginate	7.5–12.5	Heat/salt	Freeze-drying			
	—	—	Heat/salt/acid	Freeze-drying	Medical – drug delivery	2016	29
	Cellulose	—	Heat/salt/acid	Freeze-drying			
	—	1–3.5	Salt	SC-CO_2	Laboratory – catalysis/ biodetection	2016	30
Silk fibroin (SF)	—	10–17	Alcohol/salt/base	Freeze-drying	Medical – tissue engineering	2004	5
	—	4–10	Salt	Freeze-drying	Medical – tissue engineering	2005	6
	—	6–12	Heat	Freeze-drying	Medical – tissue engineering	2006	34
	Clay	0.05–5	—	Freeze-drying	Industry – not specified	2008	36
	—	5	Salt	Freeze-drying	Medical – tissue engineering	2012	17
	—	4	Heat/acid	SC-CO_2	Medical – drug delivery	2014	13
	—	4–13	Acid	SC-CO_2	Medical	2014	12
	Graphene	7–8	Base	Freeze-drying then carbonisation	Industry – capacitors	2014	14
	—	2–6	Acid/salt	SC-CO_2	Medical – tissue engineering	2015	15
	—	2–6	Acid/salt	Freeze-drying			
	—	10–15	Salt/alcohol	SC-CO_2	Medical – tissue engineering	2016	16

Table 6.1 (*Continued*)

Protein	Other	Protein %	Gelation method	Drying method	Application	Year	Ref.
Soy proteins	Clay	0.05–5		Freeze-drying	Industry – not specified	2008	36
	—	25	Chemical/heat	SC-CO_2	Industry – insulation	2012	19
	Cellulose	4–8	Heat	Freeze-drying	Industry – fluid absorbents	2013	20
	Graphene	—	Heat	Freeze-drying	Environment – water treatment	2015	21
	Graphene	—	Acid	Freeze-drying	Environment – water treatment	2015	22
	—	—	Chemical	Freeze-drying	Industry – insulation	2016	23
	Graphene	—	Acid	Freeze-drying	Environment – water treatment	2016	24
Ovalbumin	—	1.5	Heat (directly carbonised)	SC-CO_2	Environment – CO_2 absorption Laboratory – catalysis/separation	2011	9
	—	10	pH/heat	SC-CO_2	Food – microencapsulation	2015	8
	Cellulose	11	Acid then heat (directly carbonised)	Freeze-drying	Environment – water treatment	2015	10
Gluten	—	11–17	Glycerol and heat	Freeze-drying	Industry – general fluid absorbents + Medical/Cosmetics - drug delivery	2010	18
	Cellulose	11–14	Heat	Freeze-drying			
Zein	Agar	3–6	Heat	Freeze-drying	Industry – not specified	2012	25
	Silica	—	Chemical/heat/gas	SC-CO_2	Industry – thermoplastics	2014	26
Collagen	Cellulose	—	Chemical	Freeze-drying	Medical – wound dressing/ Tissue engineering	2014	11
	Alginate	3	Gas	SC-CO_2	Industry – insulation Medical – tissue engineering	2015	64

collagen, silk fibroin, gluten, soy protein, corn zein and whey protein), often alongside aerogels of each of the pure constituents for comparison. A wide range of applications from tissue engineering to absorbent materials and to capacitors have all spurred on the development of hybrid bio-aerogels due to their customisable physiochemical properties.

6.2 Processing and Fabrication of Protein-based Aerogels

An aerogel is a porous material derived from a gel, in which the liquid component of the gel has been replaced with a gas, such that an aerogel has two distinct phases of matter (*i.e.* solid and gas). The synthesis of an aerogel begins with a sol – a stable colloidal suspension of solid particles in a continuous liquid medium. The protein is the solid phase in the sol of a protein-based aerogel; or in the case of a hybrid protein aerogel, the solid phase may consist of both proteins and polymers. Gelation of the sol creates a continuous network of the solid particles that are suspended in the liquid medium (Figure 6.1). Subsequently, evaporation of the liquid medium by drying allows an aerogel to be formed as the gas (normally air) replaces the liquid medium that surrounds the gel network. The following sections will outline the role of protein chemistry in the general scheme of aerogel synthesis and production.

6.2.1 Sol–Gel Processing

In protein chemistry the route to a sol involves manipulation of certain conditions to make the protein less soluble in aqueous solution, a transformation termed *protein aggregation*. Proteins are dissolved in an aqueous solution at a pH and salt concentration near to physiological concentrations in order to mimic biological conditions. Such biomimetic approaches support natural (or native) protein conformation and activity that is useful for studies and applications of normal protein conformation and function. Generally, the formation of an aerogel does not require the protein to maintain its native conformation, and in fact more facile aerogel formation is often possible using alternative conformations induced prior to gelation.[30]

Proteins with an alternative (or non-native) conformation are described as being denatured. A sol may be obtained from a protein solution by precipitating the protein as stable aggregates that remain suspended in the liquid. Precipitation of the protein is achieved by disrupting the chemical interactions that stabilise the conformation of a protein molecule in water. The destabilised protein will exhibit different chemical moieties at the protein surface that tend to lead to protein aggregation. The newly exposed and/or activated chemical groups subsequently create new chemical interactions within the protein and with other protein molecules. Such

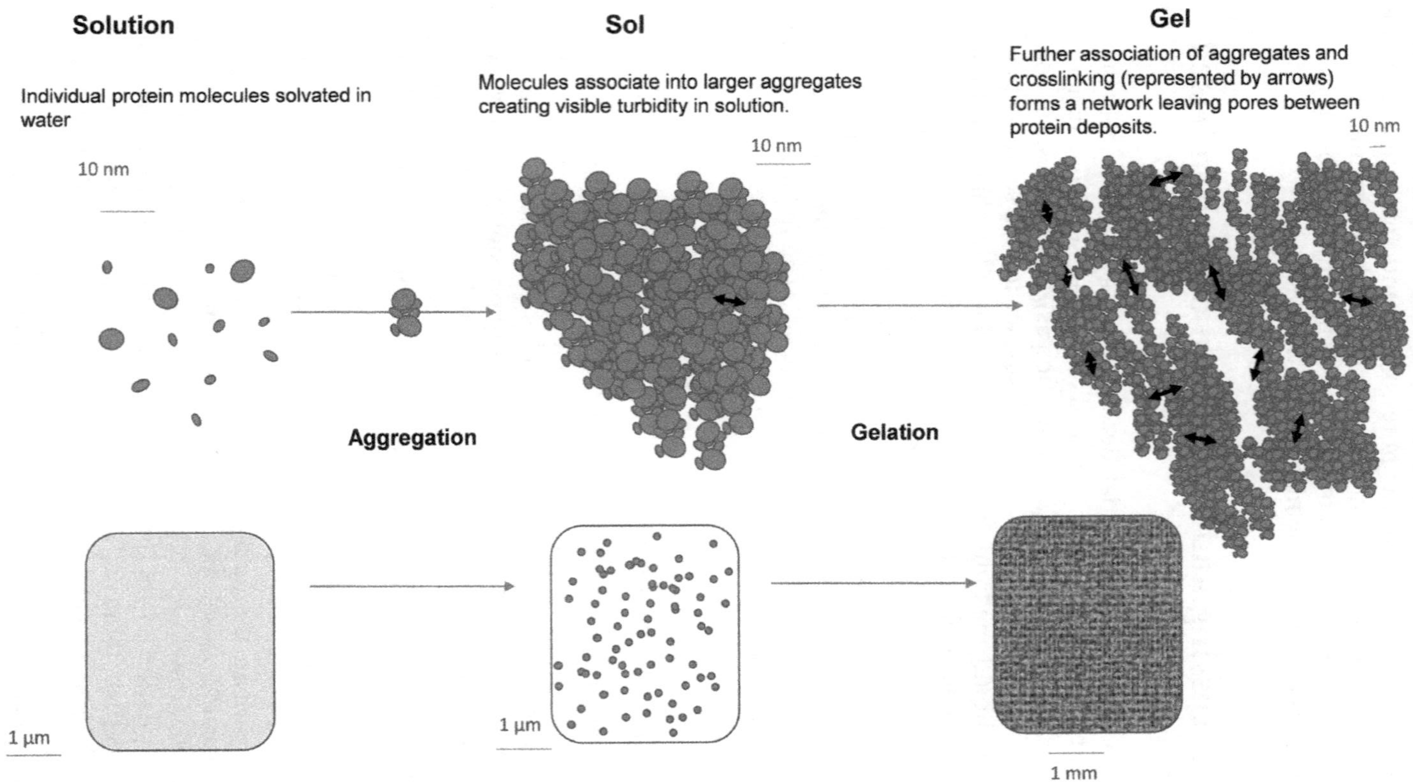

Figure 6.1 Representation of molecular and visible changes in a protein solution transformed into a hydrogel.

interactions could be hydrogen bonding between polar groups (–OH, –SH, $–NH_2$), salt bridges between charged groups ($–COO^-$, $–NH_3^+$) or hydrophobic interactions of non-polar groups ($–CH_3$, aromatic groups). Although, in general, it is the hydrophobic residues that drive protein aggregation as this reduces their interaction with the surrounding water.[37] The hydrophobic residues would normally be buried within a native protein conformation, away from the water molecules in which the protein is solvated. Once the native conformation is disrupted and hydrophobic residues are exposed, these denatured proteins develop hydrophobic interactions with neighbouring protein molecules to exclude water, leading to aggregation of the protein. The aggregated solid protein needs to be stable in the solvent, otherwise a solid mass of precipitate will form and separate completely from the water, rather than progressing to the hard gel stage.

Prevention of complete precipitation in part relies on the protein and solution conditions. However, complete precipitation can also be prevented by agitating the sol and/or rapidly progressing to the gelation stage. The aggregated protein will remain dispersed throughout the solvent if a delicate balance between attraction and repulsion of the aggregates can be maintained. Spontaneous gelation will occur without the need for further gelation induction if crosslinking and associative forces are more dominant than any electrostatic repulsion.[39]

The formation of a gel (in this case a hydrogel) from the aggregation stage is essentially a continuation of the same chemical driving forces. Gelation is partly due to the same protein–protein chemical interactions as those described above, and is partly due to irreversible chemical crosslinking (*e.g.* disulphide bridge formation).[38,39] Disulphide crosslinks are covalent chemical bonds, usually formed between thiol (–SH) groups on amino acids. The formation of irreversible chemical crosslinks mark the permanent transition from the sol to gel and can be between any functional groups on the protein able to form covalent bonds (*e.g.* $–NH_3^+$, $–S^-$, $–COO^-$, *etc.*). The importance of crosslinking was illustrated by Chen *et al.* who conducted comparative studies of whey protein isolate (WPI) aerogels using salt and heat. It was observed that combining a denaturation step (heat) with a crosslinking inducer (salt) significantly improved the mechanical properties of the aerogel when compared to those only treated with heat. The increased crosslinking due to the salt not only improved the network strength, but did so without affecting the density.[27]

The gel network formation is controlled by both the rate of the crosslinking reaction and the amount of net charge on the aggregates. Charged groups on proteins (*e.g.* $–COO^-$, $–NH_3^+$, $–S^-$) endow the overall molecule (and aggregate) with a positive or negative charge depending on the pH of the solution. Electrostatic repulsion between aggregates will slow the association of protein with greater charge on the aggregates. However, the electrostatic repulsion must be overcome by hydrophobic interactions in order to form a gel network, with the gel network then being stabilised by crosslinking. The morphology and physical characteristics of a gel are a result of the balance between the attractive and repulsive forces within the protein network and the overall degree of covalent crosslinking.[37]

The protein concentration also influences the formation of the gel. The continuous gel structure cannot be formed below a certain concentration (critical concentration) as a certain amount of protein entanglement or cross-over is required to sustain the solid network.[37] Consequently, the higher the protein concentration, the less need there is for strong driving forces to induce aggregation prior to crosslinking.

The chemistry of protein gelation can be categorised as intrinsic or extrinsic.[37] The type of protein will dictate the intrinsic properties such as the amino acid residue composition, while the molecular weight of the protein determines the likelihood of spontaneous gel formation. Additionally, there are five extrinsic factors (or solution properties) that can be manipulated to increase the likelihood of gelation, these are protein concentration, pH, temperature, pressure, and ionic strength/type of ion in solution.[40] Techniques aimed at manipulating the above properties were the focus of sol–gel chemistry in the early stages of protein aerogel processing.

Denaturation or aggregation techniques are chemically simple, being induced by either a pH change, salt addition or increase in temperature. The technique often depends on the natural tendency of the protein (or polymer) to gel. For example, Selmer *et al.* heated an ovalbumin solution to simultaneously denature and induce crosslinking in the protein, producing a hydrogel in a one-step process.[8] This is effectively identical to the process of cooking an egg, producing a white gelatinous substance that is irreversibly changed from the viscous semi-transparent raw egg white. These two stages can be combined into a single denaturation and gelation step as aggregation and gelation involve similar chemistry and one is essentially an extension of the other.

Manipulation of the extrinsic properties (pH change, salt concentration, heat, *etc.*) is the most accessible method of inducing gel formation, although gel formation may also be induced by additives. For example, aldehydes can be used to react with any protein to form covalent bonds that help to solidify the gel network.[11,19] Other additives can be used to replace water as the solvent in the solution (*e.g.* methanol[5]) so as to disrupt normal protein–water interactions. High pressure (200–500 MPa) can also be used to induce hydrophobic interactions and disulphide bonds between protein molecules, while enzymes are able to catalyse the crosslinking of protein amino acid side chains.[39] Although high pressure and enzymatic techniques are utilised in protein sol–gel chemistry, these methods have not yet been applied to the protein gels used to synthesise aerogels.

pH-, salt-, heat- and additive-induced gels are discussed in detail below, using examples from the literature on protein-based aerogels. It is also possible to employ a combination of techniques to induce gel formation (*e.g.* pH change, salt or a crosslinking agent in combination with heat) (Table 6.1).

6.2.1.1 Heat-induced Gelation

Dissolving protein precursors in an aqueous solution and inducing gelation through heating is the simplest and most common method used to produce protein hydrogels. A variety of protein types have been gelled using this method, including the previously described 'cooking' of the ovalbumin protein by Selmer *et al.*[8] Increasing the temperature of a protein encourages dehydration and unfolding of the protein conformation by destabilising the chemical interactions and exposing residues within the protein core. The exposed residues are generally hydrophobic and will contribute to protein aggregation as the denatured protein seeks to re-stabilise its conformation within the aqueous solution. Examples of proteins where straightforward heat-induced gelation has been used include whey proteins,[27–29] ovalbumin,[8–10] zein,[25] gluten[18] and soy proteins.[20,21,24] Moreover, the rate of protein aggregation, and hydrophobicity and size of the protein may influence the homogeneity of the gel (*e.g.* homogenous transparent gel *c.f.* turbid coagulum-type gel). A gel may not form if the molecular weight and concentration of the protein are below critical levels.[37]

In some proteins, the heating step can separate denaturation from the gelation step. For example, a common gelation technique with whey proteins is known as the cold-set gel procedure. The protein is first denatured *via* heating and then cooled, remaining soluble due to repulsive charges on the protein molecules. Aggregation (and subsequent gel formation) is then induced by the addition of a salt. Separating these two steps is useful for studying the effects of denaturation, aggregate size, and crosslinking on the final aerogel morphology and properties.[41,42] However, the fast and simple single-step heating could be a preferable gelation option for the industrial up-scaling of the process.

6.2.1.2 pH-induced Gelation

Varying the pH away from the isoelectric point (pI) of the protein induces a non-native protein conformation by altering the quantity and type of charged ions on the protein. Carboxylic acid ($R\text{-}COOH \rightleftharpoons R\text{-}COO^-$) and amine ($R\text{-}NH_2 \rightleftharpoons R\text{-}NH_3^+$) groups become protonated or deprotonated as the solution pH changes. Changing the protonation state of these chemical groups affects the internal electrostatic interactions of the protein and its solubility. Charged groups on the protein interact with each other (*via* ionic bonding) and with water. If the location or type of a charge on the individual protein molecule is changed (due to pH) these intramolecular protein bonds and the protein interaction with water can be disrupted, resulting in a denatured protein. In addition to intramolecular effects, the pH change can also affect the intermolecular interactions of the protein molecule with other protein molecules. The ratio of the charges (negative and positive) results in an overall molecular charge for each individual molecule. This charge helps to determine the amount of attraction that a single protein molecule will have to the neighbouring protein molecule in solution. A high degree of overall charge in one direction (*i.e.* highly negative or highly positive) will increase repulsion between individual protein molecules and counter-intuitively slow aggregation. Thus, an extreme change in pH value from the protein pI will denature protein molecules but resist aggregation through high electrostatic repulsion of molecules. To induce aggregation in this state, the charges need to be shielded by the addition of a counter ion from a salt, which will neutralise the charges and allow protein molecules to aggregate.

Most frequently, a simple hydrochloric acid or sodium hydroxide adjustment for pH manipulation is performed prior to other processing steps being employed (*e.g.* heating).[8,10,28] Alternatively, pressurised CO_2 gas is usually used to manipulate the pH in the gelation of the SF protein.[12,13,15,38] Glucono-δ-lactone is another acid used to modify the pH of protein solutions and is common in sol–gel food chemistry of whey proteins.[37] Ahmadi *et al.* reported the use of glucono-δ-lactone alongside heat and salt to induce gelation of a whey protein solution in the production of aerogels.[29]

6.2.1.3 Salt-induced Gelation

Salt additions (in excess of physiological concentrations) result in aggregation and gelation *via* an increase of ions in the solution. The ions (usually Na^+, Ca^{2+} and Cl^-) are electrostatically attracted to charged moieties ($-COO^-$, $-S^-$ and $-NH_3^+$) on the protein, thereby creating a shielding effect. The salt ions effectively prevent the charges on the proteins from interacting with water *via* hydrogen bonding. Inner hydrophobic residues are then exposed and contribute towards the protein–protein interaction that drives aggregation. In addition, the electrostatic repulsive forces between protein molecules are decreased due to the ion shielding effect. Consequently, the protein molecules tend to aggregate and a gel is formed (Figure 6.2). Salt-induced gelation is most commonly used in WPI gels in which NaCl and $CaCl_2$ are known to be gelation agents of whey proteins. WPI will gel directly upon addition of $CaCl_2$ or NaCl to the solution, without the need for prior heat denaturation.[37] Salt can be used to induce aggregation and gelation of already denatured (but still soluble) whey proteins in such solutions. In this technique, the protein is first denatured (but not aggregated) using controlled heating, subsequently salt is added for the gelation step. This gelation technique is termed cold-set gelation as salt-induced gel formation is carried out at ambient temperatures.[42]

The power of electrostatic forces within the protein dispersions (or sol) has been demonstrated by an incorporation of charged metal particles, in this case gold.[30] The gold nanoparticles nucleate on the surface of the denatured protein providing an electrostatic charge on the protein molecule. Following addition of salt as the gelation agent, the Au–Au repulsion is greatly reduced allowing the denatured protein to gel instantly.[30]

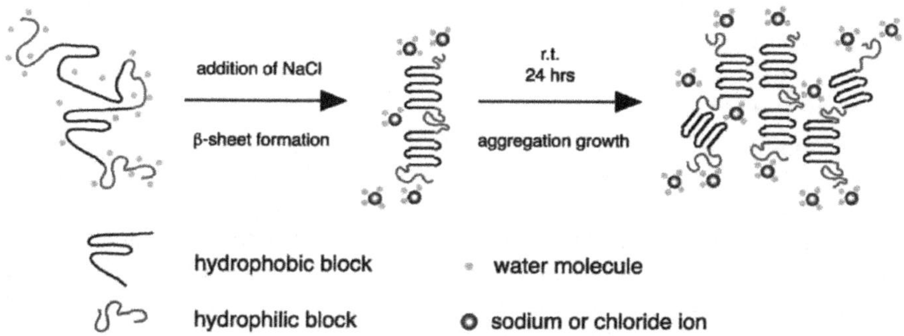

Figure 6.2 Schematic of mechanistic basis for the structural changes occurring during salt-induced protein aggregation.
Reprinted from Biomaterials, 26, U.-J. Kim, J. Park, H. J. Kim, M. Wada and D. L. Kaplan, Three-dimensional aqueous-derived biomaterial scaffolds from silk fibroin, 2775–2785, Copyright 2005, with permission from Elsevier.

Salt-induced gelation has also been shown to occur with SF proteins.[5,6] A salt leaching method was employed where salt granules are used as porogens, a technique that has been adapted from the production of porous scaffolds for use in tissue engineering.[5,6] The porogen acts as a physical mould, allowing the protein to precipitate in the spaces around it to create a scaffold. The use of a salt porogen effectively saturates the aqueous solution with dissolved ions and initiates the salt-induced gelation described above. An investigation of the structural changes induced by the salt porogens concludes that salt-induced dehydration exposes hydrophobic residues, initiating β-sheet formation in the secondary structure, a type of protein conformational change.[6] The salt is then removed by a leaching process with water followed by a solvent exchange with alcohol which is eventually evaporated, leaving behind the porous structure. The resulting porous structure can be further dried using freeze-drying to ensure that all of the solvent is removed from the resulting aerogel.

6.2.1.4 Chemical Additives

A less commonly used technique to achieve gelation is the introduction of other molecules to the solution that induce gelation *via* solvent removal or chemical crosslinking. Chemical crosslinking between protein molecules drives aggregation and stabilises the network within the hydrogel. Although covalent crosslinking occurs naturally in some proteins, the degree and rate of crosslinking may be increased by chemical additives. Crosslinking can be induced by adding a reactive molecule, such as an aldehyde, that forms covalent bonds with reactive groups on protein amino acids.[23] An example of this technique is the synthesis of a collagen-cellulose hybrid aerogel by prior oxidation of the cellulose fibres to create dialdehyde functionality on the glucose molecules of the cellulose chain. The dialdehyde-functionalised cellulose reacts with amine groups on the collagen protein, thereby crosslinking the two biopolymers.[11] Thus, the gelation of the collagen and cellulose is spontaneously driven by a crosslinking reaction of the aldehyde groups on the cellulose and amines of the protein. A crosslinking agent that forms covalent bonds with two protein functional groups can be utilised to achieve gelation in non-hybrid cases (*i.e.* where only the protein contributes towards the gel network structure). Repetition of the protein + crosslinking agent reaction results in two new covalent bonds which allow the crosslinking molecule to act like a bridge between two points on the protein molecule or molecules. An example of this bridging mechanism was observed for soy protein, formaldehyde and tannin aerogels as reported by Amaral-Labat *et al.* (Figure 6.3).[19] Formaldehyde reacts with both the tannin and protein molecules and can undergo two reduction reactions, allowing the formation of $-CH_2-$ bridges between molecules. Thus, a complex network of protein–protein, protein–tannin and tannin–tannin crosslinks are formed to create the gel.

Chemical crosslinking techniques take advantage of the existing chemistry within the proteins. The logical next step is to attempt an even wider range of chemically-induced crosslinking or other chemical interactions between protein aggregates by exploiting the chemistry of existing amino acid residues on the protein. Although this is slightly more complicated than a simple heat or pH induced gelation, this kind of chemical manipulation during the formation of the gel networks could be the key to tailoring the resulting aerogel properties such as the elastic modulus, strength, density and homogeneity.

The importance of these covalent crosslinks has been demonstrated in ovalbumin wet-gels by Van Kleef *et al.* The resulting mechanical properties of the hydrogel are attributed directly to the number of thiol groups and therefore the disulphide covalent crosslinks in the protein network.[39] Although

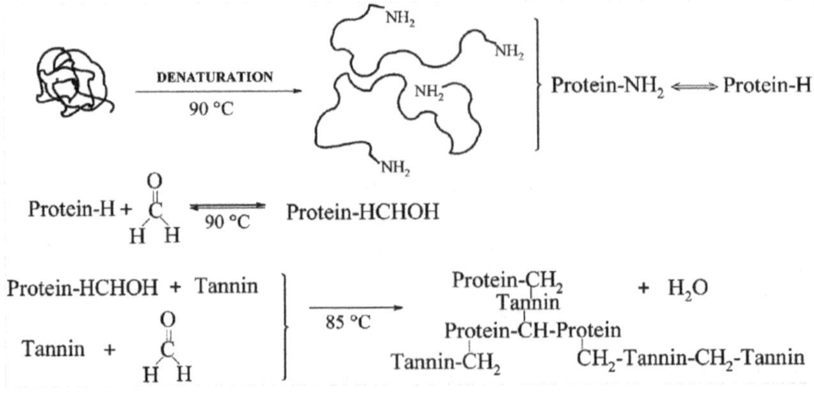

Figure 6.3 Supposed crosslinking reactions of soy protein, tannin and formaldehyde. Adapted from ref. 19 with permission from The Royal Society of Chemistry.

these studies were completed on the hydrogel rather than the corresponding aerogel, this work provides clear evidence that covalent crosslinking (*e.g.* disulphide bonds) in a protein network is an important molecular feature in describing the processing–structure–property paradigm of protein-based gels.

Molecules that interfere with protein solvation, specifically a dehydration agent, may also induce gelation. For example, the use of alcohol to dehydrate the protein and encourage aggregation has been applied to the processing of protein aerogels.[5] Alcohol-induced gelation is similar in its mechanisms to that of heat-induced gelation. Alcohol destabilises the native protein conformation by reducing the protein–solvent interactions due to the alcohol being significantly less polar than water. Alcohol dehydrates the protein molecules, and with a lower solubility of proteins in alcohol compared to water, the proteins are forced to aggregate. β-sheet structures form due to water removal during the aggregation. β-sheets are a secondary protein structure in which the backbone of the protein takes on a pleated conformation. β-sheets may be part of the normal protein structure or induced in a protein because other backbone geometries have been destabilised. The chemistry of protein folding is complicated, but it is necessary to note here that the β-sheet structure often stabilises hydrophobic amino acid residues that are exposed in denatured proteins prior to the aggregation. It is the formation of non-native β-sheet structures that causes the globular proteins in the neuronal tissue to denature and adopt the dangerous plaque-forming fibrous conformations, known as amyloid fibres, in those suffering from Alzheimer's disease. This mechanism of protein denaturation has been exploited by Nyström *et al.* to produce extremely low density aerogels from β-lactoglobulin, a normally globular whey protein.[30]

Therefore, it is clear that various parameters related to the protein solution and subsequent processing methods may affect the formation of a protein-based gel network (Figure 6.4). The network is composed of crosslinked protein aggregates that are in turn composed of denatured protein molecules. The gel network can range from a loosely connected agglomeration of large aggregates (in which conditions favour large globular protein aggregation) to a tightly crosslinked network of smaller, fibrous aggregates separated by smaller spacing. The aggregation and gelation factors are influenced by the protein type, protein concentration, heat, pH, salt and the presence of crosslinking molecules.

6.2.2 Gel–Aerogel Processing

The term "aerogel" is used to refer to a "microporous solid in which the dispersed phase is a gas".[2] Thus, the wet-gel prepared in aqueous solution requires conversion to a dry state where air is dispersed throughout the gel network. The drying step aims to eliminate or minimise the capillary forces that are present during the evaporation of the solvent. Capillary forces are responsible for the structural collapse and subsequent material shrinkage seen with drying at atmospheric pressure and room temperature. The drying

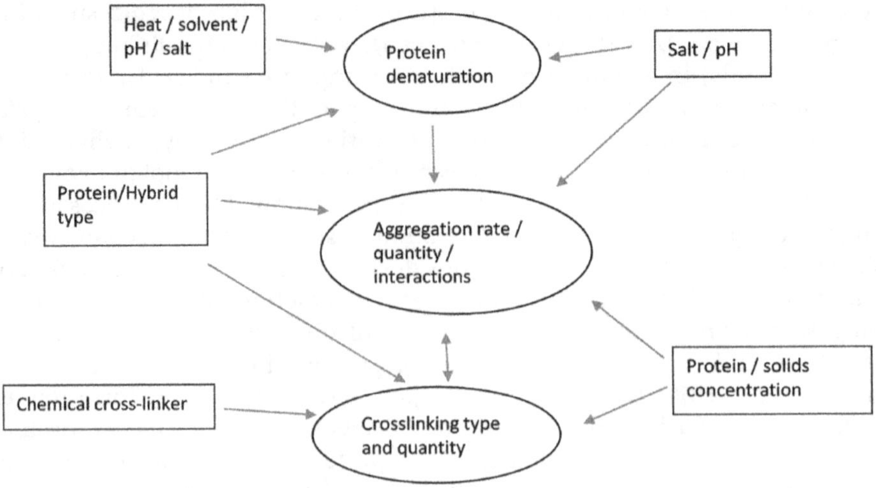

Figure 6.4 Schematic showing the effect of variables in a protein solution on the gelation process.

methods, freeze-drying or supercritical carbon dioxide drying, employed at this stage are designed to avoid directly crossing the liquid–gas phase boundary. Freeze-drying involves the freezing and subsequent sublimation of the solvent (*e.g.* water) at low pressure. Supercritical carbon dioxide (SC-CO_2) drying involves replacing the liquid in the gel with supercritically heated and compressed CO_2 that is then evaporated by reducing the pressure. Thus, the gas phase is achieved *via* a supercritical phase and collapse of the gel network is avoided. As seen in sol–gel processing, the choice of solvent for protein solutions is always water, but a solvent that may be dissolved in SC-CO_2 is required for this process. Therefore, water is solvent-exchanged with ethanol, or a similar non-polar solvent, that is more soluble in the SC-CO_2 and can be flushed out and replaced with the SC-CO_2.

Interestingly, the selection of the drying technique in the production of protein-based aerogels has not been discussed in the short history of these materials (please refer to Table 6.1). Thus, there is no apparent preference of one technique over the other as a function of the protein or application. At these early stages in the protein aerogel field, researchers may be non-selective about drying techniques or may even experiment with both[15,28] due to a lack of compelling evidence for the selection of one technique over the other. Cost, accessibility, ease of use, environmental impact and effect on resulting aerogel properties will all be factors taken into consideration when a drying technique is selected. The exact effects of the drying process on aerogel properties may have high importance for the future of aerogel processing beyond the laboratory. Researchers may simply opt for the cheapest or most accessible option. Basic morphological and other physical properties such as strength, are not always reported in the literature with novel protein aerogels. The various applications of these gels mean that researchers are interested in investigating very different properties of the aerogels, making it harder to collate information and draw basic conclusions regarding the processing–property relationships of protein-based aerogels. Nevertheless, this chapter will investigate what relationships there may be between drying technique (among other processing differences) and the resulting aerogel morphologies and properties.

6.3 Morphology of Aerogels

The morphological structure of aerogels is characterised in terms of the quantity, shape, size and connectivity of its pores. A detailed knowledge of the pore structure is central to understanding the performance of an aerogel in a specific application. Low density and high porosity are two of the defining features of all aerogels. Density and porosity are interrelated meaning that lower density aerogels will generally have a greater percentage of porosity and total pore volume. Density is measured using straightforward laboratory assessments and can be used to calculate the quantity of air, or percentage of porosity, that the aerogel contains. Highly accurate mass and dimension measurements are taken of the entire aerogel to determine the *bulk* density of the intact aerogel structure. Measurements of the solid component in the aerogel can be taken by grinding it down and measuring the volume of the powdered aerogel in a pycnometer. A pyncometer is a device that detects the volume of a solid sample through displacement of a gas, usually helium. The sample is introduced into a pressurised chamber and the pressure change of the gas held in the chamber is used to calculate the volume that the sample occupies (pressure and volume are related by the ideal gas law). Combining this measurement with the mass of the sample, the *skeletal* density of the solid component can be calculated. Bulk and skeletal density can then be used to calculate the porosity. Dividing the bulk density by the skeletal density gives the fraction of the aerogel volume that is due to the solid phase. Subtracting this value from one then yields the fraction of the aerogel that is air, thus indicating the porosity.[19] Alternatively, the total volume of the pores (measured in cm^3 of pore space per gram of aerogel) can be calculated by subtracting the volume of the skeletal structure (solids) from the bulk volume.

Porosity can be further elucidated by measurements using nitrogen adsorption–desorption techniques. Nitrogen gas (or any appropriate gas) can be introduced in a controlled manner to a chamber containing the sample. The gas will adsorb onto the surface of the material and upon subsequent release at increasing temperatures, result in a measurable increase in gas pressure. An isotherm tracking these pressure changes can be plotted and, using appropriate mathematical tools, the volume of gas adsorbed on the material surface can be calculated. This technology allows the specific surface area (measured in meters squared per gram of aerogel structure) of the aerogel to be calculated from the quantity of gas (nitrogen) absorbed on the surface. The most popular analysis used to calculate the surface area from nitrogen adsorption–desorption information is based on the Brunauer–Emmett–Teller (BET) equation.[43] Other theories used to analyse the nitrogen adsorption–desorption information include the Dubinin–Radushkevich method for determining micropore volume[44] and the Barrett–Joyner–Halenda (BJH) method for determining pore size distribution.[45] Pore size and shape can also be crudely assessed using microscopy such as scanning electron microscopy (SEM), which can be useful when nitrogen absorption–desorption approaches cannot accurately assess the pore size of interest.[28] The International Union of Pure and Applied Chemistry (IUPAC) defines pore sizes in three categories: <2 nanometres (nm) pores are 'microporous', 2–50 nm pores are mesoporous and >50 nm are macroporous.[46] Limitations of the BET equation come into effect when samples exhibit a wide range of pore sizes. If adsorption is limited to a single layer of gas molecules (termed monolayer adsorption) then the results are best converted to surface area data using a different model, the Langmuir equation.[47] However, if the adsorption occurring is of more than one layer of gas molecules (termed multilayer adsorption) then results are best interpreted using the BET equation.[48] Other adsorption phenomenon such as pore filling (gas molecules completely fill micropores) and capillary condensation (gas molecules condensing to liquid in pore corners) further complicate analysis and require other modifications to these models (*e.g.* the Dubinin–Radushkevich method). Complications in analysis occur when these types of adsorption happen simultaneously in the sample. Generally, this can be prevented in flat surface adsorption by controlling the vapour pressure as lower pressures are not conducive to multilayer adsorption and thus the Langmuir model can be used. The BET model is most appropriate if multilayer and condensation adsorption is assumed in the porous material.[47] However, neither of the above models is an accurate analysis of the resulting isotherms when these two types of adsorption occur simultaneously, requiring a choice to be made as to which model represents the majority of the adsorption. Thus, BET data reflects only the surface area in the mesoporous range.

The morphology of macropores can extend well into the micrometre (μm) scale and can be very different to that of the smaller meso- and micro-pores that are restricted to the nanometre (nm) scale. Different morphologies based on these ranges can emerge when the aerogel structure contains sheet-like solid scaffolds.[11,21,22] These sheets or walls can be deposited in layers or web-like structures with micron-sized slits or spherical macropores between walls. Upon closer inspection these walls or sheets can also contain pores of the meso- and micro-pore range within the wall structure itself (Figure 6.5). One drawback of the nitrogen adsorption–desorption BET technology is its inaccuracy when determining surface areas of matrices with larger macropores (>100 nm).[28] Thus, the pore distribution should be discussed alongside the surface area measurements where possible. The reported surface area values should therefore be considered as an indication of the surface area of the meso- and micro-porous range, rather than of the total surface area of the aerogel skeleton.

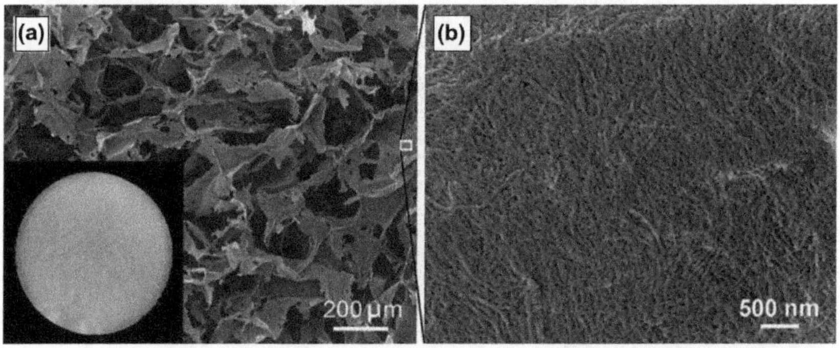

Figure 6.5 Macroscopic view and SEM images of collagen-cellulose aerogel. Large macropores between scaffold walls can be seen at >200 micrometers (a) while nanometer sized micropores can be seen within the wall of the scaffold at higher magnification (b).
Reprinted from Composites Science and Technology, **94**, T. Lu, Q. Li, W. Chen and H. Yu, Composite aerogels based on dialdehyde nanocellulose and collagen for potential applications as wound dressing and tissue engineering scaffold, 132–138, Copyright 2014, with permission from Elsevier.

Another technique used for measuring porosity is mercury intrusion porosimetry (MIP), in which mercury is forced into the evacuated sample while the intrusion pressure and quantity of mercury introduced are recorded. The pore sizes and volumes can be calculated from the intrusion as a specific pressure of mercury will correspond with intrusion into a specific pore size. Measuring the amount of mercury introduced at a known pressure, allows calculation of the pore volume at a particular pore size.[49] Increasing the pressure will introduce an additional mercury quantity from which the same calculation of pore volume can be made but for a slightly smaller pore size, and so on.

The methods used to measure density and porosity demonstrate that they are intimately linked and thus basic density measurements can be used to crudely assess porosity. The typical density of a silica aerogel is approximately 0.1 $g\,cm^{-3}$, although this may range from 0.00153 to 0.9 $g\,cm^{-3}$.[50,51] The average protein aerogel density reported is approximately 0.2 $g\,cm^{-3}$, while the porosity averages approximately 92%.[5,6,8,9,12,13,15–18,27–30,34] Although 0.2 $g\,cm^{-3}$ is slightly heavier than the typical densities of other aerogels, in general the protein aerogels exhibit densities, porosities and even mechanical properties that rival those of the non-protein aerogels. Notably, early protein aerogels have been reported with densities as low as 0.03 $g\,cm^{-3}$.[5] Since these early reports, lower density protein aerogels have only been achieved by amyloid fibril aerogels, having a phenomenally low density of 0.003 $g\,cm^{-3}$.[30]

The summary of morphology information given in Table 6.2 attempts to place protein aerogels and their hybrids in context with other aerogel types from the literature.[52–59] It should be noted that the variety of protein types,

Table 6.2 A comparison of morphological data from different aerogel types.

Aerogel type	Density ($g\,cm^{-3}$)	% porosity	SA ($m^2\,g^{-1}$)	Pore size distribution	Average mesopore diameter (nm)	Ref.
Silica	0.003–0.5	90–99.8	500–1200	Mostly mesoporous	20–40	51,53,57,61
Synthetic organic	0.02–0.2	>80	400–1200	Mostly lower mesoporous range	3–20	51,52,54–56,58
Carbonised	0.001–0.5	99.9	600–800	Micro and low mesopore range	7–20	41,55
Polysaccharide	0.008–0.46	52–99	30–850	Mesoporous/macroporous	2–11	50, 59, 60
Proteins	0.003–0.5	75–99.9	0.12–478	Mesoporous/macroporous	2–30	5,6,8,9,12,13,15,18,19,23,27–30,34
Hybrid proteins	0.02–0.12	90–95	1.72–451	Mesoporous/macroporous	3–40	10,11,14,18,20–22,24–27,29,36,64

gelation methods, drying methods and hybrid material types can have a profound influence on the aerogel morphologies and properties. This very wide scope of variables means that the average protein aerogel morphologies and properties are less useful for a comparison with the more widely studied silica and other organic aerogel fields. However, dissecting the resulting morphological information with respect to these processing variables will reveal more specific relationships between certain protein aerogel processes and the resulting morphologies and properties.

6.3.1 Composition

The type and amount of protein is the first stage in controlling protein aerogel morphology. A clear trend in increasing density and a decrease in porosity with increasing protein concentration can be seen across all protein types. However, the effect on density becomes less pronounced beyond a certain protein concentration and is independent of protein type and gelation methods.[6,18,27,30,34]

Protein (or total solids) concentration has an immediate effect on the rate of protein solution aggregation and the rate of gel network formation. More aggregates contribute more solids to the scaffold of the aerogel resulting in denser network formation. SEM micrographs can demonstrate the decrease in pore sizes with increasing aerogel density (Figure 6.6). Interestingly, increased protein concentrations also result in a marked increase in the BET surface area.[15] It might be expected that aerogels with reduced porosity should have correspondingly lower surface areas as reduced pore numbers result in reduced scaffold walls. However, higher density structures formed at higher protein concentrations demonstrate the opposite trend. The effect may be related to the decrease in average pore size that is seen with increased concentrations and densities. A larger pore will have reduced surface area when compared to many smaller pores of equivalent volume. Even if the combined pore volume of these smaller pores is slightly less than the single larger pore, surface area can still be higher. Moreover, BET surface area measurements are usually restricted to the mesoporous range and thus a reduction in pore size to the mesoporous range will result in a greater number of pores detectable by the technique.

The type of protein chosen to form the aerogel also dictates the nature of the gel network and ultimately the aerogel morphology. Certain proteins are chosen based on their structural and chemical compositions with known influences on the resulting gel viscosities, strengths and functionalities. This aerogel morphological dependency on protein type can be clearly seen when average structures from the various protein types are compared, for example SF tends to produce largely macroporous structures of lower densities, while whey and soy proteins tend to be denser yet more mesoporous (Table 6.3). An obvious exception is that of Nyström *et al.* who produced a phenomenally light 0.003 g cm^{-3} protein aerogel from a whey protein using the unique amyloid fibril induction as discussed earlier.[30] The amyloid fibril structures

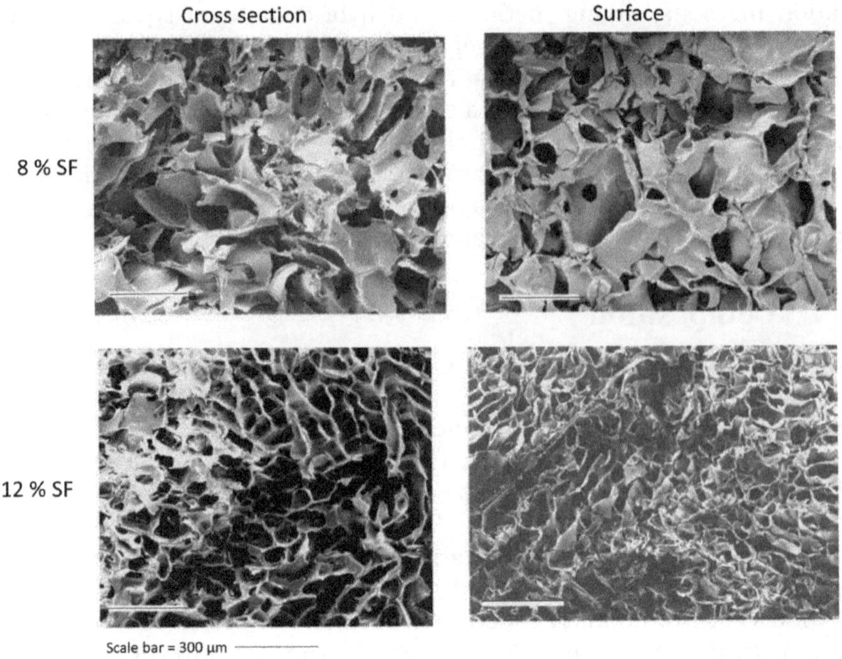

Figure 6.6 Cross section (left) and surface (right) morphologies of silk fibroin aerogels prepared from 8% protein solution (top) and 12% protein solution (bottom).
Reproduced from Journal of Materials Science: Materials in Medicine, Preparation of 3-D regenerated fibroin scaffolds with freeze drying method and freeze drying/foaming technique, **17**, 2006, 1353, Q. Lv, © Springer Science + Business Media, LLC 2006, with permission of Springer.

were induced conformations of β-lactoglobulin, a normally globular (non-amyloid) protein, which is one of the two main proteins found in whey. Whey protein aerogels reach their lowest density of 0.06 $g\,cm^{-3}$ in the hybrid alginate-whey aerogels formulated by Chen *et al.*,[27] but are usually around 0.1–0.2 $g\,cm^{-3}$. Thus, the fibrillar nature of the amyloid version must be contributing towards the incredible gelling ability in these aerogels, which is a distinct advantage of using a protein whose fibrous aggregates can maintain a less dense network. Nyström's β-lactoalbumin amyloid aerogels surpass even the lightest polysaccharide aerogels that have a density of 0.008 $g\,cm^{-1}$.[60] This incredible achievement highlights the manipulation of the protein structure as a key advancement and provides researchers with tantalising prospects for future aerogel research.

6.3.2 Methods

Once the protein type is selected, the density and porosity can be further influenced by the method of aerogel production. This can be categorised

Table 6.3 A comparison of drying method and reported morphological data from protein aerogels.

Protein	Drying method	Density g cm^{-3}	% porosity	Pore volume cm^3 g^{-1}	Pore distribution			Average mesopore width	SA m^2 g^{-1} (<100 nm pore size range)	Year	Ref.
					Micro	Meso	Macro				
Whey proteins (WP)	SC-CO_2	0.30–0.43	—	1.34–1.72	Mesoporous and macroporous			12–27 nm	14–447	2012	28
	Freeze-drying	0.23–0.30	—	—	Macroporous			—	<5		
	Freeze-drying	0.11–0.26	—	—	—			—	—	2013	27
	Freeze-drying	0.13	—	0.0061	Mostly meso and micro			—	2.08	2016	29
β-lactoglobulin amyloid (WP)	SC-CO_2	0.003–0.003	98–99.9	2.7	—	10%	Mostly macro	—	325	2016	30
Silk fibroin	Freeze-drying	0.03–0.12	87–99	—	Macroporous – 15–155 μm			—	—	2004	5
	Freeze-drying	—	85–96	—	Mostly large macropores			—	—	2005	6
	Freeze-drying	0.09–0.20	75–97	—	Large macropores up to 150 μm: 1000 fold greater than mesopores			—	—	2006	33
	Freeze-drying	—	—	—	Macroporous (~400 μm)			—	—	2012	17
	Freeze-drying	—	—	0.04–0.05	Shift towards both macroporous and microporous ends of spectrum from SC-CO_2 results			3 nm	45–59	2015	15
	SC-CO_2	0.058	—	—	None	Mostly mesopores	Up to 130 nm	20 nm	424	2014	13
	SC-CO_2	—	—	—	Mostly mesoporous			—	—	2014	12
	SC-CO_2	—	—	1.6–1.8	Mostly mesoporous			17 nm	260–308	2015	15
	SC-CO_2	—	88–93	—	Macroporous			—	—	2016	16
Soy proteins	SC-CO_2	0.19–0.25	84–88	3.31–4.67	3–4%	40–50%	40–60%	30–40 nm	384–478	2012	19
	Freeze-drying	—	—	0.001–0.009	Smallest	>50%	>Micro	8–40 nm	0.12–4.49	2016	23
Ovalbumin	SC-CO_2	0.07	95–97	0.38–0.7	8–24%	85–92%	—	—	247–476	2011	9
	SC-CO_2	0.22–0.47	—	0.2–2.4	Mesoporous and macroporous			—	16–380	2015	8
Gluten	Freeze-drying	0.08–0.20	49–93	—	Mostly macro (30–73 μm)			—	—	2010	18

into two key stages namely, gel network formation and drying. At the gel network stage, the amount and type of protein, the solubility conditions and additional crosslinking chemical affect protein denaturation, aggregation and crosslinking. Thus, it is expected that these variables will influence the porosity and density of the resulting gel network. The second stage of the process involves drying and can also have a significant effect on the resulting porosity and density, as the solvent is removed from this gel network and replaced with air.

Factors that influence the gel network during the gelation process are the next important variables to be considered after protein type as they control how the protein will aggregate and gel. These gelation techniques depend intimately on the type of protein being gelled. As discussed, pH changes, ionic strength adjustments and heat-induced denaturation involve destabilisation of existing protein chemical bonds and interactions. Some proteins can be gelled at acidic pH values,[29] while others can be gelled at basic or neutral pH values[8,14] depending on the pI. The effect of pH is best seen when compared against the pI of the protein as this indicates how much charge each protein molecule carries and therefore how fast aggregation (and gel formation) can occur. When a lower density is observed, the surface area of the aerogel structure is reduced and the size of the protein aggregates is increased when the pH is close to the pI of the protein (Figure 6.7). These effects can be described by considering how the rate of protein aggregation and subsequent network crosslinking are altered with changes in the protein electrostatic charge and the level of denaturation. At pH values close to the pI of the protein there is little to no repulsion between protein molecules once they are denatured by other means (*e.g.* heat) and thus aggregation (driven by hydrophobic interactions) is fast, creating large globular aggregates of protein before gelation and any crosslinking sets in. As a result, the gel network is comprised of very large aggregates which subsequently create larger pores and less dense aerogels due to limited crosslinking. At pH values far from the pI of the protein, aggregation is slower due to ionic charges on the protein molecules resulting in smaller aggregates where protein denaturation has had more time to allow for unfolding of proteins.[8] These aggregates contain more exposed reactive groups and can then crosslink more, becoming a more filamentous network before aggregate size is increased excessively.[28] Thus, the resulting gels are denser and more homogenous, and pore size is shifted to a more micro- and/or mesoporous range.

In a similar manner, the use of salt in the protein gelation procedure can affect the morphologies of the gels. The best examples of this are whey protein aerogels that are gelled using pH, salt or heat methods. The cold-set gelation method involves the use of both salt and heat, allowing a similar control over the aggregation and gelation rates observed with pH change methods. The addition of salt decreases the surface charge in proteins, thereby increasing aggregation. This is similar to the control of the aggregation rate seen in pH manipulation.[8] It is also possible to denature the

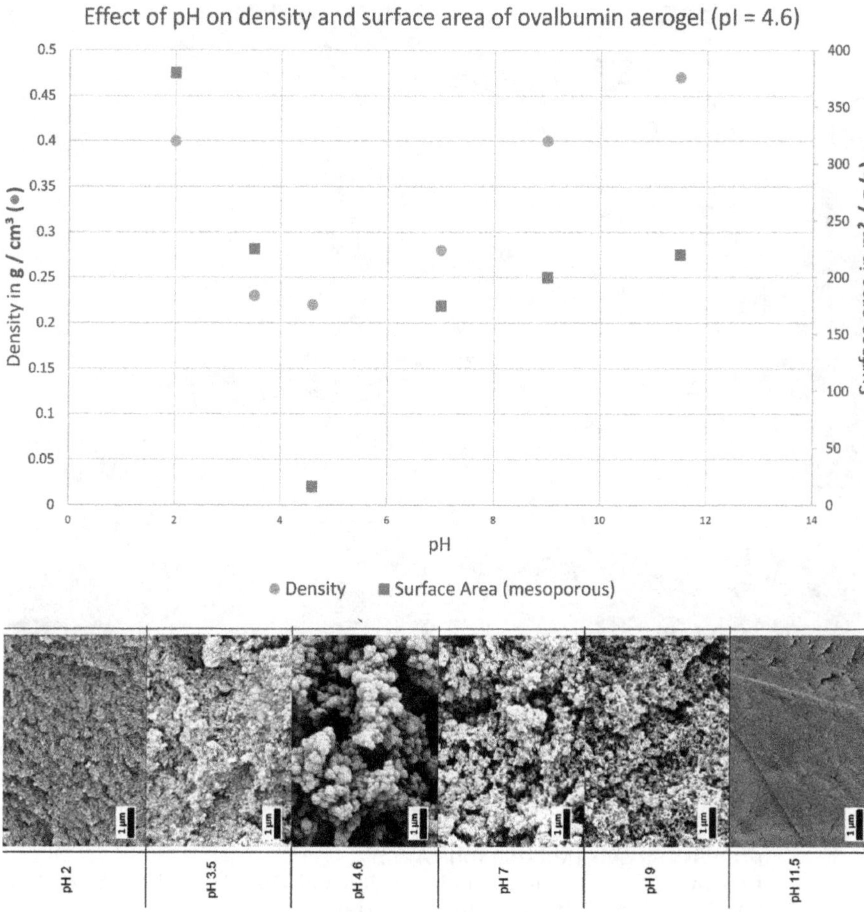

Figure 6.7 (Top) Plot of ovalbumin aerogel densities and BET surface areas with varying pH. (Bottom) SEM images of ovalbumin aerogels showing aggregate size increase as the pH approaches the pI (4.6) of the protein. Both the data for the plot and the SEM images are obtained from ref. 8.
Reprinted from The Journal of Supercritical Fluids, **106**, I. Selmer, C. Kleemann, U. Kulozik, S. Heinrich and I. Smirnova, Development of egg white protein aerogels as new matrix material for microencapsulation in food, 42–49, Copyright 2015, with permission from Elsevier.

protein with heat prior to the salt treatment. When aggregation is subsequently induced, due to the shielding nature of salt ions on the protein charges, repulsion between the protein molecules is reduced resulting in a gel network with a high degree of crosslinking. This results in a stronger network that is mesoporous, rather than the sheet-like and macroporous gels induced by heating only (Figure 6.8).[27,29] Chemical crosslinks also influence the gel network homogeneity and average pore size. The crosslinks, such as the $-CH_2-$ bridges formed between amine groups in the

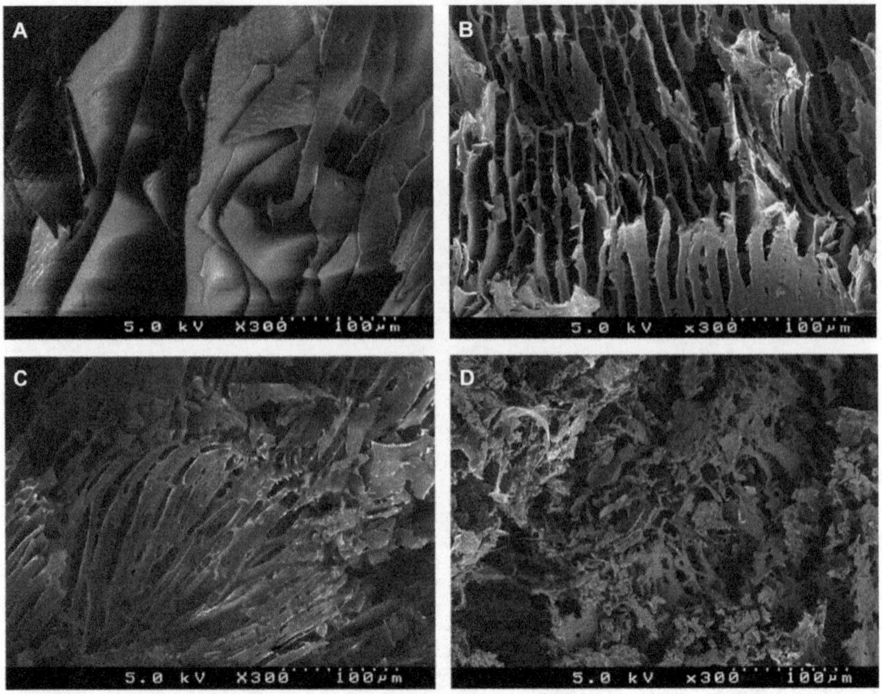

Figure 6.8 SEM micrographs of whey protein isolate (WPI) aerogels. Micrograph B (WPI denatured by salt and heat treatment) demonstrates the effect of dual gelation method over simple heat treatment (micrograph A). Micrographs C (WPI and alginate) and D (WPI, alginate and clay) demonstrate the effect of hybridization with other polymers as compared to pure protein aerogels (Micrographs A & B).
Reprinted from European Polymer Journal, **49**, H.-B. Chen, Y.-Z. Wang and D. A. Schiraldi, Foam-like materials based on whey protein isolate, 3387–3391, Copyright 2013, with permission from Elsevier.

formaldehyde crosslinking reaction, create extensive networks. The resulting aerogels have less distinguishable aggregates, but are more homogeneous with reduced pore sizes (Figure 6.9) and increased BET surface areas.[23]

Following the gelation stage, the morphology of aerogels can be further influenced by the drying process. During SC-CO_2 drying, gels are expected to experience the least amount of surface tension within the pores and therefore are less prone to collapse of the internal structure.[61] Consequently, the aerogels are expected to have the best preserved gel network structures in the final dry scaffold. Freeze-drying techniques are expected to involve a small amount of structural breakdown as the solvent undergoes freezing before it is sublimed at low pressures. Interestingly, the freeze-drying technique tends to produce aerogels with lower densities and higher porosities, compared with those produced by SC-CO_2 drying.[28] The average density of protein aerogels obtained by SC-CO_2 drying is 0.19 g cm^{-3} [8,9,13,18,19,27] while that *via* freeze-drying is 0.12 g cm^{-3}.[5,11,17,19,24,26–29,34] This surprising combination of

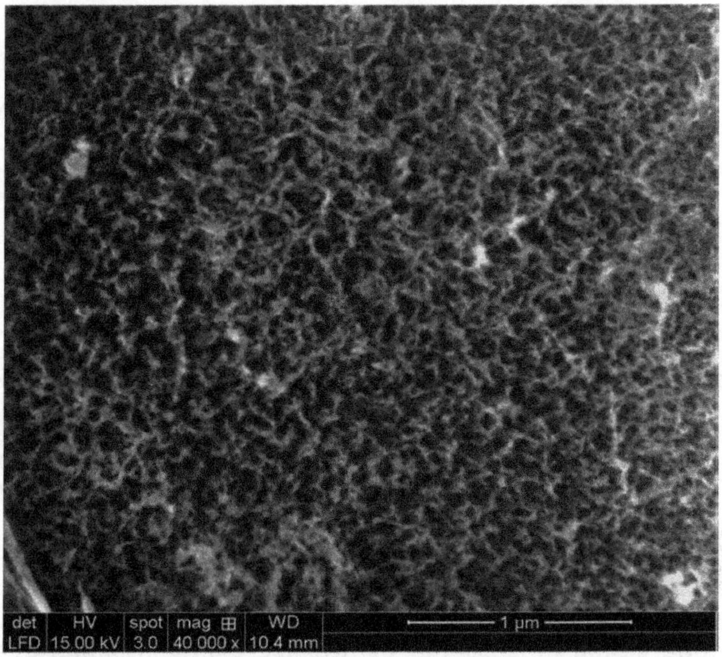

Figure 6.9 SEM micrograph of soy protein based aerogel with tannin and formaldehyde crosslinking. Gels tend to be homogenised and more mesoporous (pore size <100 nm) when crosslinking is strong.
SEM image Adapted from ref. 19 with permission from The Royal Society of Chemistry.

results has been elucidated by Betz *et al.* who concluded that the use of freeze-drying produces a shift in pore size towards a more macroporous structure (pore diameters >50 nm) in which surface area and pore volume readings are harder to determine by BET N_2 adsorption–desorption.[28] Larger ice crystals form as the freezing stage of the process is extended, meaning that growth of the ice porogens increases the pore size over those that were initially present.[23] Thus, aerogels obtained by freeze-drying have lower densities and lower BET surface areas due to the larger pore sizes created *via* large ice-crystal formation.[28] Differences in density and pore size caused by these drying techniques can be observed macroscopically or at higher magnifications using SEM (Figure 6.10).[15]

6.3.3 Hybrid Protein-based Aerogels

Hybrid protein-based aerogel production is driven by the idea of improving or tailoring the resulting properties of the aerogel. In general, proteins are not fibrous in nature like most carbohydrates, which limits their mechanical properties in the wet-gels and aerogels. A comparison of protein gels and

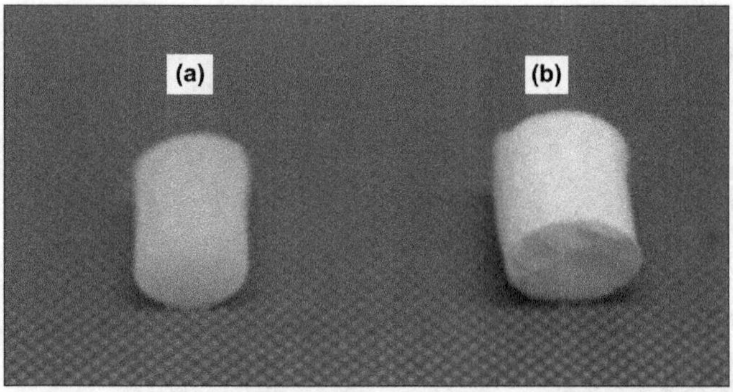

Figure 6.10 Whey protein aerogels prepared by supercritical carbon dioxide drying (a) or by freeze-drying (b).
Reprinted from The Journal of Supercritical Fluids, 72, M. Betz, C.A. García-González, R.P. Subrahmanyam, I. Smirnova and U. Kulozik, Preparation of novel whey protein-based aerogels as drug carriers for life science applications, 111–119, Copyright 2012, with permission from Elsevier.

their hybrids demonstrates a preference for globular (non-fibrous) proteins such as those found in whey and soy isolates, zein and gluten to be supplemented with other non-proteinaceous constituents. Often these second constituents are polysaccharides, which in addition to being biodegradable and biocompatible, can also provide additional strength and absorbency to the aerogels, due to a more fibrous nature.[19] The naturally fibrous (SF) or induced fibrous proteins (amyloid fibrils of β-lactoglobulin) tend to be used in hybrids less often, although this may also be due to the intended application of the aerogel. The obvious exception to this apparent trend is collagen (gelatin) which, with the exception of Kistler's work, has only been produced in hybrids thus far, but is naturally fibrous.[11,35,62–64]

When hybrid protein aerogels are analysed alongside their pure protein counterparts, in general their densities and porosities are more desirable (Table 6.4). The densities are significantly reduced to an average value of approximately 0.08 g cm^{-3}.[11,14,18,20,21,24,26,27,29,36,64] The hybrid aerogels represented are mostly based on polysaccharide–protein aerogels and clay–protein aerogels, having average densities of 0.076 and 0.085 g cm^{-3}, respectively.

Chen *et al.* demonstrated that the density of a WPI aerogel (0.165 g cm^{-3}) reduces as the whey protein is mixed with first alginate (0.098 g cm^{-3}) and then with both clay and alginate (0.082 g cm^{-3}). However, these hybrid densities are still higher than the corresponding alginate aerogel density (0.047 g cm^{-3}). Interestingly, the incorporation of alginate and clay appear to reduce the pore size, which should result in increased density (Figure 6.8). However, Chen *et al.* observed that the alginate hybrids appear to have thinner layered structures when compared to the pure whey protein aerogels, which explains the reduced pore size and density.[27]

Table 6.4 Summary of morphological data and important findings of hybrid protein aerogels.

Protein	Hybrid type	Density g cm^{-3}	Pore volume cm^3 g^{-1}	Pore distribution: Micro	Pore distribution: Meso	Pore distribution: Macro	Average mesopore diameter	SA m^2 g^{-1} (limited to mesopore range)	Findings	Ref.
Whey protein	Alginate	0.06–0.1	—		—		—	—		26
	Cellulose	0.11–0.16	0.0025–0.0078	Micro and mesoporous			—	1.72–3.04	Nano-crystaline cellulose increases pore size while micro-crystalline cellulose reduces it	28
Silk Fibroin	Graphene	—	—	Between graphene sheets: macroporous Within graphene sheets: 1–160 nm pores			11 nm	180.7	Higher silk to graphene ratio results in denser structures	14
	Clay	0.08–0.12	—	—			—	—		36
Soy Protein	Clay	0.06–0.1	—	—			—	—		36
	Cellulose	0.111–0.115	—	Macroporous			—	1.9	Increasing soy content decreases porosity and surface area	19
	Graphene	—	0.02–0.6	Large number of ~2 nm			3–5 nm	30–156	Graphene has slit shaped small pores between layers (with macropores)	20
	Graphene	—	0.843	14%	85%	1%	10	30–119		21

Table 6.4 (*Continued*)

Protein	Hybrid type	Density g cm^{-3}	Pore volume cm^3 g^{-1}	Pore distribution (Micro / Meso / Macro)	Average mesopore diameter	SA m^2 g^{-1} (limited to mesopore range)	Findings	Ref.
Ovalbumin	Cellulose	—	0.18	Mostly mesoporous with 34% microporous volume	—	38		10
Gluten	Cellulose	0.08–0.20	—	Mostly macro (30–73 μm) – pore size reduces as density increases	—	—		17
Zein	Agar	0.03–0.04	—	Mostly macroporous but significant mesopore presence	43 nm	451	Addition of protein to agar reduced pore size, increasing SA	24
	Silica	—	—	Macroporous	—	—		25
Collagen	Cellulose	0.02–0.03	—	—	—	—	Highly absorbent	11
	Alginate	0.043	3.24	—	—	208	Collagen increases pore size, reducing SA and pore volume compared to alginate alone	64

Various other hybrid combinations of proteins and polysaccharides reveal that incorporation of polysaccharides generally leads to reduced densities and pore size shifts towards mesoporosity as compared to the 100% protein aerogels. However, some studies show that the average gel density of protein aerogels is unaffected by the addition of polysaccharides, although they may improve other properties such as the viscosity, homogeneity and strength of the gels.[18]

6.4 Processing–Property Relationships

Looking beyond the morphology, it is possible to draw correlations between processing techniques and other properties. Properties such as strength and absorbency can be influenced by the morphology of the aerogels, which is in turn influenced by the processing as previously discussed. The morphology of the aerogel and the properties it may influence are ultimately connected to the intended application of the aerogel. In silica aerogel technology, high porosity, smaller pore sizes and lower densities are preferable when the aerogels are to be used in thermal insulation.[51,65] Targeted morphologies that are specified by the application are also a focus of protein aerogel development.

In industry, materials must have higher mechanical and insulating properties, while processing must be achieved in a cost effective manner. A technical challenge for aerogel insulating materials is the combination of both low density and small pore size that is useful for increasing the thermal insulating properties of the material.[19] Thus, many protein aerogels that are targeted at general industry material applications tend to use hybrids due to the combined effect of protein and polysaccharide in increasing the strength, and reducing the density and pore size, while remaining economically viable. Mechanical properties are useful measurements for industrial based applications such as packaging and construction, but are also important for medical applications such as tissue culture. A useful comparison for bio-aerogel mechanical performance are the compressive moduli of silica aerogels which range from 50–400 kPa and are current competitors for the medical applications of aerogels.[57,58] Protein aerogels demonstrate compressive moduli ranging from 19 kPa[15] to 18.2 MPa[27] and with an average of 2.5 MPa[5,6,15,27] they can offer a competitive edge over silica aerogels. However, polysaccharides are by comparison a much better mechanically performing aerogel option and thus the polysaccharide-protein hybrid aerogels offer the best compromise between strength, biodegradability and biocompatibility.

Whey protein aerogels produced by the cold-set gelation method (combination of heat denaturation and salt gelation) have significantly improved mechanical properties, with compressive moduli up to 20 times higher than those prepared by a simple heat gelation process of equivalent density.[27] This effect is undoubtedly due to the crosslinking and protein association in the gel network and could become a useful insight for future protein aerogel research into tailored mechanical properties. Mechanical properties are improved in hybrid protein aerogels with an average compressive modulus of 3320 kPa (calculated using all reported compressive moduli from ref. 20, 27, 36). Protein aerogels are in general not this strong without a corresponding loss in density. The work of Arboleda *et al.* extensively examines how the cellulose content of soy protein–cellulose hybrid aerogels affects the mechanical properties of the gels. Not only does the addition of the nano-fibrillar cellulose markedly improve the compressive modulus of the soy aerogel, but interestingly, the effect is not linear. The improved compressive modulus can be seen at just a 1 : 5 ratio of cellulose to soy protein. Aerogels with just a 1 : 3 ratio of cellulose to soy protein demonstrate equivalent strength to pure nano-fibrillar cellulose gels. Arboleda *et al.* concluded that the inclusion of the cheaper soy protein in these cellulose aerogels *"endows the system with interesting chemical features while maintaining a high compression modulus"*.[20] Clearly the hybrid has advantages over both types of parent aerogels. Such studies are a useful indication of how cheaper protein products, such as soy, can be incorporated into the stronger polysaccharide aerogels for viable industrial applications.

Tissue engineering applications are at the centre of SF aerogel research as silk is already a popular material for *in vivo* studies and medical applications. The unusual amino acid composition and subsequent secondary protein structure makes it less vulnerable to degradation *in vitro* and *in vivo* and its natural fibrous structure and mechanical strength make it ideal as a scaffold.[38] Additionally, in medical applications larger aerogel pores are more suitable for supporting cell growth in tissue cultures as cell sizes approach micrometre scales rather than nanometre scales.[15] Thus, the SF aerogel field is suited for salt leaching, CO_2 acidification and freeze-drying techniques that lead to the larger pore sizes required for cell culture scaffolds.

In laboratory technology, high pore volume and surface areas at the nanometre scale indicate increased efficiency of aerogels as absorbents and filters.[46] Absorbency is crucial for applications in drug delivery, waste-water treatments and bio-sensor applications.[66] Thus, many aerogels intended as absorbent materials in medical or laboratory applications have processes optimised for more homogenous gels with reduced pore sizes and larger surface areas.

6.5 Protein Encapsulation

Proteins are also used in silica aerogel technology in a hybrid-like structure resulting in *biologically functional* encapsulated proteins. This technology, although not strictly 'bio' in its origin due to the use of silica aerogels, can be considered a type of hybrid material with proteins as one component. Essentially a silica aerogel matrix is impregnated with protein (more specifically, an enzyme) which is immobilised inside the pores of the structure, rather than being a structural element of the gel network itself. The purpose of this protein encapsulation is to create a bio-catalytic or bio-specific filtration scaffold, a material which can detect certain molecules in liquid or air. Extending this concept out of the laboratory allows these bio-catalytic scaffolds to be used in applications such as forensics and waste-water treatment in addition to laboratory assays. Enzyme catalysis and bio-detection offer an unparalleled level of chemical specificity which is exploited in these types of applications and aerogels and can help them become more practical.

Enzymes are utilized in laboratories for a myriad of catalytic identification and isolation assays of biological fluids. These assays are usually conducted in solution, in a single vessel. Immobilizing enzymes on a solid scaffold such as an aerogel adds utility and practicability to these techniques. The key to immobilising an enzyme on a surface is being able to fix the enzyme without disturbing its biological functionality. Rather than a chemical immobilization, encapsulation within the pores of an aerogel physically restrains the enzyme (Figure 6.11), allowing it to retain its normal chemical function. Furthermore, the open porous nature of the aerogel allows the testing fluid to permeate the structure and filter through the enzyme laden matrix, resulting in a chemically selective (*via* the enzyme) filtration, catalysis or bio-detection of that fluid. In fact, the very first report of enzyme encapsulation in aerogels indicates that the matrix can even protect the immobilised enzyme from external degradation by factors such as heat.[67] Since this first report, research in this area has included various processing and solvent techniques,[68] application in biodiesel processing,[69] and the ability of these aerogel-enzymes to be bioactive in air.[70–73] Although this technology produces aerogels with a different structural matrix to the protein aerogels discussed thus far, the processing techniques are remarkably similar in a practical sense. Preparing the gel is performed using the same two-phase process of gelation and drying. The enzyme is simply added at the wet-gel stage prior to drying. The technique allows the enzyme to permeate the silica gel network and once dry remains immobilised within the pores of the silica matrix.

The incredible feature of this technology is its ability to retain the chemical functionality of the enzyme after the drying phase. Maury *et al.* were able to demonstrate that encapsulation within an aerogel matrix can even enhance the performance and the storage stability of the enzyme.[74] Manipulation of the aerogel properties, such as pore size, has also allowed this type of bio-incorporated aerogel technology to extend to bacterial cells. Bacterial cells have sizes on the micrometer scale rather than nanometre (as for proteins) and so encapsulation requires a macroporous aerogel structure. Power *et al.* demonstrated this novel concept by encapsulating bacterial cells in a silica aerogel and using it to detect the presence of a virus in air.[66] This work demonstrates the huge potential of aerogels and their unique properties within the field of bio-detection and bio-sensors.

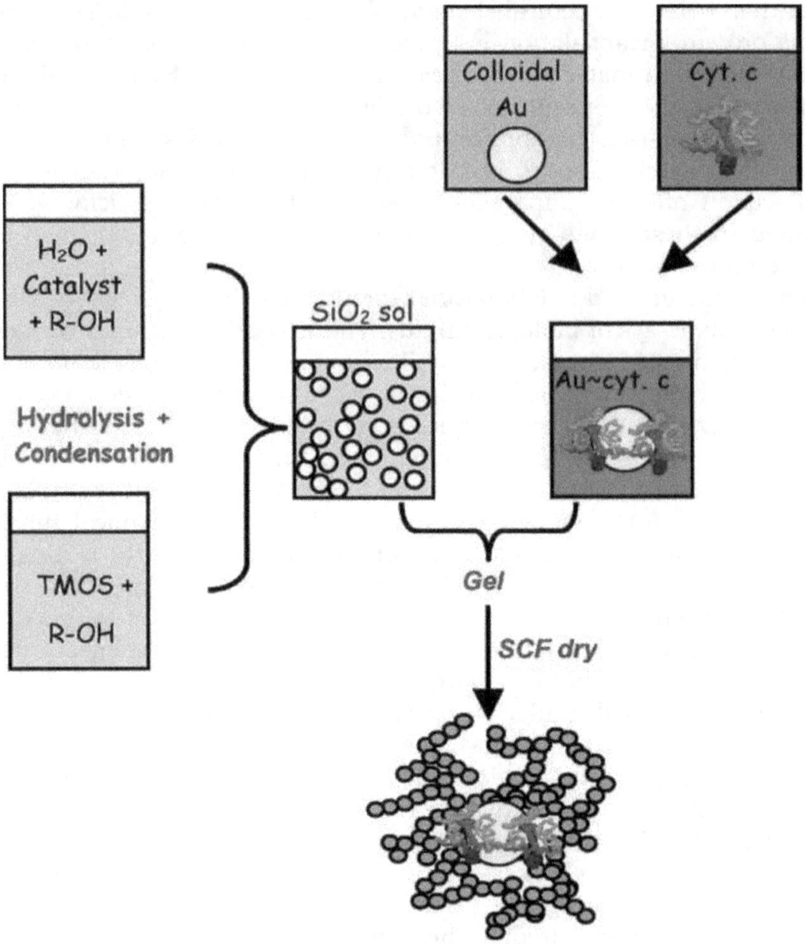

Figure 6.11 Schematic of the process used to encapsulate a cytochrome c protein within a silica aerogel.
Reprinted from Journal of Non-Crystalline Solids, 350, J. M. Wallace, R. M. Stroud, J. J. Pietron, J. W. Long and D. R. Rolison, The effect of particle size and protein content on nanoparticle-gold-nucleated cytochrome c superstructures encapsulated in silica nanoarchitectures, 31–38, Copyright 2004, with permission from Elsevier.

6.6 Conclusions

Protein aerogels are able to produce equivalent morphological features to other aerogels. The most exciting prospect is the vast chemical potential of protein molecules and the unique chemical and biochemical specificity they may offer, both during processing and in the resulting aerogel. They can obtain densities rivalling those of the silica and synthetic carbon aerogels and also generate variety in pore size. Their inherent biocompatibility and biodegradability is retained throughout the aerogel processing, which can rely solely on simple aqueous chemistry and freeze-drying techniques. In addition, proteins obtained as a by-product from industry, in particular from the food processing industry, can be a relatively cheap and an accessible source of biopolymers for aerogels. The morphology and strength can be improved by protein choice, process parameters and hybridisation with polysaccharides and clay. Manipulation of the process parameters is generally chemically simple (pH or salt concentration changes) and most improvements in morphology are achieved using the cheaper and easier freeze-drying technique rather than supercritical CO_2 drying. The catalytic potential of proteins has been exploited by encapsulating enzymes in aerogel matrices, a unique use of proteins in aerogel technology. Less desirable qualities of the protein aerogels are generally higher densities and lower porosities along with relatively weaker mechanical properties when compared to polysaccharide-based bio-aerogels. However, protein-based bio-aerogels already offer improved chemical functionality, cost and control of morphologies. The scope of the research in this field is large and promising, and much more specific research is needed to further elucidate the connections between process and morphology. A greater understanding of these connections will lead to better customisation of the protein aerogels to a particular application. The increasing focus on this field should produce further process improvements and help achieve similar standards in density, strength and porosity as the polysaccharide-based aerogels.

References

1. S. S. Kistler, *Nature*, 1931, **127**, 741.
2. A. Du, B. Zhou, Z. Zhang and J. Shen, *Materials*, 2013, **6**, 941.
3. L. M. Hair, R. W. Pekala, R. E. Stone, C. Chen and S. R. Buckley, *J. Vac. Sci. Technol., A*, 1988, **6**, 2559–2563.
4. G. M. Glenn and D. W. Irving, *Cereal Chem.*, 1995, **72**, 155–161.
5. R. Nazarov, H.-J. Jin and D. L. Kaplan, *Biomacromolecules*, 2004, **5**, 718-726.
6. U.-J. Kim, J. Park, H. Joo Kim, M. Wada and D. L. Kaplan, *Biomaterials*, 2005, **26**, 2775–2785.
7. S. Kistler, *J. Phys. Chem.*, 1932, **36**, 52–64.
8. I. Selmer, C. Kleemann, U. Kulozik, S. Heinrich and I. Smirnova, *J. Supercrit. Fluids*, 2015, **106**, 42–49.
9. R. J. White, N. Yoshizawa, M. Antonietti and M.-M. Titirici, *Green Chem.*, 2011, **13**, 2428–2434.
10. S.-M. Alatalo, F. Pileidis, E. Mäkilä, M. Sevilla, E. Repo, J. Salonen, M. Sillanpää and M.-M. Titirici, *ACS Appl. Mater. Interfaces*, 2015, 7, 25875–25883.
11. T. Lu, Q. Li, W. Chen and H. Yu, *Compos. Sci. Technol.*, 2014, **94**, 132–138.
12. R. R. Mallepally, M. A. Marin and M. A. McHugh, *Acta Biomater.*, 2014, **10**, 4419–4424.

13. M. A. Marin, R. R. Mallepally and M. A. McHugh, *J. Supercrit. Fluids*, 2014, **91**, 84–89.
14. Y. S. Yun, S. Y. Cho and H.-J. Jin, *Macromol. Res.*, 2014, **22**, 509–514.
15. R. R. Mallepally, M. A. Marin, V. Surampudi, B. Subia, R. R. Rao, S. C. Kundu and M. A. McHugh, *Biomed. Mater.*, 2015, **10**, 035002.
16. L. Baldino, S. Cardea and E. Reverchona, *Chem. Eng.*, 2016, 49.
17. D. Yao, S. Dong, Q. Lu, X. Hu, D. L. Kaplan, B. Zhang and H. Zhu, *Biomacromolecules*, 2012, **13**, 3723–3729.
18. T. O. J. Blomfeldt, R. T. Olsson, M. Menon, D. Plackett, E. Johansson and M. S. Hedenqvist, *Macromol. Mater. Eng.*, 2010, **295**, 796–801.
19. G. Amaral-Labat, L. Grishechko, A. Szczurek, V. Fierro, A. Pizzi, B. Kuznetsov and A. Celzard, *Green Chem.*, 2012, **14**, 3099–3106.
20. J. C. Arboleda, M. Hughes, L. A. Lucia, J. Laine, K. Ekman and O. J. Rojas, *Cellulose*, 2013, **20**, 2417–2426.
21. Y. Zhuang, F. Yu, J. Ma and J. Chen, *New J. Chem.*, 2015, **39**, 3333–3336.
22. Y. Zhuang, F. Yu, J. Ma and J. Chen, *RSC Adv.*, 2015, **5**, 27964–27969.
23. B. Bi, Q. Lin, P. Zheng, F. Li, T. Ouyang and N. Chen, *Trans. Chin. Soc. Agric. Eng.*, 2016, **32**, 309–314.
24. Y. Zhuang, F. Yu, J. Ma and J. Chen, *Desalin. Water Treat.*, 2016, **57**, 9510–9519.
25. A. A. Shamsuri, D. K. Abdullah and R. Daik, *Cellul. Chem. Technol.*, 2012, **46**, 45–52.
26. L. Verdolotti, M. Oliviero, M. Lavorgna, V. Iozzino, D. Larobina and S. Iannace, *J. Cell. Plast.*, 2014, 0021955X14529138.
27. H.-B. Chen, Y.-Z. Wang and D. A. Schiraldi, *Eur. Polym. J.*, 2013, **49**, 3387–3391.
28. M. Betz, C. García-González, R. Subrahmanyam, I. Smirnova and U. Kulozik, *J. Supercrit. Fluids*, 2012, **72**, 111–119.
29. M. Ahmadi, A. Madadlou and A. A. Saboury, *Food Chem.*, 2016, **196**, 1016–1022.
30. G. Nyström, M. P. Fernández-Ronco, S. Bolisetty, M. Mazzotti and R. Mezzenga, *Adv. Mater.*, 2016, **28**, 472–478.
31. T. Pojanavaraphan, R. Magaraphan, B.-S. Chiou and D. A. Schiraldi, *Biomacromolecules*, 2010, **11**, 2640–2646.
32. H.-B. Chen, B.-S. Chiou, Y.-Z. Wang and D. A. Schiraldi, *ACS Appl. Mater. Interfaces*, 2013, **5**, 1715–1721.
33. G. H. Altman, F. Diaz, C. Jakuba, T. Calabro, R. L. Horan, J. Chen, H. Lu, J. Richmond and D. L. Kaplan, *Biomaterials*, 2003, **24**, 401–416.
34. Q. Lv and Q. Feng, *J. Mater. Sci.: Mater. Med.*, 2006, **17**, 1349–1356.
35. G. R. Bardajee, A. Pourjavadi and R. Soleyman, *J. Polym. Res.*, 2011, **18**, 337–346.
36. K. Finlay, M. D. Gawryla and D. A. Schiraldi, *Ind. Eng. Chem. Res.*, 2008, **47**, 615–619.
37. A. Totosaus, J. G. Montejano, J. A. Salazar and I. Guerrero, *Int. J. Food Sci. Technol.*, 2002, **37**, 589–601.

38. A. Matsumoto, J. Chen, A. L. Collette, U.-J. Kim, G. H. Altman, P. Cebe and D. L. Kaplan, *J. Phys. Chem. B*, 2006, **110**, 21630–21638.
39. F. Van Kleef, J. Boskamp and M. Van den Tempel, *Biopolymers*, 1978, **17**, 225–235.
40. A. Totosaus, J. G. Montejano, J. A. Salazar and I. Guerrero, *Int. J. Food Sci. Technol.*, 2002, **37**, 589–601.
41. M. Ahmadi, A. Madadlou and A. A. Sabouri, *Food Chem.*, 2015, **174**, 97–103.
42. A. C. Alting, R. J. Hamer, C. G. de Kruif and R. W. Visschers, *J. Agric. Food Chem.*, 2003, **51**, 3150–3156.
43. S. Brunauer, P. H. Emmett and E. Teller, *J. Am. Chem. Soc.*, 1938, **60**, 309–319.
44. M. Dubinin, *Carbon*, 1989, **27**, 457–467.
45. E. P. Barrett, L. G. Joyner and P. P. Halenda, *J. Am. Chem. Soc.*, 1951, **73**, 373–380.
46. Z. Ulker and C. Erkey, *J. Controlled Release*, 2014, **177**, 51–63.
47. K. Sing, *Colloids Surf., A*, 2001, **187**, 3–9.
48. S. Brunauer, P. H. Emmett and E. Teller, *J. Am. Chem. Soc.*, 1938, **60**, 309–319.
49. S. Westermarck, Use of mercury porosimetry and nitrogen adsorption in characterisation of the pore structure of mannitol and microcrystalline cellulose powders, granules and tablets, 2000.
50. X. Chang, D. Chen and X. Jiao, *J. Phys. Chem. B*, 2008, **112**, 7721–7725.
51. A. S. Dorcheh and M. Abbasi, *J. Mater. Process. Technol.*, 2008, **199**, 10–26.
52. S. A. Al-Muhtaseb and J. A. Ritter, *Adv. Mater.*, 2003, **15**, 101–114.
53. F. Kirkbir, H. Murata, D. Meyers and S. R. Chaudhuri, *J. Non-Cryst. Solids*, 1998, **225**, 14–18.
54. R. W. Pekala and D. W. Schaefer, *Macromolecules*, 1993, **26**, 5487–5493.
55. H. Sun, Z. Xu and C. Gao, *Adv. Mater.*, 2013, **25**, 2554–2560.
56. F. J. Maldonado-Hódar, M. A. Ferro-Garcıa, J. Rivera-Utrilla and C. Moreno-Castilla, *Carbon*, 1999, **37**, 1199–1205.
57. T. M. Tillotson and L. W. Hrubesh, *J. Non-Cryst. Solids*, 1992, **145**, 44–50.
58. R. Pekala, C. Alviso, F. Kong and S. Hulsey, *J. Non-Cryst. Solids*, 1992, **145**, 90–98.
59. C. Rudaz, R. Courson, L. Bonnet, S. Calas-Etienne, H. Sallée and T. Budtova, *Biomacromolecules*, 2014, **15**, 2188–2195.
60. C. A. García-González, M. Alnaief and I. Smirnova, *Carbohydr. Polym.*, 2011, **86**, 1425–1438.
61. H. D. Gesser and P. C. Goswami, *Chem. Rev.*, 1989, **89**, 765–788.
62. L. Baia, M. Baia, V. Danciu, M. G. Albu, V. Coşoveanu, D. Iordăchescu and V. Trandafir, *J. Optoelectron. Adv. Mater.*, 2008, **10**, 933–936.
63. M. Sachithanadam and S. C. Joshi, *Procedia Eng.*, 2014, **75**, 51–55.
64. S. P. Raman, P. Gurikov and I. Smirnova, *J. Supercrit. Fluids*, 2015, **106**, 23–33.
65. A. C. Pierre and G. M. Pajonk, *Chem. Rev.*, 2002, **102**, 4243–4266.

66. M. Power, B. Hosticka, E. Black, C. Daitch and P. Norris, *J. Non-Cryst. Solids*, 2001, **285**, 303–308.
67. S. Braun, S. Rappoport, R. Zusman, D. Avnir and M. Ottolenghi, *Mater. Lett.*, 1990, **10**, 1–5.
68. Y. K. Li, M. J. Chou, T. Y. Wu, T. R. Jinn and Y. W. Chen-Yang, *Acta Biomater.*, 2008, **4**, 725–732.
69. O. Orçaire, P. Buisson and A. C. Pierre, *J. Mol. Catal. B: Enzym.*, 2006, **42**, 106–113.
70. A. S. Harper-Leatherman, M. Iftikhar, A. Ndoi, S. J. Scappaticci, G. P. Lisi, K. L. Buzard and E. M. Garvey, *Langmuir*, 2012, **28**, 14756–14765.
71. J. M. Wallace, J. K. Rice, J. J. Pietron, R. M. Stroud, J. W. Long and D. R. Rolison, *Nano Lett.*, 2003, **3**, 1463–1467.
72. A. S. Harper-Leatherman, E. R. Pacer and N. D. Kosciuszek, *J. Visualized Exp.*, 2016, **109**, e53802.
73. J. M. Wallace, R. M. Stroud, J. J. Pietron, J. W. Long and D. R. Rolison, *J. Non-Cryst. Solids*, 2004, **350**, 31–38.
74. S. Maury, P. Buisson, A. Perrard and A. C. Pierre, *J. Mol. Catal. B: Enzym.*, 2005, **32**, 193–203.

CHAPTER 7

Hybrid Green Aerogels: Processing and Morphology

KOW KIEN-WOH*[a] AND ROZITA YUSOFF[b]

[a] Department of Chemical and Environmental Engineering, Faculty of Science and Engineering, The University of Nottingham Ningbo China, Ningbo 315100, China; [b] Department of Chemical Engineering, Faculty of Engineering, University of Malaya, 50603 Kuala Lumpur, Malaysia
*Email: Kien-Woh.Kow@nottingham.edu.cn; kowkw@outlook.com

7.1 Types of Hybrid Bio-based Aerogels

Hybrid bio-based aerogels are formed by incorporating other phases of materials into bio-based aerogels. The rationale for such hybridization can be one of the following, or a combination of both:

(i) Synergistic effect
   Pure aerogels usually exhibit undesirable properties that might weaken their performance. Incorporating other materials can have a remarkable positive impact on the properties of the hybrid aerogels, resulting in better stability (reduced agglomeration of particles, oxidation rate), mechanical properties (reduced fragility[1–11]), adsorption performance (adsorption capacity,[12–16] selectivity[17–21]), thermal properties,[3–7,9,22–29] electrical properties,[2,30] magnetism[31–34] and biocompatibility (scaffold for tissue engineering,[35–38] drug delivery[39–42]).

(ii) Reaction template
   The porous structure of aerogels is suitable to act as a template for the synthesis of nanoparticles. This template can nucleate and control

Green Chemistry Series No. 58
Biobased Aerogels: Polysaccharide and Protein-based Materials
Edited by Sabu Thomas, Laly A. Pothan and Rubie Mavelil-Sam
© The Royal Society of Chemistry 2018
Published by the Royal Society of Chemistry, www.rsc.org

the growth of nanoparticles in three ways. The simplest way is by acting as an inert support to control the uniform size and/or shape growth.[24,31,43–45] In addition, particle size, shape and pore restriction can also immobilize particles from agglomeration. Such immobilization is crucial to fully utilize the functionality of nanoparticles, which otherwise may be severely deteriorated upon agglomeration. However, this physical immobilization may still be dislodged from the template when it is used under vigorous agitation. The second way for the template to hold the nanoparticles in place is by electrostatic interaction and hydrogen bonding originating from the hydroxyl and oxygen-rich groups.[33,46] These functional groups can serve as nano-reacting sites, whereby inorganic nanoparticles can be anchored tightly to avoid particle agglomeration and hence achieve controlled growth. Finally, such functional groups can also act as reducing sites that react with nanoparticle precursors to nucleate nanoparticles.[10,34,47–57] The resulting hybrid bio-based aerogel may[24,31,33,34,43–45,48–50] or may not[46] exhibit a synergistic effect. In some cases, cellulose aerogel was calcined[58] to recover the nanoparticles, as cellulose can easily be burned.

In the early 2000s, hybrid aerogels were first reported separately by two research teams led by Hunt[41] and Risen[59] respectively. Hunt's team tried to improve the biocompatibility of a silica aerogel by adding chitosan, whereas Risen's team focused on using a chitosan-silica aerogel as a template to synthesize metallic nanoparticles. Figure 7.1 shows that during the period from 2000 to 2005, this area was mainly dominated by Risen's research team, while research on bio-based hybrid aerogels was mainly silent during this period. Bio-based hybrid aerogels regained attention amongst the research community a few years after the success of a cellulose aerogel, which was synthesized by Josef's research group in 2006.[60] This cellulose aerogel offered a new bio-based precursor to synthesize hybrid aerogels. The high biocompatibility of cellulose aerogels that could potentially be applied to tissue engineering and drug delivery caused research on cellulosic hybrid aerogels to boom after 2011.

It is generally accepted that bio-based hybrid aerogels contain two or more phases with various chemical components. However, there is no classification for the type of structures formed. Hence, it is crucial to define some of the common terms that describe different types of hybrid structures. The essence of an aerogel structure is the continuous backbone that gives rise to its rigidity and porosity. Throughout this chapter, the term "skeletal phase" is used to describe this continuous backbone of the aerogel. This term is equivalent to "primary phase", "primary network", "host" and "backbone" in the literature. Many terms have been used to describe other phase/s that were added to synthesize the hybrid aerogel. All these terms are related to how these phases are being incorporated into the skeletal phase, which includes the secondary phase, incorporated phase, embedded phase,

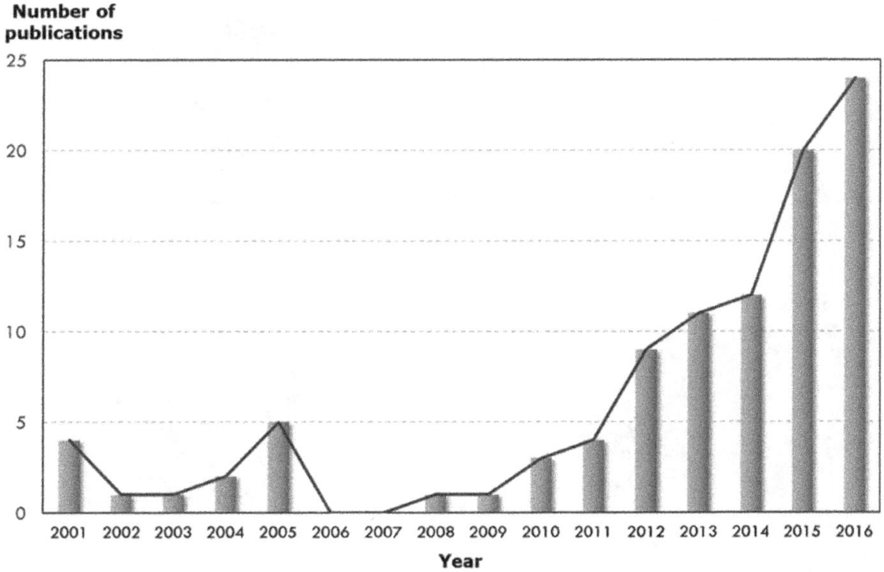

Figure 7.1 Number of publications related to bio-based hybrid aerogels from 2000–2016 (based on the bibliography list in this chapter).

anchored phase, coated phase, dopant, shell structure, encapsulation agent, filling agent and the cross-linked agent. Hence, in general, the term "incorporated phase" is used to describe any material added to the skeletal phase in the hybrid aerogel throughout this chapter. The incorporated phase can be further divided into the following categories, depending on the hybrid structure formed:

(a) Coating agent
The incorporated phase forming a monolayer or multilayer particles on the surface of the skeletal phase throughout the gel network as shown in Figure 7.2b.

(b) Encapsulation
Similar to the coating agent, the incorporated phase forms a monolayer or multilayer particles on the surface of the skeletal phase. However, the encapsulation only involves the outer surface of the skeletal phase. It is the outer surface of the skeletal phase that is either encapsulating (shell) other particles (please see Figure 7.2d) or being encapsulated (core) by other particles (see Figure 7.2c). In either case, the inner pores of the skeletal phase are not in contact with other particles. Hence, it is less homogenous as compared with the coating agent.

(c) Anchored phase
The incorporated phase is grafted onto the skeletal phase as shown in Figure 7.2e. One end of the anchored phase is usually bonded to the active sites of the skeletal phase, which is spaced spatially in the

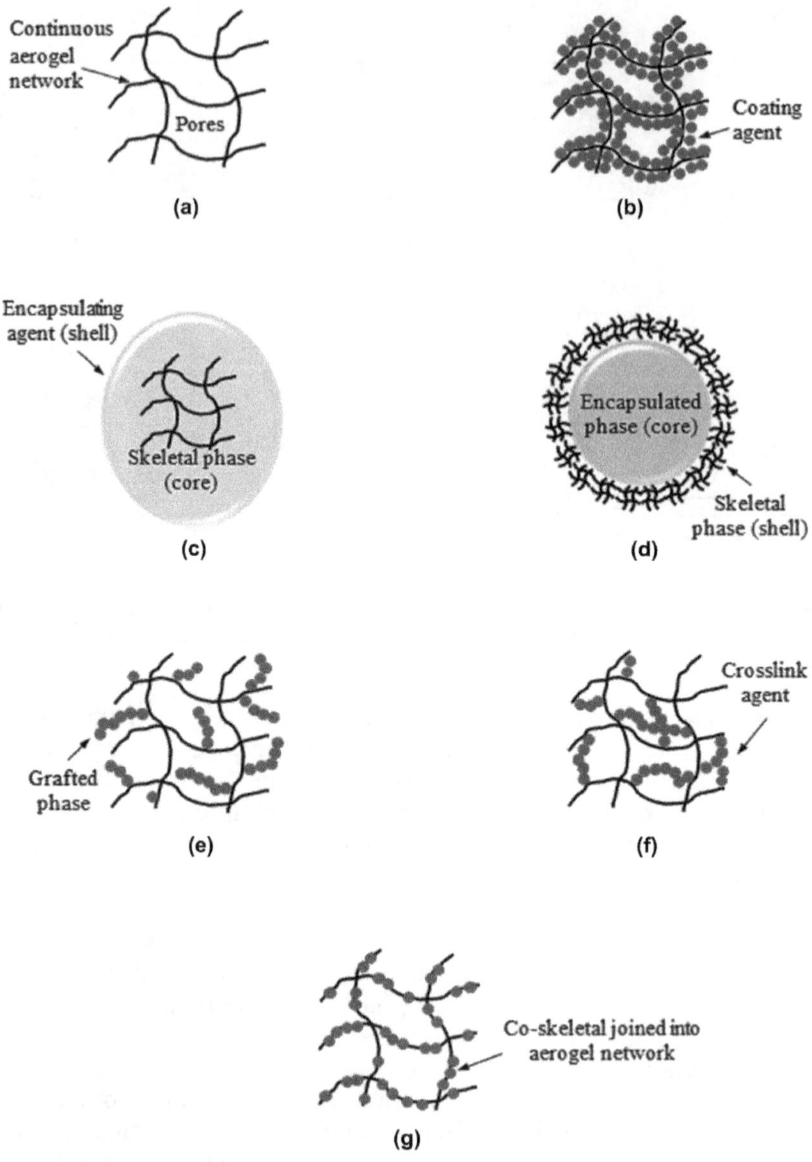

Figure 7.2 Structures in the hybrid aerogel: (a) skeletal phase; (b) coating; (c) encapsulation (skeletal phase as core); (d) encapsulation (skeletal phase as shell); (e) anchor phase; (f) crosslink agent and (g) co-skeletal phase.

network of the skeletal phase, whereas the other end of the incorporated phase is "free" and not bonded to the skeletal phase.

(d) Crosslink agent

Unlike the anchored phase, the crosslink agent binds the ends of its particles to at least two sides of the skeletal phase (see Figure 7.2f).

Hence, two or more sides of the skeletal phase are linked together by the crosslink agent.

(e) Co-skeletal phase

The incorporated phase is joined into the skeletal phase to form part of the network as shown in Figure 7.2g. Since both phases form the network, the phase having a smaller composition is termed the "co-skeletal" phase.

7.1.1 Organo-hybrid

The organic compounds that are incorporated into bio-based aerogels are mostly aimed toward drugs delivery purposes, as shown in Table 7.1. Less emphasis has been placed on improving the mechanical or thermal properties, as in the case of inorgano-hybrid aerogels. This may be due to the nature of the organic compound, commonly having weak bonding and thus not showing

Table 7.1 Types of organo-hybrid bio-based aerogel.

Bio-based aerogel	Incorporated phase	Purpose
Protein	(i) Alginate	Reduce brittleness[1]
	(ii) Ketoprofen	Drug delivery[39]
	(iii) Ibuprofen	Drug delivery[40]
	(iv) Nano-crystalline cellulose	Nutraceuticals[61]
Nano-crystalline cellulose (NCC)	(i) Bleached cellulose fibers	Mechanical properties[5]
Nano-fibrillated cellulose (NFC)	(i) Polyvinyl-alcohols	Superabsorbent[20,21]
	(ii) Xyloglucan	Mechanical properties[62]
	(iii) Bleached cellulose fibers	Mechanical properties[5]
	(iv) Methyl trichlorosilane	Oil absorption[63]
	(v) Collagen	Biocompatibility[64]
	(vi) Poly(*N*-isopropyl-acrylamide)	Thermo-responsiveness[65]
	(vii) Various polymers	Biocompatibility[66]
Chitosan	(i) Gelatin	Mechanical properties[35]
	(ii) Polybenzoxazine	CO_2 adsorption[67]
Alginate	(i) Pectin and carrageenan	Drug delivery,[42] porous structure[68]
	(ii) Starch	Mechanical properties,[13] porous structure[68]
	(iii) Gelatin	Mechanical properties,[35] porous structure[68]
	(iv) Lignin	Non-cytotoxicity,[38] porous structure[68]
Agar	(i) Cellulose/rice starch/zein protein	Dissolution and crosslink agent[69]
Pectin	(i) Aniline	Conductive polymer[70]

much of the synergistic effect in terms of mechanical and thermal properties. On the other hand, organo-hybrid bio-based aerogels that are totally made up of organic compounds may have the advantage of being biocompatible and biodegradable. In this sense, organo-hybrid aerogels are considered to be more environmental friendly as compared to the inorgano-hybrid aerogels.

7.1.2 Inorgano-hybrid

Incorporating inorganic compounds into bio-based aerogels is usually targeted towards improving the physical properties such as mechanical, thermal, magnetic and adsorption properties. The inorganic incorporated compounds consist mainly of two categories of materials, these are nanomaterials and ceramics. While ceramics are incorporated to improve the thermal and adsorption properties of aerogels, nanomaterials are used to add on a wide range of functionalities such as antibacterial activity, photocatalytic activity, photo-luminescence, magnetic properties, mechano-responsiveness and so forth, as stated in Table 7.2.

7.2 Synthesis

Many routes were proposed to synthesize hybrid bio-based aerogels. In general the synthesis involves the following steps: (i) formation of sol from raw materials, (ii) gelation of sol to form the skeletal phase, (iii) doping incorporated phase to the skeletal phase, and (iv) drying of the hybrid gel.

7.2.1 Formation of Sol

Raw materials are dissolved to form the sol which contains small clusters (monomers, dimers, oligomers) of atoms prior to gelation. The dissolution methods of hybrid bio-based aerogels are mainly adapted from the pure bio-based aerogels such as cellulose, chitosan, pectin, alginate, and so forth. Among the bio-based precursors, those with limited solubility in water pose a major challenge in the dissolution step. The insolubility of precursors such as cellulose and chitin is attributed to the tight hydrogen bonding between the polymer chains. Hence, both chemical and physical pre-treatments are required to weaken the hydrogen bonding prior to dissolution. Hydrogen bonding cleavage can be accomplished using either one of the following treatments:

(a) Polar solvent systems such as N-methylmorpholine oxide (NMMO).[22,60,90]
(b) TEMPO (2,2,6,6-tetramethylpiperidine-1-oxylradical) mediated oxidation and related hybrid solvents.[5,20,74,91]
(c) Alkaline/urea aqueous solutions.[6,22,24,31,32,34,46,49,58,60,63,87,90]
(d) Ionic liquids (IL).[22,44,90,92]
(e) Deep eutectic solvents (DES).[90]

These chemical pre-treatments are usually accompanied by other physical pre-treatments such as freezing-thawing, mechanical shearing and/or ultrasonic impingement to further weaken the hydrogen bonding.

Table 7.2 Types of inorgano-hybrid bio-based aerogel.

Bio-based aerogel	Incorporated phase	Purpose
Nano-crystalline cellulose (NCC)	(i) Graphene oxide	Adsorption[18]
Nano-fibrillated cellulose (NFC)	(i) Silver	Mechanical properties[71]
	(ii) Silica	Thermal stability,[72] mechanical properties,[36,73] heat insulation,[6,9,23] acoustic insulation[23]
	(iii) Zeolite	Heat insulation[74]
	(iv) Aluminum hydroxide	Flame retardancy[28]
	(v) Zinc oxide	Antibacterial activity[46]
	(vi) Polypyrrole and silver	Antibacterial activity,[49] electrical conductive and pressure responsive properties[75]
	(vii) Titanium oxide	Adsorption,[76,77] mechanical properties,[78] photocatalytic activity[58,79,80]
	(viii) Carbon nanotube	Mechanoresponsive conductivity,[33,34] electrical fatigue behaviour[81]
	(ix) Cobalt ferrites	Magnetic behaviour,[31,33] mechanical properties[82]
	(x) Graphene oxide	Electromagnetic shielding,[32] heat insulation[27]
	(xi) Silver, gold, and platinum nanoparticles	Supporting medium[43,45]
	(xii) $Ag–Ag_2S$	Catalysis[47]
Chitosan	(i) Graphene oxide	CO_2 capture[12]
	(ii) Graphene	Capacitance[30]
	(iii) Silica	Porous structure,[11,59] mechanical properties,[8] composite structure,[83] biocompatibility,[41] absorption,[84] adsorption[24,46,49,85]
	(iv) Zeolite	Adsorption[16]
	(v) Titanium oxide	Surface area[11]
	(vi) Silica–Au(III)	Photo-formation of gold nanoparticles[55]
	(vii) Montmorillonite	Adsorption[14]
	(viii) Magnesium hydroxide	Flame retardancy[24]
	(ix) Zn–In–Cu alloy quantum dot	Photo-luminiscence[44]
	(x) Aniline–silica	Template to synthesize silver and gold nanoparticles[48]
	(xi) Metallic nanoparticles-silica	Functionalized composite[86]
Chitin	(i) Halloysite nanotubes	Adsorption[87]
Pectin	(i) Silica	Heat insulation[29]
Alginate	(i) Halloysite nanotubes	Cytocompatibility[37]
	(ii) Silica	Porosity[88]
	(iii) Carbon nanotubes/silver	Pressure sensing[89]
Xanthan gum	(i) Sodium montmorillonite	Flame retardancy[26]
Gum Arabic	(i) Sodium montmorillonite	Heat insulation[7]

In addition to the bio-based components, there are hybrid gels with other precursors (such as silica) that need to be prepared as a sol. Two or more-sol systems can either be gelled separately to be used later or gelled directly together (co-gelation) to form the hybrid.

7.2.2 Gelation

Gelation of the dissolved precursor/s occurs when these atomic clusters are condensed and polymerized. The process is heavily dependent on the electric potential of these atomic clusters, which in turn reflect on the pH of the sol. With appropriate pH adjustment, which is usually done by addition of acids and bases to the sol, rapid gelation can occur when the iso-electric point is achieved among these clusters. An alternative to the network formation is to crosslink these atomic clusters with a crosslinking agent. This method is especially crucial when the sol concentration is too low to form a rigid network.

Interestingly, besides pH adjustment and crosslinking, some have reported accelerated gelation by (i) pressurized carbon dioxide and (ii) addition of graphene oxide sheet (GOS). In the pressurized CO_2 method, gelation is induced by the weak acidity of CO_2 in the water containing samples of alginate, cellulose and chitosan hybrid aerogels.[38,68,93] As for graphene oxide sheet, the accelerated gelation in the cellulose hybrid aerogel was confirmed by Zhang *et al.*[94] and it is attributed to the interaction of GOS with the cellulose chain as shown in Figure 7.3.

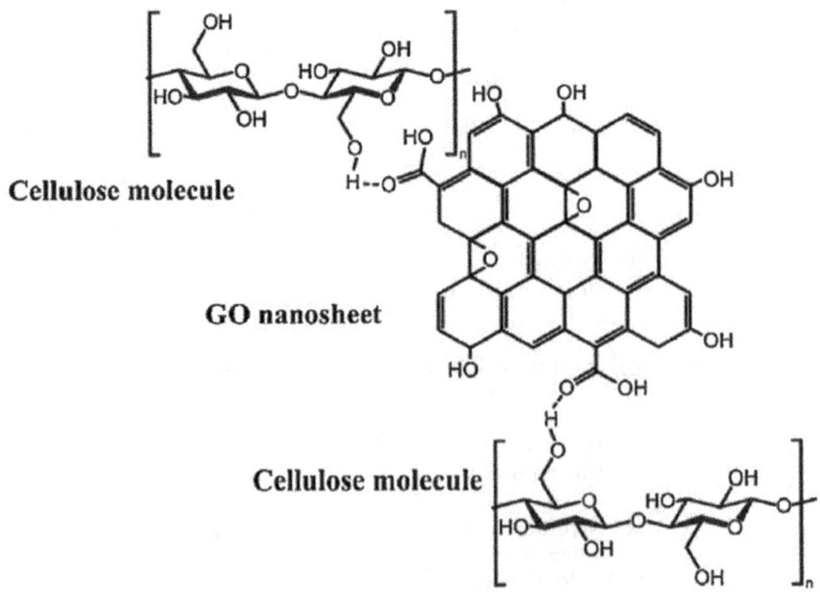

Figure 7.3 Interaction of hydrogen bonds between cellulose and GOS.[32] Reprinted with permission from *Carbohydrate Polymers*, **150**, C. Wan and J. Li, Graphene oxide/cellulose aerogels nanocomposite: Preparation, pyrolysis, and application for electromagnetic interference shielding, 172–179, Copyright 2016, with permission from Elsevier.

7.2.3 Doping

The addition of an incorporated phase can either be performed (i) after the gelation of the skeletal phase or (ii) before the gelation in a co-gelation manner. In the former method, the incorporated phase is attached to the precast skeletal phase by:

(a) Deposition/encapsulation: The incorporated phase is readily deposited onto the surface of the precast skeletal phase in the form of coating[5,7,13,23,26,37,62,81] or encapsulation[37] as shown in Figure 7.2. There are also incorporated phases that can only be formed after its precursor is reacted in the presence of the skeletal phase. Such a reaction may be initiated by adding other reagents,[6,24,31,33,34,36,44,46,48,58,83,95,96] or be photo-catalytic[55] and so forth. For encapsulation, the incorporated phase is pre-mixed with the precursor of the skeletal phase. By adding chemical reagents, the skeletal phase evolves and encapsulates the incorporated phase within itself.[14,16,28,32,59,68,72,87,97]
(b) Silylation: This is mainly for hybrid gels containing silica where the alkyl-silanes are coated or grafted onto the skeletal phase[5,63] as illustrated in Figure 7.2.
(c) Chemical vapor deposition (CVD): This is a thin layer of coating in nanometres, such as titania[76,77] that can be functionalized. However, CVD usually requires high energy consumption and toxic materials.
(d) Active groups reduction: These are the reactive functional groups on the skeletal phase which react with the incorporated phase containing metal oxides and cause it to be reduced it to metal nanoparticles[31,45–47,71,86] which can then be attached to the skeletal phase.
(e) Crosslinked agent: This is the incorporated phase, which is commonly another polymeric species reacted with the skeletal network on its ends to join these networks together.[37,38,68,87,98]

Besides the above mentioned conventional methods, the co-gelation method recently received considerable attention within the research community. By having two or more miscible sol systems, co-gelation can be carried out and the incorporated phase is likely to form simultaneously with the skeletal phase. In some cases, the incorporated phase can also act as a co-skeletal phase as shown in Figure 7.2. This so called "one-pot" synthesis was carried out by mixing the precursors of the skeletal phase and the incorporated phase to form a homogenous solution, as shown in Figure 7.4.

Gelation of a mixture may be self-initiated or it can be initiated by pH adjustment. The following advantages can be associated with this co-gelation method:

(i) *Solvent reduction*: In the conventional methods, the pores of the precast skeletal phase is filled with another solvent before adding

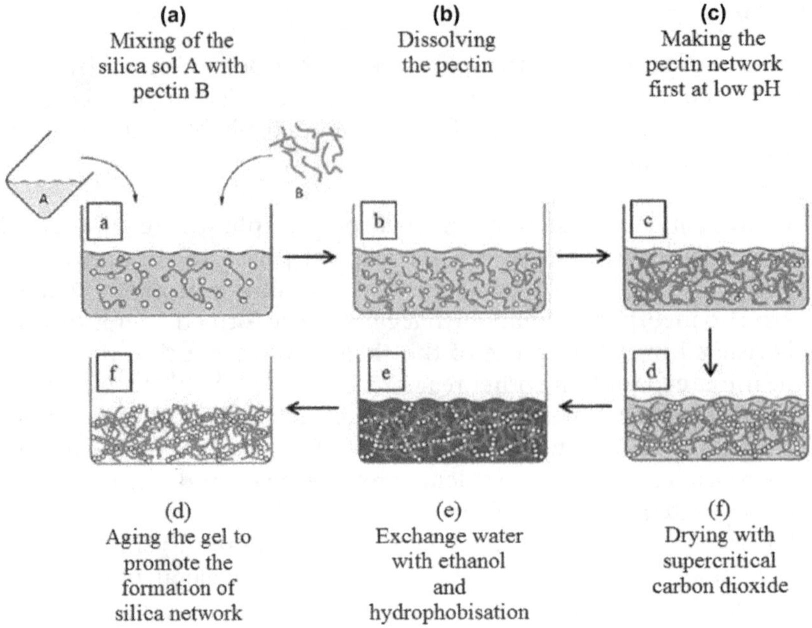

Figure 7.4 Schematic diagram of the "one-pot" synthesis of a pectin-silica hybrid aerogel.[29]
Adapted from ref. 29 with permission from John Wiley and Sons, Copyright © 2015 Wiley-VCH Verlag GmbH & Co. KGaA, Weinheim.

the incorporated phase. The equilibrium for mass transfer of the incorporated phase into these porous networks is easily reached due to the dilute action of the pore fluid. Thus, the extra amount of the incorporated phase is required to generate a sufficient concentration gradient for the incorporated phase to diffuse into the pores. Whereas in the co-gelation method, the incorporated phase is embedded in the pores during the formation of the skeletal phase. Hence, less of the incorporated phase is needed.[99]

(ii) *Time saving*: In the conventional method, the mass transfer of the incorporated phase into the pores of the skeletal phase is diffusion limited. This process is very time consuming and in addition a large amount of the incorporated phase is required. The duration for such a process can be largely reduced by embedding the incorporated phase prior to co-gelation.[22,73,99]

(iii) *Alteration of the microstructure*: The presence of the incorporated phase can cause an ionic or steric effect during co-gelation. As a result, evolution of the microstructure may be altered by these effects.[8,22,29,41,73] In addition, there is a possibility for the incorporated phase to join into the skeletal phase and form a co-skeletal phase[69] as illustrated in Figure 7.2.

7.2.4 Drying

As in a conventional aerogel, the drying of a hybrid aerogel can be accomplished *via* freeze drying, supercritical drying and ambient pressure drying. A discussion of each is provided below:

(a) Supercritical drying: This is by far the best technique to preserve the pore structures of the gel. The aerogel yield contains an isotropic hierarchical structure of micro-, meso- and macropores. The products formed also suffer a lower degree of shrinkage as compared to those formed by freeze drying and ambient pressure drying.
(b) Freeze drying: This technique is well known for its ease of operation and low cost. However, the expansion of the pore fluid upon freezing can induce micro-cracks in the sample. In addition, hierarchical structures can be eliminated in the process leaving mainly macropores. Another interesting aspect of freezing is the fact that it tends to exhibit anisotropic morphology. Such a phenomenon is usually attributed to the anisotropic growth of ice crystals during freezing.[7,27,62]
(c) Ambient pressure drying: This is done by attaching other agents onto the surface of the gel pores, which can strengthen the skeletal phase and also reduce the surface tension of the pore fluid. The method has a great potential to be scaled up for mass production due to the simplicity and low energy consumption. However, the hybrid aerogels formed usually suffer from a large degree of shrinkage.

7.3 Morphology

The morphology of hybrid aerogels are usually visualized using conventional imaging techniques such as scanning electron microscopy (SEM) or field emission scanning electron microscopy (FESEM). However, transmission electron microscope (TEM) is used to confirm the presence of a crystallite, typically from metallic nanoparticles that are doped into the skeletal phase. Under certain configurations, elemental mapping can also be carried out on both the SEM and TEM images. In this section, various nanostructures of bio-based hybrid aerogels captured by SEM and TEM will be discussed. In addition, a new synchrotron X-ray tomography will be introduced. This technique is able to probe the degree of anisotropy of the incorporated phase within the skeletal phase.

7.3.1 SEM Images

Most of the neat bio-based aerogels usually display a 3D open structure as shown in Figure 7.5. Nevertheless, the hybrid version normally exhibits a 2D porous sheet-like nanopaper structure (Figure 7.6).

The formation of a 2D porous sheet may be attributed to the growth of doped materials on the skeletal phase with a 3D open structure; especially

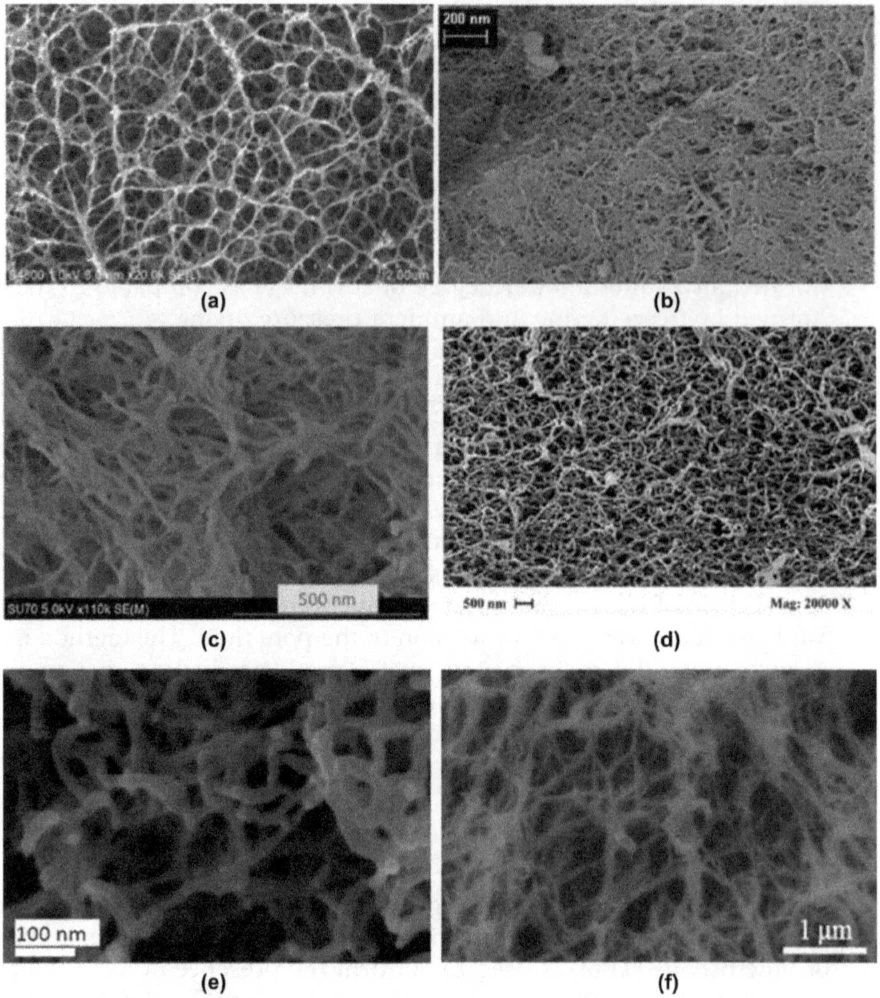

Figure 7.5 SEM images of neat (a) cellulose aerogel,[34] (b) alginate aerogel,[13] (c) silk fibroin,[40] (d) chitosan,[100] (e) pectin,[22] and (f) chitin.[101] Part A reprinted from *Carbohydrate Polymers*, **89**, S. Liu, Q. Yan, D. Tao, T.Yu and X. Liu, Highly flexible magnetic composite aerogels prepared by using cellulose nanofibril networks as templates, 551–557, Copyright 2012, with permission from Elsevier. Part B reprinted from *Powder Technology*, **285**, S. Antonyuk, S. Heinrich, P. Gurikov, S. Raman and I. Smirnova, Influence of coating and wetting on the mechanical behaviour of highly porous cylindrical aerogel particles, 34–43, Copyright 2015, with permission from Elsevier. Part C reprinted from *The Journal of Supercritical Fluids*, **91**, M. A. Marin, R. R. Mallepally and M. A. McHugh, Silk fibroin aerogels for drug delivery applications, 84–89, Copyright 2014, with permission from Elsevier. Part D reprinted from *The Journal of Supercritical Fluids*, **103**, L. Baldino, S. Concilio, S. Cardea, I. De Marco, E. Reverchon, Complete glutaraldehyde elimination during chitosan hydrogel drying by SC-CO2 processing, 70–76, Copyright 2015, with permission from Elsevier. Part F reproduced from ref. 101 with permission from John Wiley and Sons, © 2013 Wiley-VCH Verlag GmbH & Co. KGaA, Weinheim.

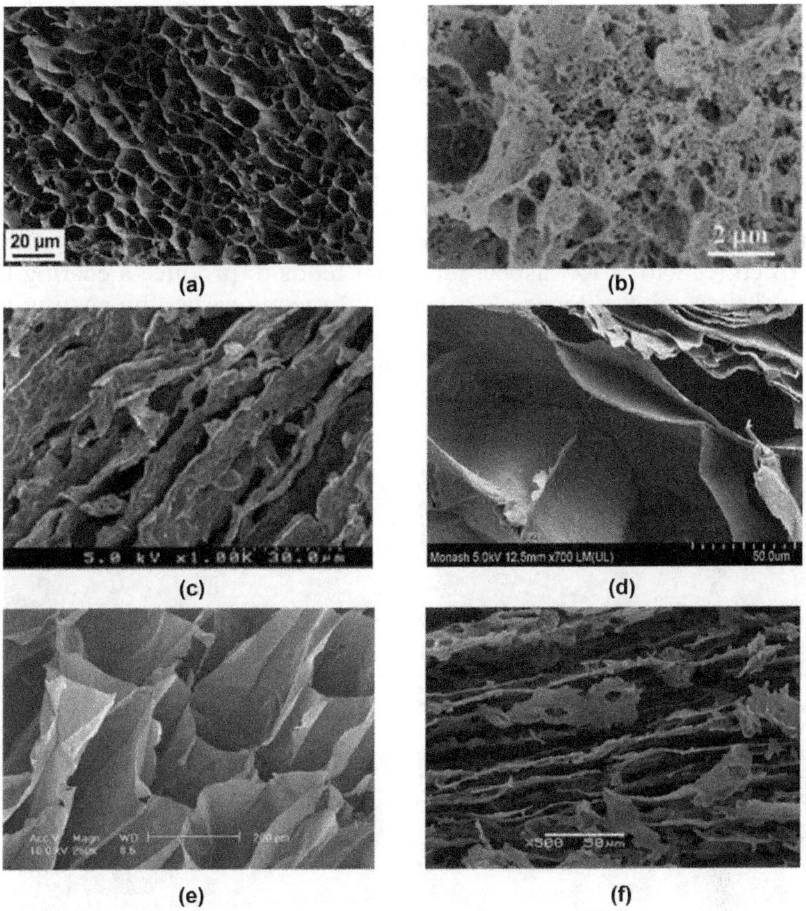

Figure 7.6 SEM images of hybrid (a) cellulose-functionalized carbon nanotubes aerogel,[81] (b) cellulose-graphene oxide aerogel,[32] (c) chitosan-graphene oxide aerogels,[12] (d) alginate-halloysite aerogel,[102] (e) chitin-halloysite nanotubes aerogel,[87] and (f) gum Arabic-clay aerogel.[7] Parts A and C reproduced from ref. 81 and 12 with permission from the Royal Society of Chemistry. Part B reprinted from *Carbohydrate Polymers*, **146**, C. Wan and J. Li, Cellulose aerogels functionalized with polypyrrole and silver nanoparticles: In-situ synthesis, characterization and antibacterial activity, 362–367, Copyright 2016, with permission from Elsevier. Part D reprinted from *Applied Clay Science*, **101**, C. S. C. Chiew P. E. Poh, P. Pasbakhsh, B. T. Tey, H. K. Yeoh, E. S. Chan, Physicochemical characterization of halloysite/alginate bionanocomposite hydrogel, 444–454, Copyright 2014, with permission from Elsevier. Part E reprinted from *International Journal of Biological Macromolecules*, **58**, M. Liu, Y. Zhang, J. Li, C. Zhou, Chitin-natural clay nanotubes hybrid hydrogel, 23–30, Copyright 2013, with permission from Elsevier. Part F reprinted from *Industrial Crops and Products*, **91**, L. Wang, M. Sánchez-Soto and T. Abt, Properties of bio-based gum Arabic/clay aerogels, 15–21, Copyright 2016, with permission from Elsevier.

with sheet-like or needle-like dopants. The process might be driven by a diffusion limited aggregate (DLA) phenomenon, whereby the incorporated phase is constantly diffused and immobilized by the aggregates formed along the network of the skeletal phase. As a result, the 2D porous sheet structure is evolved, which resembles a coral patterns.[3] In addition, the anisotropic growth of ice crystals during free drying can also create a continuous gap between the microstructures of the hybrid aerogel.

Unlike the sheet-like or needle-like dopants, these dopants have particle-like structures with a low aspect ratio that tends to produce a coating layer on the skeletal phase (Figure 7.7a and b). In some cases, the coating layer covers the entire surface of the skeletal phase forming a hollow tube structure. It is believed that the addition of metallic nanoparticles usually results in the formation of such structures.

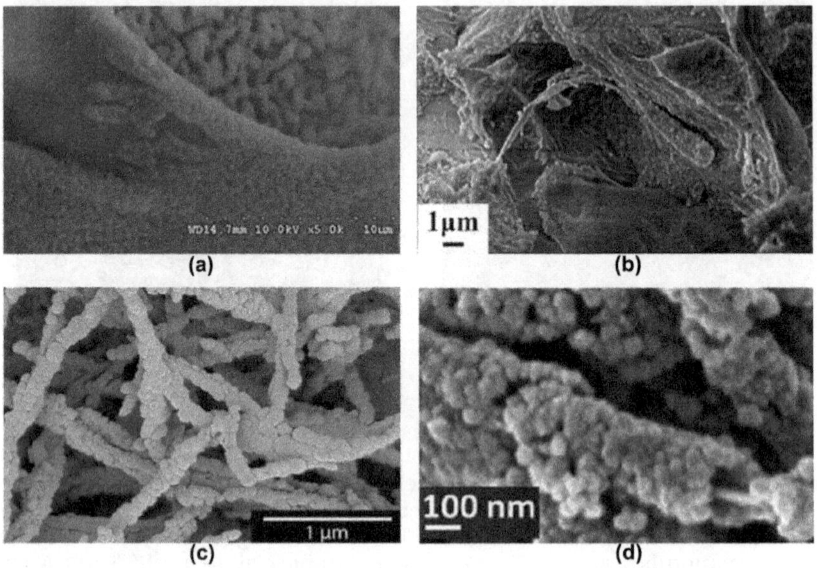

Figure 7.7 SEM images of a hybrid aerogel with an incorporated phase coated on the skeletal phase; (a) and (b) cellulose coated with silica,[6,36] hollow tube structure of (c) nanocellulose fibrils coated with silica,[72] and (d) bacterial cellulose templated with cobalt-ferrite nanoparticles.[33] Part A reprinted from *Carbohydrate Polymers*, **98**, J. Shi, L. Lu, W. Guo, J. Zhang and Y. Cao, Heat insulation performance, mechanics and hydrophobic modification of cellulose–SiO2 composite aerogels, 282–289, Copyright 2013, with permission from Elsevier. Part B reprinted from *Carbohydrate Polymers*, **147**, J. Fu, S. Wang, C. He, Z. Lu, J. Huang, Z. Chen, Facilitated fabrication of high strength silica aerogels using cellulose nanofibrils as scaffold, 89–96, Copyright 2016, with permission from Elsevier. Part C reproduced from ref. 72 with permission from the Royal Society of Chemistry. Part D reproduced from *Carbohydrate Polymers*, **137**, S. Menchaca-Nal, C. L. Londoño-Calderón, P. Cerrutti, M. L. Foresti, L. Pampillo, V. Bilovol, R. Candal and R. Martínez-García, Facile synthesis of cobalt ferrite nanotubes using bacterial nanocellulose as template, 726–731, Copyright 2016, with permission from Elsevier.

Nanoparticles may also be anchored/grafted onto the skeletal phase without coating the entire surface (as shown in Figure 7.8). This anchoring action results from the active group reduction of the precursors of the nanoparticles as previously elucidated in Section 7.2.3(d). This makes the bio-based aerogels suitable to be used as a support or template for the synthesis, and prevents agglomeration of nanoparticles.

In addition to the method of incorporation and the aspect ratio of the dopant, the morphology of hybrid aerogels is also affected by the drying method employed. Freeze dried hybrid aerogels tend to display anisotropic microstructures (Figure 7.9) due to the anisotropic growth of ice within the pores. Liu *et al.* proposed a mechanism for the formation of such anisotropic structures,[103] which is illustrated in Figure 7.10.

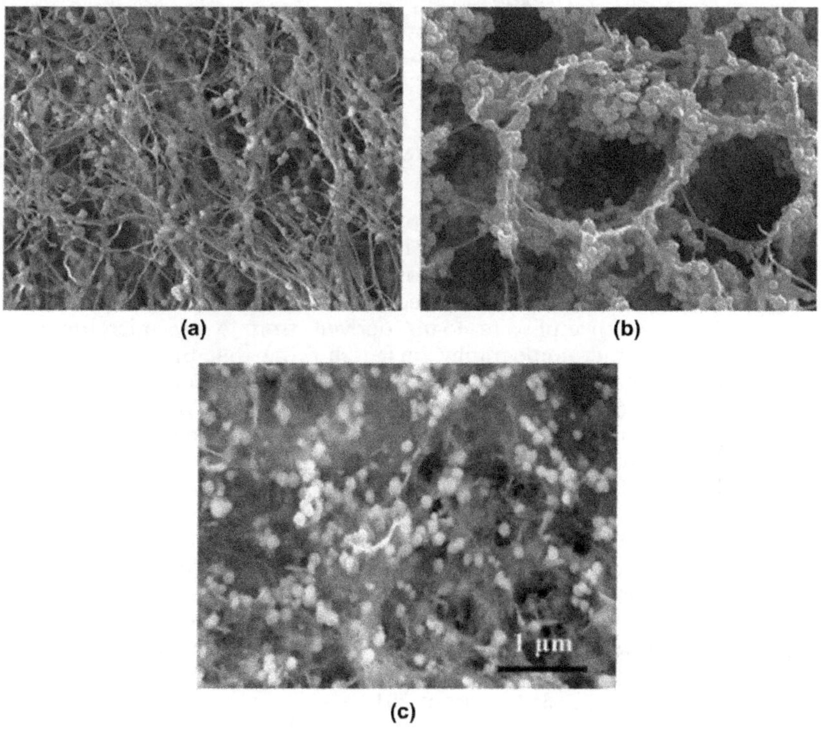

Figure 7.8 SEM images of hybrid aerogel with incorporated phase anchored/grafted onto skeletal phase; (a) and (b) bacterial nanofibril cellulose grafted with cobalt-ferrite,[82] and (c) wheat straw cellulose grafted with cobalt ferrite.[31] Part A and B reproduced from *Nature Nanotechnology*, Making flexible magnetic aerogels and stiff magnetic nanopaper using cellulose nanofibrils as templates, **5**, 2010, 584–588, R. T. Olsson, M. A. S. Azizi Samir, G. Salazar-Alvarez, L. Belova, V. Ström, L. A. Berglund, O. Ikkala, J. Nogués and U. W. Gedde, Copyright 2010, with permission of Springer. Part C reprinted from *Carbohydrate Polymers*, **134**, C. Wan and J. Li, Synthesis of well-dispersed magnetic CoFe2O4 nanoparticles in cellulose aerogels *via* a facile oxidative co-precipitation method, 144–150, Copyright 2015, with permission from Elsevier.

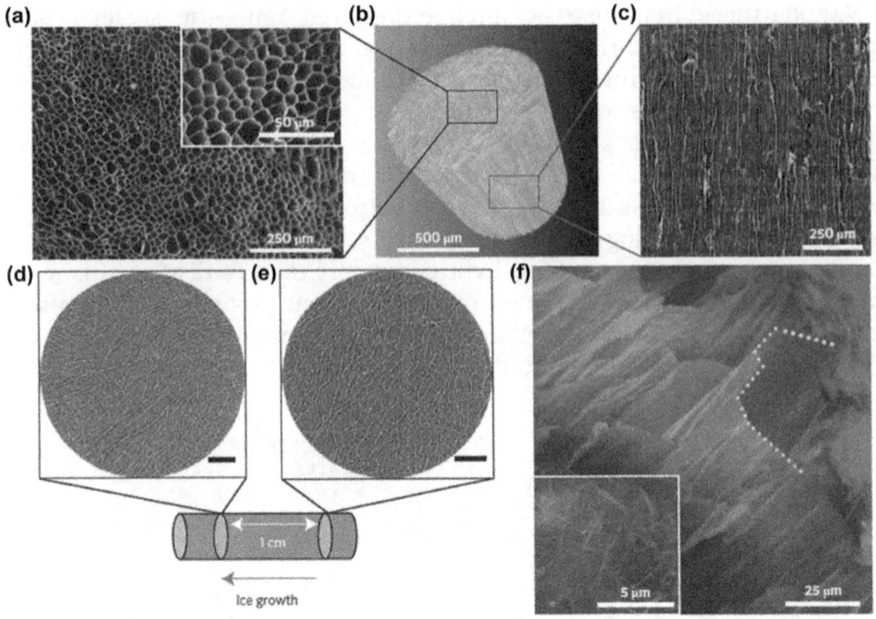

Figure 7.9 Microstructure of freeze-cast nanocomposite foams. (a) SEM cross-section image of a freeze-cast nanocomposite foam containing cellulose nanofibres (CNF), graphene oxide (GO), sepiolite (SEP) and boric acid (BA). (b) Three-dimensional reconstruction of the tubular pore structure of the nanocomposite foam derived from X-ray microtomography. (c) X-ray microtomography image showing that the tubular pores are straight and several millimetres long in the nanocomposite foam with a composition of 77% CNF/10% GO/10% SEP/3% BA (in wt%). (d) and (e) X-ray microtomography cross-sections of a nanocomposite foam, taken through the upper and lower parts, respectively (scale bars, 100 µm). (f) HRSEM image of a foam wall, where the dotted line indicates a section of the tubular cell. Inset: Distributed SEP nanorods within the cell wall. The nanomaterials are homogeneously distributed in the cell walls, forming an anisotropic tubular pore structure as a result of the unidirectional freeze-casting process.
Reprinted by permission from Macmillan Publishers Limited: *Nature Nanotechnology* (B. Wicklein, A. Kocjan, G. Salazar-Alvarez, F. Carosio, G. Camino, M. Antonietti and L. Bergström, *Nature Nanotechnology*, 2015, **10**, 277–283) Copyright 2015.

However, the presence of such anisotropic structures is heavily dependent on the cooling rate during freezing and the types of pore fluid it contains. This is evidenced from the cellulose-graphene oxide hybrid aerogels that were reported by two research groups using different freezing temperatures.[27,32] The results are shown in Figures 7.6(b) and 7.9(a), where both of these hybrid aerogels were freeze dried at different cooling rates. Sehaqui *et al.* also reported similar evidence, whereby microstructures formed near the cooling source can differ from that formed in the middle fractured section of the sample (Figure 7.11).[62] The higher degree of anisotropy in the

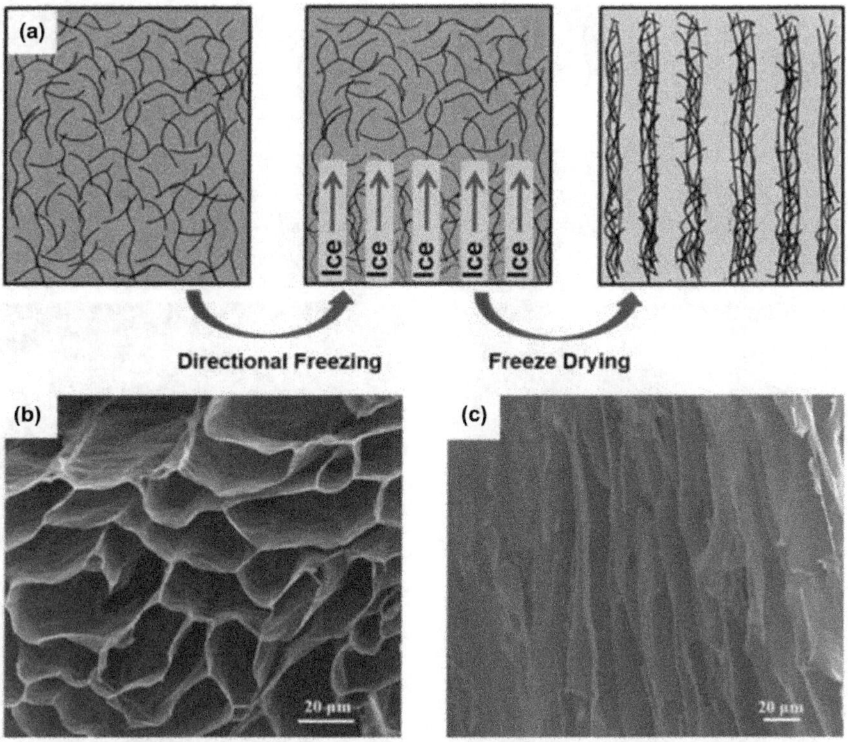

Figure 7.10 (a) A schematic illustrating the directional-freezing and freeze–drying; (b) top-view and (c) side-view SEM images of the anisotropic porous structure of AGA-8.7.
Reprinted with permission from *Carbon*, **100**, T. Liu, M. Huang, X. Li, C. Wang, C.-X. Gui and Z.-Z. Yu, Highly compressible anisotropic graphene aerogels fabricated by directional freezing for efficient absorption of organic liquids, 456–464, Copyright 2016, with permission from Elsevier.

microstructure observed at the fractured section is caused by the anisotropic growth of ice crystals along the vertical direction. Whereas, at the surface next to the cooling source that served as the nucleation sites for ice growth, the anisotropic structure of the aerogel is less obvious. Besides the cooling rate, the nature of the pore fluid plays a critical role in creating anisotropic structures in the hybrid aerogel. Anisotropic growth of crystals from the pore fluids is favoured in the case of a polar solvent such as water, whereas when a non-polar solvent, such as liquid propane is used as a pore fluid, the 3D isotropic porous network of aerogel can be preserved.[62]

7.3.2 TEM Images

TEM plays an essential role in imaging hybrid aerogels, especially for the metal-hybrid type. TEM is used to probe the crystal structure of metallic nanoparticles doped to bio-based gels. Figures 7.12c and 7.13b illustrate

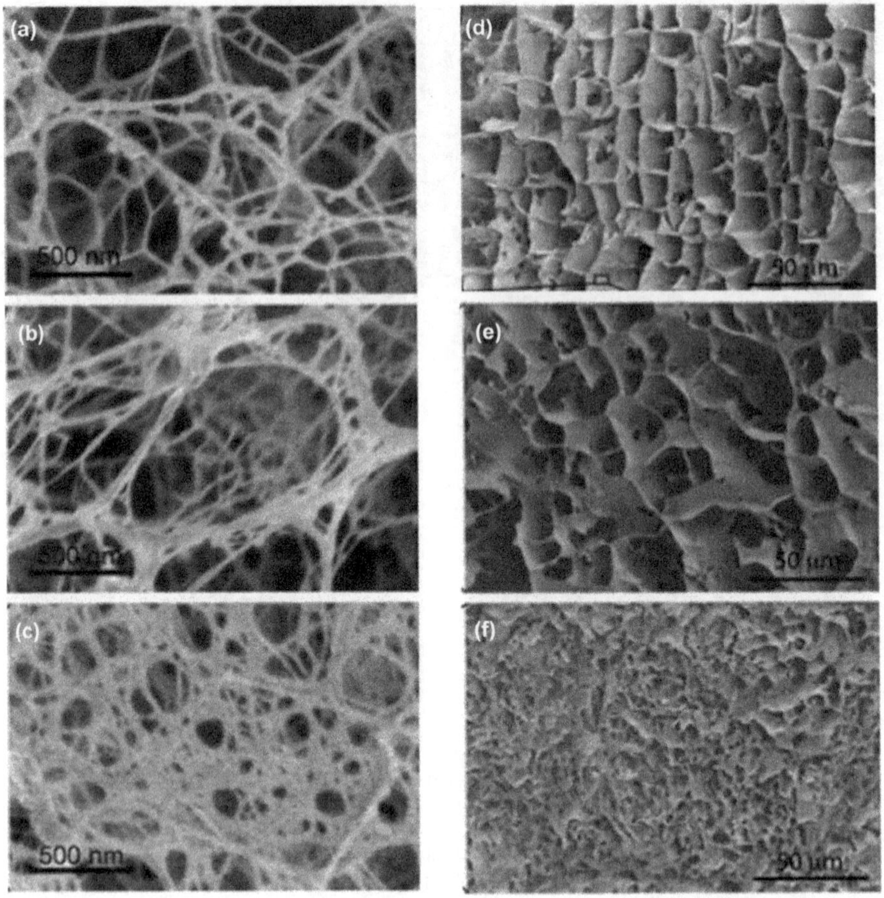

Figure 7.11 Microstructure of microfibrillated cellulose (MFC) – xyloglucan (XG) hybrid aerogels cross section at (a–c) the end of the sample next to the cooling source, (d–f) middle part of fractured sample. All pairs are compared from the lowest to highest density, which corresponds to images from top to bottom.
Reproduced from ref. 62 with permission from the Royal Society of Chemistry.

some of the crystallites in hybrid aerogels. Furthermore, TEM can give a clear picture of the distribution of nanoparticles that are anchored/grafted onto the skeletal phase. Figures 7.14 and 7.15 are examples of TEM micrographs of the cellulose skeletal phase grafted with inorganic nanoparticles.

7.3.3 Synchrotron X-ray Tomography

It is worth mentioning that the properties of hybrid aerogels can be greatly influenced by the degree of anisotropy in the microstructure. However, SEM and TEM images that focus on a tiny region of the composite are only able to

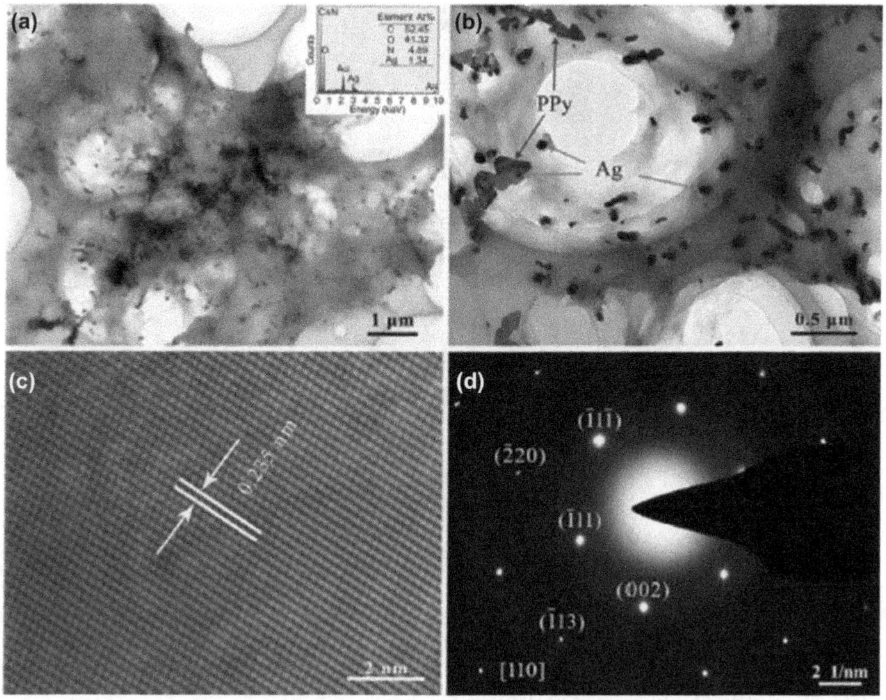

Figure 7.12 (a) Low-magnification TEM image, (b) high-magnification TEM image, (c) HRTEM image and (d) SAED pattern of Polypyrrole/silver/cellulose aerogel, respectively. C.
Adapted from *Carbohydrate Polymers*, **146**, C. Wan and J. Li, Cellulose aerogels functionalized with polypyrrole and silver nanoparticles: In-situ synthesis, characterization and antibacterial activity, 362–367, Copyright 2016, with permission form Elsevier.

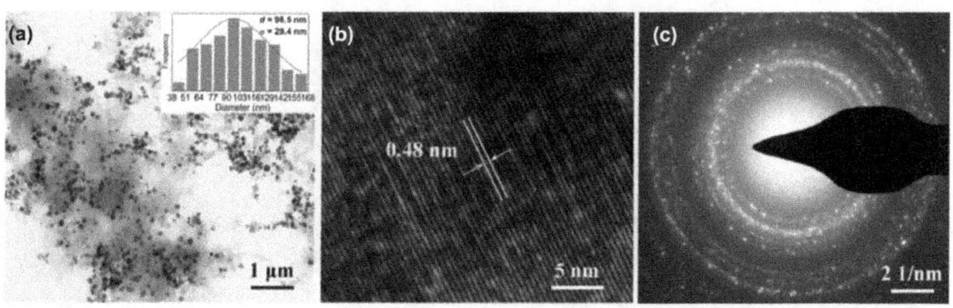

Figure 7.13 (a) TEM, (b) high-resolution TEM, and (c) SAED images of $CoFe_2O_4$@ Cellulose aerogel. Inset of (a) presents the size distribution of the $CoFe_2O_4$ nanoparticles in a $CoFe_2O_4$@ Cellulose aerogel.
Reprinted from *Carbohydrate Polymers*, **134**, C. Wan and J. Li, Synthesis of well-dispersed magnetic $CoFe_2O_4$ nanoparticles in cellulose aerogels *via* a facile oxidative co-precipitation method, 144–150, Copyright 2017, with permission from Elsevier.

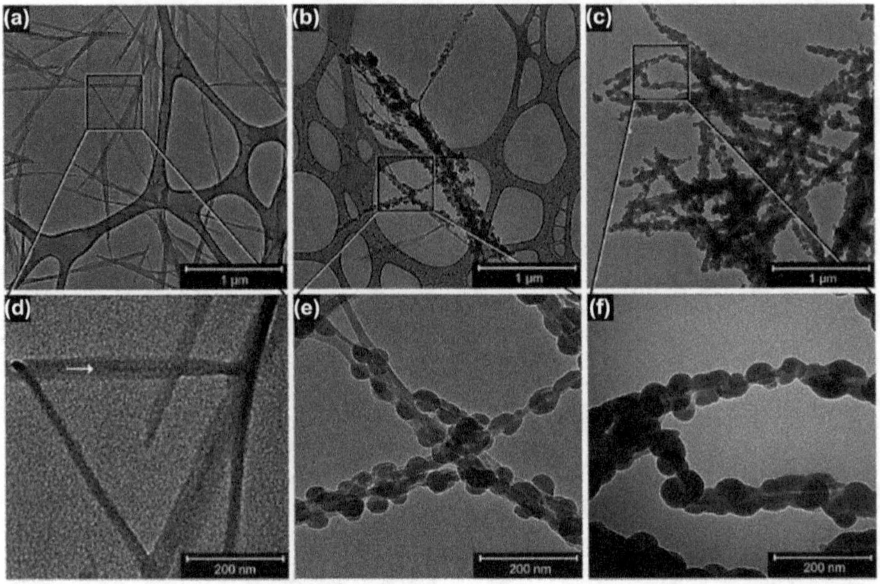

Figure 7.14 TEM images of (a) uncoated cellulose nanofibrils (CNF), (b) silica embedded with CNFs coated in 2-propanol at different CNF concentrations: 3 mg mL^{-1} and (c) 1 mg mL^{-1}. (d–f) are highlighted areas from images (a–c) respectively.
Reproduced from ref. 72 with permission from the Royal Society of Chemistry.

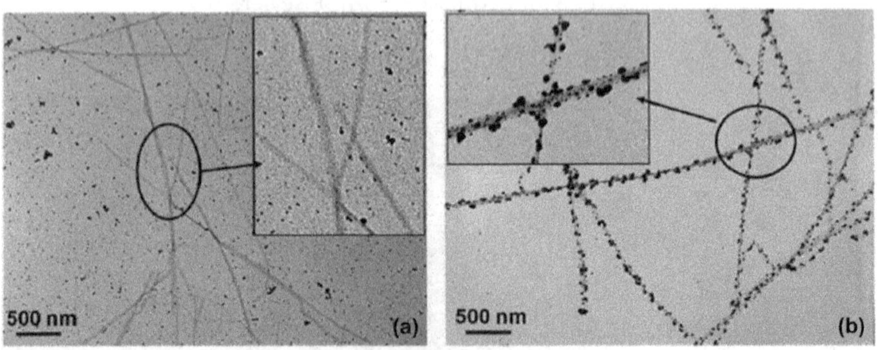

Figure 7.15 TEM images of silver nanoparticles synthesis on tunicate cellulose nanocrystals (a) without cetyltrimethylammonium bromide (CTAB) and (b) with CTAB.
Reprinted with permission from S. Padalkar, J. R. Capadona, S. J. Rowan, C. Weder, Y.-H. Won, L. A. Stanciu and R. J. Moon, Natural Biopolymers: Novel Templates for the Synthesis of Nanostructures, *Langmuir*, 2010, **26**(11), 8497–8502, Copyright 2014 American Chemical Society.

probe the degree of anisotropy locally, rather than giving the global picture. Amongst the research conducted in this area, Sedighi Gilani and her research group have pioneered the use of synchrotron X-ray tomography in the imaging of nano-fibrillated cellulose (NFC) incorporated into the silica aerogel.[83] Figure 7.16 shows some of the tomography images reported by

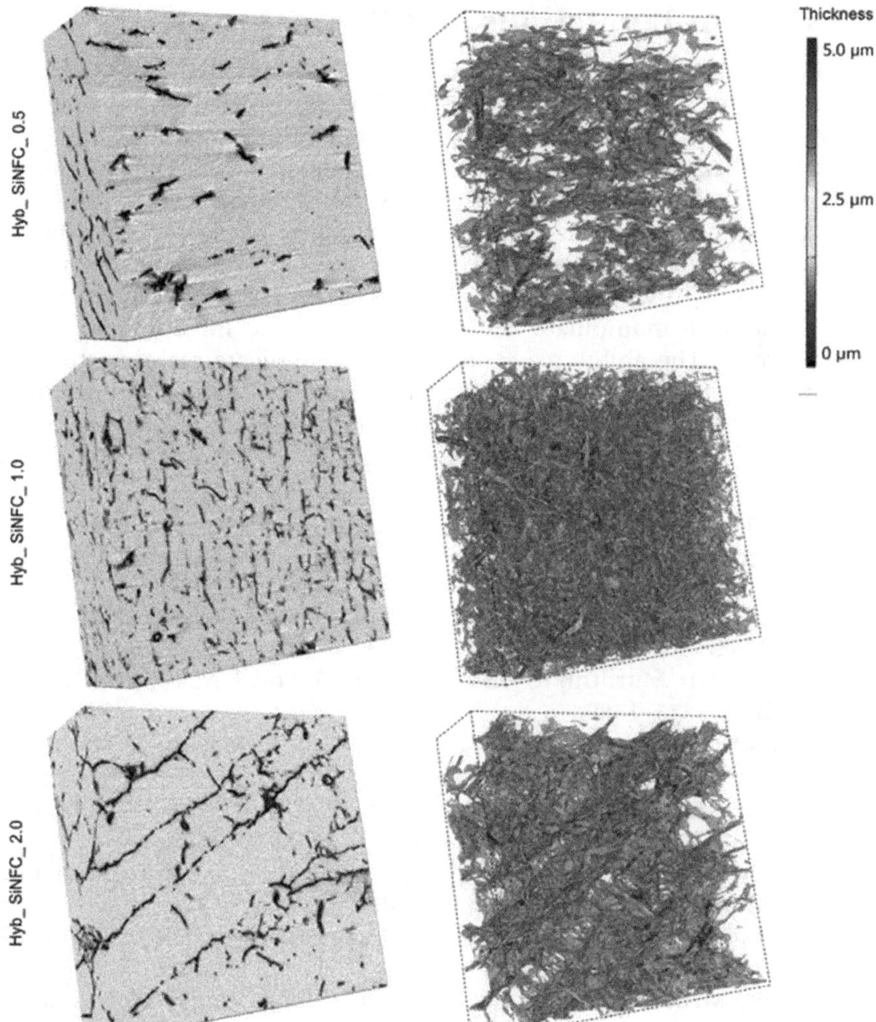

Figure 7.16 (a) 3D volume rendering of (a) silica aerogel and (b) thickness map of the respective silylated NFC scaffold.
Reprinted from *Composites Science and Technology*, **124**, S. G. Marjan, M. N. Boone, J. L Fife, S. Zhao, M. M. Koebel, T. Zimmermann and P. Tingaut, Structure of cellulose-silica hybrid aerogel at sub-micron scale, studied by synchrotron X-ray tomographic microscopy, 71–80, Copyright 2016, with permission from Elsevier.

this group. This technique can be very useful for hybrid aerogels, whereby the degree of anisotropy of the incorporated phase inside the skeletal phase can be visualized. This should be popularized in the study of hybrid aerogels, especially with bio-based aerogels which usually contain components with a high aspect ratio embedded in another matrix.

7.4 Current Limitations and Future Prospects

The commercial potential of bio-based hybrid aerogels is still limited by the same constraints faced in pure bio-based aerogels, represented by extensive solvent exchange and drying processes which are energy consuming or difficult to be scaled-up. Therefore, more research efforts should be devoted to minimizing solvent usage and to finding viable alternatives for the current drying processes. From an environmental point of view, very limited discussions have been carried out on separation and recycling CO_2 in the supercritical CO_2 drying processes. Furthermore, there is very scarce research on how to manipulate the orientation of the incorporated phase into the matrix. The ability to do so would that indicate real-time tunable properties of the hybrid aerogel have formed. Such achievements would definitely revolutionize the use of hybrid aerogels, which in return would benefit many applications, especially in tissue engineering and pharmaceuticals.

References

1. H.-B. Chen, Y.-Z. Wang and D. A. Schiraldi, *Eur. Polym. J.*, 2013, **49**, 3387–3391.
2. D. J. Boday, B. Muriithi, R. J. Stover and D. A. Loy, *J. Non-Crystal. Solids*, 2012, **358**, 1575–1580.
3. G. Hayase, K. Kanamori, K. Abe, H. Yano, A. Maeno, H. Kaji and K. Nakanishi, *ACS Appl. Mater. Interfaces*, 2014, **6**, 9466–9471.
4. A. Javadi, Q. Zheng, F. Payen, A. Javadi, Y. Altin, Z. Cai, R. Sabo and S. Gong, *ACS Appl. Mater. Interfaces*, 2013, **5**, 5969–5975.
5. B. Seantier, D. Bendahou, A. Bendahou, Y. Grohens and H. Kaddami, *Carbohydr. Polym.*, 2016, **138**, 335–348.
6. J. Shi, L. Lu, W. Guo, J. Zhang and Y. Cao, *Carbohydr. Polym.*, 2013, **98**, 282–289.
7. L. Wang, M. Sánchez-Soto and T. Abt, *Ind. Crops Prod.*, 2016, **91**, 15–21.
8. S. Zhao, W. J. Malfait, E. Jeong, B. Fischer, Y. Zhang, H. Xu, E. Angelica, W. M. Risen, J. W. Suggs and M. M. Koebel, *ACS Sustainable Chem. Eng.*, 2016, **4**, 5674–5683.
9. S. Zhao, Z. Zhang, G. Sèbe, R. Wu, R. V. Rivera Virtudazo, P. Tingaut and M. M. Koebel, *Adv. Funct. Mater.*, 2015, **25**, 2326–2334.
10. X. Liu and W. M. Risen, in *Natural Fibers, Plastics and Composites*, ed. T. W. Frederick and E. W. Norman, Kluwer, Boston, 2004.
11. A. El Kadib and M. Bousmina, *Chemistry*, 2012, **18**, 8264–8277.

12. A. A. Alhwaige, T. Agag, H. Ishida and S. Qutubuddin, *RSC Adv.*, 2013, **3**, 16011–16020.
13. S. Antonyuk, S. Heinrich, P. Gurikov, S. Raman and I. Smirnova, *Powder Technol.*, 2015, **285**, 34–43.
14. H. Ennajih, R. Bouhfid, E. M. Essassi, M. Bousmina and A. El Kadib, *Microporous Mesoporous Mater.*, 2012, **152**, 208–213.
15. M. S. Sajab, C. H. Chia, C. H. Chan, S. Zakaria, H. Kaco, S. W. Chook, S. X. Chin and A. A. M. Noor, *RSC Adv.*, 2016, **6**, 19819–19825.
16. A. Djelad, A. Morsli, M. Robitzer, A. Bengueddach, F. di Renzo and F. Quignard, *Molecules*, 2016, **21**, E109.
17. R.-L. Liu, X.-Q. Li, H.-Q. Liu, Z.-M. Luo, J. Ma, Z.-Q. Zhang and Q. Fu, *RSC Adv.*, 2016, **6**, 30301–30310.
18. Y. Wang, S. Yadav, T. Heinlein, V. Konjik, H. Breitzke, G. Buntkowsky, J. J. Schneider and K. Zhang, *RSC Adv.*, 2014, **4**, 21553–21558.
19. X. Yang, Thesis, McMaster University, 2014.
20. T. Zhai, Q. Zheng, Z. Cai, H. Xia and S. Gong, *Carbohydr. Polym.*, 2016, **148**, 300–308.
21. Q. Zheng, Z. Cai and S. Gong, *J. Mater. Chem. A*, 2014, **2**, 3110–3118.
22. A. Demilecamps, Thesis, Ecole Nationale Supérieure des Mines de Paris, 2015.
23. J. Feng, D. Le, S. T. Nguyen, V. Tan Chin Nien, D. Jewell and H. M. Duong, *Colloids Surf., A*, 2016, **506**, 298–305.
24. Y. Han, X. Zhang, X. Wu and C. Lu, *ACS Sustainable Chem. Eng.*, 2015, **3**, 1853–1859.
25. N. Leventis, A. Sadekar, N. Chandrasekaran and C. Sotiriou-Leventis, *Chem. Mater.*, 2010, **22**, 2790–2803.
26. L. Wang, D. A. Schiraldi and M. Sánchez-Soto, *Ind. Eng. Chem. Res.*, 2014, **53**, 7680–7687.
27. B. Wicklein, A. Kocjan, G. Salazar-Alvarez, F. Carosio, G. Camino, M. Antonietti and L. Bergström, *Nat. Nanotechnol.*, 2015, **10**, 277–283.
28. B. Yuan, J. Zhang, J. Yu, R. Song, Q. Mi, J. He and J. Zhang, *Sci. China: Chem.*, 2016, **59**, 1335–1341.
29. S. Zhao, W. J. Malfait, A. Demilecamps, Y. Zhang, S. Brunner, L. Huber, P. Tingaut, A. Rigacci, T. Budtova and M. M. Koebel, *Angew. Chem., Int. Ed.*, 2015, **54**, 14282–14286.
30. P. Hao, Z. Zhao, Y. Leng, J. Tian, Y. Sang, R. I. Boughton, C. P. Wong, H. Liu and B. Yang, *Nano Energy*, 2015, **15**, 9–23.
31. C. Wan and J. Li, *Carbohydr. Polym.*, 2015, **134**, 144–150.
32. C. Wan and J. Li, *Carbohydr. Polym.*, 2016, **150**, 172–179.
33. S. Menchaca-Nal, C. L. Londoño-Calderón, P. Cerrutti, M. L. Foresti, L. Pampillo, V. Bilovol, R. Candal and R. Martínez-García, *Carbohydr. Polym.*, 2016, **137**, 726–731.
34. S. Liu, Q. Yan, D. Tao, T. Yu and X. Liu, *Carbohydr. Polym.*, 2012, **89**, 551–557.
35. L. Baldino, S. Cardea and E. Reverchon, *Chem. Eng. Trans.*, 2015, **43**, 739–744.

36. J. Fu, S. Wang, C. He, Z. Lu, J. Huang and Z. Chen, *Carbohydr. Polym.*, 2016, **147**, 89–96.
37. M. Liu, L. Dai, H. Shi, S. Xiong and C. Zhou, *Mater. Sci. Eng., C*, 2015, **49**, 700–712.
38. S. Quraishi, M. Martins, A. A. Barros, P. Gurikov, S. P. Raman, I. Smirnova, A. R. C. Duarte and R. L. Reis, *J. Supercrit. Fluids*, 2015, **105**, 1–8.
39. M. Betz, C. A. García-González, R. P. Subrahmanyam, I. Smirnova and U. Kulozik, *J. Supercrit. Fluids*, 2012, **72**, 111–119.
40. M. A. Marin, R. R. Mallepally and M. A. McHugh, *J. Supercrit. Fluids*, 2014, **91**, 84–89.
41. M. R. Ayers and A. J. Hunt, *J. Non-Cryst. Solids*, 2001, **285**, 123–127.
42. V. S. S. Gonçalves, P. Gurikov, J. Poejo, A. A. Matias, S. Heinrich, C. M. M. Duarte and I. Smirnova, *Eur. J. Pharm. Biopharm.*, 2016, **107**, 160–170.
43. J. Rooke, C. de Matos Passos, M. Chatenet, R. Sescousse, T. Budtova, S. Berthon-Fabry, R. Mosdale and F. Maillard, *J. Electrochem. Soc.*, 2011, **158**, B779–B789.
44. H. Wang, Z. Shao, M. Bacher, F. Liebner and T. Rosenau, *Cellulose*, 2013, **20**, 3007–3024.
45. J. Cai, S. Kimura, M. Wada and S. Kuga, *Biomacromolecules*, 2009, **10**, 87–94.
46. C. Wan and J. Li, *Mater. Des.*, 2015, **83**, 620–625.
47. D. Dutta, R. Hazarika, P. D. Dutta, T. Goswami, P. Sengupta and D. K. Dutta, *RSC Adv.*, 2016, **6**, 85173–85181.
48. S. Zhao, H. Xu, L. Wang, P. Zhu, W. M. Risen Jr and J. William Suggs, *Microporous Mesoporous Mater.*, 2013, **171**, 147–155.
49. C. Wan and J. Li, *Carbohydr. Polym.*, 2016, **146**, 362–367.
50. C. Yao, X. Liu and M. R. William, in *Aerogels Handbook*, ed. M. A. Aegerter, N. Leventis and M. M. Koebel, Springer, New York, 2011.
51. X. Hu, S. Ji and W. M. Risen, *MRS Proc.*, 2001, **702**, U3.4.1.
52. X. Liu, Y. Zhu, C. Yao and W. M. Risen, *MRS Proc.*, 2003, **788**, L2.1.
53. C. Yao and M. R. J. William, *MRS Proc.*, 2005, **847**, 515–520.
54. C. Yao, X. Liu and W. M. Risen, *MRS Proc.*, 2005, **888**, 0888-V05-07.
55. N. Kuthirummal, A. Dean, C. Yao and W. Risen Jr, *Spectrochim. Acta, Part A*, 2008, **70**, 700–703.
56. X. Liu, Thesis, Brown University, 2005.
57. X. Liu, C. Yao and W. M. Risen, *MRS Proc.*, 2005, **899**, 0899-N06-06.
58. Z. Liu, P. Wu, S. Yang, H. Wang and C. Jin, *Int. J. Polym. Sci.*, 2016, **2016**, 8.
59. X. Hu, K. Littrel, S. Ji, D. G. Pickles and W. M. Risen Jr, *J. Non-Cryst. Solids*, 2001, **288**, 184–190.
60. I. Josef, K. W. Hedda and K. Gregor, *Lenzinger Ber.*, 2006, **86**, 137–143.
61. M. Ahmadi, A. Madadlou and A. A. Saboury, *Food Chem.*, 2016, **196**, 1016–1022.

62. H. Sehaqui, M. Salajkova, Q. Zhou and L. A. Berglund, *Soft Matter*, 2010, **6**, 1824–1832.
63. C. Wan, Y. Lu, Y. Jiao, C. Jin, Q. Sun and J. Li, *J. Appl. Polym. Sci.*, 2015, **132**, 42037–42044.
64. T. Lu, Q. Li, W. Chen and H. Yu, *Compos. Sci. Technol.*, 2014, **94**, 132–138.
65. X. Zhang, Y. Wang, J. Zhao, M. Xiao, W. Zhang and C. Lu, *ACS Sustain. Chem. Eng.*, 2016, **4**, 4321–4327.
66. N. Pircher, S. Veigel, N. Aigner, J. M. Nedelec, T. Rosenau and F. Liebner, *Carbohydr. Polym.*, 2014, **111**, 505–513.
67. A. A. Alhwaige, Case Western Reserve University, 2014.
68. S. P. Raman, P. Gurikov and I. Smirnova, *J. Supercrit. Fluids*, 2015, **106**, 23–33.
69. A. S. Ahmad, K. A. Dzulkefly and D. Rusli, *Cellulose Chem. Technol.*, 2012, **46**, 45–52.
70. H.-B. Zhao, L. Yuan, Z.-B. Fu, C.-Y. Wang, X. Yang, J.-Y. Zhu, J. Qu, H.-B. Chen and D. A. Schiraldi, *ACS Appl. Mater. Interfaces*, 2016, **8**, 9917–9924.
71. H. Dong, J. F. Snyder, D. T. Tran and J. L. Leadore, *Carbohydr. Polym.*, 2013, **95**, 760–767.
72. D. Liu, Q. Wu, R. L. Andersson, M. S. Hedenqvist, S. Farris and R. T. Olsson, *J. Mater. Chem. A*, 2015, **3**, 15745–15754.
73. J. Cai, S. Liu, J. Feng, S. Kimura, M. Wada, S. Kuga and L. Zhang, *Angew. Chem. Int. Ed.*, 2012, **51**, 2076–2079.
74. D. Bendahou, A. Bendahou, B. Seantier, Y. Grohens and H. Kaddami, *Ind. Crops Prod.*, 2015, **65**, 374–382.
75. S. Zhou, M. Wang, X. Chen and F. Xu, *ACS Sustain. Chem. Eng.*, 2015, **3**, 3346–3354.
76. M. Kettunen, R. J. Silvennoinen, N. Houbenov, A. Nykänen, J. Ruokolainen, J. Sainio, V. Pore, M. Kemell, M. Ankerfors and T. Lindström, *Adv. Funct. Mater.*, 2011, **21**, 510–517.
77. J. T. Korhonen, M. Kettunen, R. H. A. Ras and O. Ikkala, *ACS Appl. Mater. Interfaces*, 2011, **3**, 1813–1816.
78. C. Schütz, J. Sort, Z. Bacsik, V. Oliynyk, E. Pellicer, A. Fall, L. Wågberg, L. Berglund, L. Bergström and G. Salazar-Alvarez, *PLoS One*, 2012, 7, e45828.
79. M. Wang, I. V. Anoshkin, A. G. Nasibulin, J. T. Korhonen, J. Seitsonen, J. Pere, E. I. Kauppinen, R. H. A. Ras and O. Ikkala, *Adv. Mater.*, 2013, **25**, 2428–2432.
80. L. Melone, L. Altomare, I. Alfieri, A. Lorenzi, L. De Nardo and C. Punta, *J. Photochem. Photobiol. A: Chem.*, 2013, **261**, 53–60.
81. M. Wang, I. V. Anoshkin, A. G. Nasibulin, R. H. A. Ras, Nonappa, J. Laine, E. I. Kauppinen and O. Ikkala, *RSC Adv.*, 2016, **6**, 89051–89056.
82. R. T. Olsson, M. A. S. Azizi Samir, G. Salazar Alvarez, L. Belova, V. Strom, L. A. Berglund, O. Ikkala, J. Nogues and U. W. Gedde, *Nat. Nano*, 2010, **5**, 584–588.

83. M. Sedighi Gilani, M. N. Boone, J. L. Fife, S. Zhao, M. M. Koebel, T. Zimmermann and P. Tingaut, *Compos. Sci. Technol.*, 2016, **124**, 71–80.
84. Q. Ma, Y. Liu, Z. Dong, J. Wang and X. Hou, *J. Appl. Polym. Sci.*, 2015, **132**, 41770–41781.
85. J. Wang, Q. Zhou, D. Song, B. Qi, Y. Zhang, Y. Shao and Z. Shao, *J. Sol-Gel Sci. Technol.*, 2015, **76**, 501–509.
86. C. J. Yao, X. Liu and W. M. Risen, in *Aerogels Handbook*, ed. M. A. Aegerter, N. Leventis and M. M. Koebel, Springer New York, New York, 2011, pp. 385–401.
87. M. Liu, Y. Zhang, J. Li and C. Zhou, *Int. J. Biol. Macromol.*, 2013, **58**, 23–30.
88. Z. Ulker and C. Erkey, *RSC Adv.*, 2014, **4**, 62362–62366.
89. S. Zhao, G. Zhang, Y. Gao, L. Deng, J. Li, R. Sun and C.-P. Wong, *ACS Appl. Mater. Interfaces*, 2014, **6**, 22823–22829.
90. X. Shen, J. L. Shamshina, P. Berton, G. Gurau and R. D. Rogers, *Green Chem.*, 2016, **18**, 53–75.
91. H. Wei, K. Rodriguez, S. Renneckar and P. J. Vikesland, *Environ. Sci.: Nano*, 2014, **1**, 302–316.
92. M. Deng, Q. Zhou, A. Du, J. van Kasteren and Y. Wang, *Mater. Lett.*, 2009, **63**, 1851–1854.
93. P. Gurikov, S. P. Raman, D. Weinrich, M. Fricke and I. Smirnova, *RSC Adv.*, 2015, **5**, 7812–7818.
94. J. Zhang, Y. Cao, J. Feng and P. Wu, *J. Phys. Chem. C*, 2012, **116**, 8063–8068.
95. R. J. B. Pinto, M. C. Neves, C. P. Neto and T. Trindade, *Compos. Cellulose Met. Nanopart.*, 2012.
96. M. Ahmadi, A. Madadlou and A. A. Sabouri, *Food Chem.*, 2015, **174**, 97–103.
97. Q. Peng, M. Liu, J. Zheng and C. Zhou, *Microporous Mesoporous Mater.*, 2015, **201**, 190–201.
98. H.-B. Chen, B.-S. Chiou, Y.-Z. Wang and D. A. Schiraldi, *ACS Appl. Mater. Interfaces*, 2013, **5**, 1715–1721.
99. M. M. Koebel, L. Huber, S. Zhao and W. J. Malfait, *J. Sol-Gel Sci. Technol.*, 2016, **79**, 308–318.
100. B. Lucia, C. Simona, C. Stefano, M. Iolanda De and R. Ernesto, *J. Supercrit. Fluids*, 2015, **103**, 70–76.
101. H. Lindy, Z. Lifan and T. Wim, *ChemSusChem*, 2013, 1–9.
102. C. S. C. Chiew, P. E. Poh, P. Pasbakhsh, B. T. Tey, H. K. Yeoh and E. S. Chan, *Appl. Clay Sci.*, 2014, **101**, 444–454.
103. T. Liu, M. Huang, X. Li, C. Wang, C.-X. Gui and Z.-Z. Yu, *Carbon*, 2016, **100**, 456–464.

CHAPTER 8

Modelling and Simulations of Polysaccharide and Protein Based Aerogels

AMEYA REGE,*[a] LORENZ RATKE[b] AND MIKHAIL ITSKOV[a]

[a] RWTH Aachen University, Department of Continuum Mechanics, Kackertstraße 9, 52072 Aachen, Germany; [b] German Aerospace Center, Institute of Materials Research, Linder Höhe, 51147 Cologne, Germany
*Email: rege@km.rwth-aachen.de

8.1 Introduction

Aerogels are highly nano-porous open cellular materials exhibiting many excellent material properties, such as a low bulk density, low thermal conductivity, low dielectric permittivity, high specific surface area, and excellent shock adsorption properties.[1] Aerogel literature in the 20th century has primarily focused on silica aerogels and metal oxide aerogels. Native silica aerogels have been applied to a wide spectrum of applications ranging from building insulation to high end space applications.[2] Despite their extraordinary properties, these aerogels are very fragile and tend to collapse and disintegrate easily when subjected to small stresses. This makes their application difficult in environments where the material is required to carry a structural load, or at least maintain its own structural integrity. To overcome these drawbacks, polymer and fibre reinforcement of these aerogels have been extensively explored with many positive outcomes.[3] On the other hand, an alternate synthesis approach has proven even more successful towards finding a solution to overcome the fragile properties of native silica aerogels,

Green Chemistry Series No. 58
Biobased Aerogels: Polysaccharide and Protein-based Materials
Edited by Sabu Thomas, Laly A. Pothan and Rubie Mavelil-Sam
© The Royal Society of Chemistry 2018
Published by the Royal Society of Chemistry, www.rsc.org

that is synthesis and development of organic aerogels. To this end, polysaccharide based aerogels and resorcinol-formaldehyde aerogels have been developed and demonstrate high strength and flexibility. They can also be converted into carbon aerogels upon pyrolysis. Polysaccharidic aerogels are highly porous open cellular networks of naturally functionalized hydrocarbon materials. Kistler first described the preparation of such aerogels based on polysaccharides such as agar, nitrocellulose, and cellulose, in the very first article on aerogels in 1931.[4,5] Since then, aerogels have been prepared using various polysaccharides such as cellulose, alginate, pectin, chitin, chitosan, carrageenan, agar, starch.[6] Such polysaccharidic aerogels have very high porosities, about 90–99%,[7] and surface areas as high as 600 $m^2 g^{-1}$.[8] Such aerogels and their properties make them very attractive to a wide spectrum of potential applications. The most cited application of polysaccharidic aerogels in the literature is as carriers for drug delivery.[7,9,10] Polysaccharide cellulose is an almost inexhaustible polymeric raw material with fascinating structure and properties. Several attempts to prepare cellulose-based aerogels have been reported, most of which use either cellulose derivatives or cellulose-dissolving solvents, followed by a regeneration step.[11–14] The structure of cellulose aerogels can be described as a type of nanofleece, which means that the elementary fibrils of cellulose are arranged in a random three-dimensional structure.[15] Studies addressing other polysaccharide based aerogels are relatively few. However, the ongoing interest makes them very desirable to the aerogel community. A route to prepare monolithic κ-carrageenan aerogels was reported by Ganesan and Ratke[16] discussing their meso- and macro-porous properties. Preparation of monolithic alginate aerogels *via* pH induced gelation[10] and *via* carbon dioxide induced gelation[17] is discussed. Other polysaccharides, such as chitin and chitosan have been used to prepare aerogels,[18–21] and the preparation of pectin aerogels in powdered form, and as monoliths, has also been discussed.[7,22,23] Protein based aerogels are also being developed for applications in the food industry.[24] However, studies reporting on protein aerogels remain very scarce. The details on these materials and their synthesis procedure can be found in the respectively cited articles above. In addition, they are also discussed in detail in the preceding chapters.

Answers to several outstanding questions concerning the aerogel structure and its mechanical properties could be achieved through experimental characterization and computational modelling. Both areas have their own merits and limitations. Accurate characterization at the micro-structural range is limited by the state of the art of experimental tools. Although novel experimental procedures such as nanoholotomography allowing the 3D microstructure to be resolved at a sub 50 nm level[25] are being developed at a faster pace, interpretation of the results to describe the properties of the materials is a challenging task. On the other hand, modelling can be categorized into many kinds. Depending on the chosen method, which ranges from atomistic *ab-initio* methods at the atomic scale to the molecular dynamics simulations at the molecular level, and to the mechanical models at the meso-macro level, the materials properties at the respective scales can be determined. Each method faces challenges which are directly in correlation to either the availability of the computational power and/or the availability of the required corresponding experimental data. Alternative and upcoming modelling approaches include models based on tomographic reconstructions, which have recently been applied to aerogels,[25] and they are in good agreement with standard state-of-the-art experimental findings. Measures such as fractal dimensions, pore-size distributions, and modulus of elasticity of the aerogel clusters/fibrils can be accurately obtained using molecular modelling procedures. On the other hand, the mechanical constitutive response under different loading conditions can be described using micro-mechanical modelling. It is of course known, that a greater accuracy of results is directly proportional to the decrease of the length scale in simulations. For example, quantum mechanical calculations can accurately calculate the potentials describing the bonding between atoms and molecules forming the material. However, such simulations are limited to only a very few molecules because of the increasing computational costs and power needed. Classical molecular dynamics simulations on the other hand can be performed up to atomistic systems with a total number of atoms of the order 10^6, which is still at the nano-scale regime. Meso and macro-scale mechanical models are based on simplified assumptions and strain energy functions, which although proven to be micro-mechanically motivated, often remain idealized. Thus, the current need is to theoretically understand the sub-structural evolution of the aerogels under deformation and to be able to tailor their mechanical properties, these needs could be fulfilled by a multi-scale modelling framework. Most of the above-mentioned approaches ranging from the atomistic ones to the mechanical ones have been applied on aerogels providing insightful information on the aerogel structure and mechanical properties. Other tools such as stochastic reconstructions[26] and dynamic Monte-Carlo methods[27] have also been applied on aerogels to generate structures and study the gel formation, respectively.

However, molecular modelling of polysaccharide and protein based aerogels has so far not been reported in the literature. Our recently proposed micro-mechanical models describing the constitutive response of cellulose aerogels under compression[28] and tension[29] are the only published models describing the mechanical response of these aerogels. This is because most of the focus in the literature on modelling of aerogels has been placed on silica aerogels. Hence, in the following chapter, we first discuss this literature on the modelling of aerogels in Section 8.2 and then the modelling of polysaccharides in Section 8.3. Models discussing only the mechanical properties and behaviour are elucidated. Then the recently proposed mechanical models of the polysaccharidic aerogels are discussed in detail in Section 8.4. To this end, the mechanical characterisation of these aerogels is first addressed and is followed by a brief overview on the microcell-based models for cellulose aerogels. Model validations against cellulose and κ-carrageenan aerogels are then demonstrated against corresponding experimental data.

8.2 Overview on Modelling of Aerogels

The amount of literature published on the modelling of aerogels has been rising in the past decade, but it is nevertheless rather limited. The existing modelling approaches that have been applied to aerogels can be characterised into three types:

1. Atomistic models
2. Coarse-grained models
3. Micro-mechanical models

An excellent review on modelling methods for aerogels was published in 2011 by Gelb.[30] In this subsection, we briefly overview some of these models along with some newer studies. Atomistic models are referred, using the terminology determined by Gelb, as models that simulate the hydrolysis/condensation chemistry underlying the silica sol-gel process, and dynamical simulations of the formation of sol particles and gels. These procedures simulate the formation of aerogels and do not describe the final product itself. Hence, they have almost no impact on understanding the mechanical behaviour of aerogels under deformation. However, coarse-grained models have been proven to clarify certain linear-elastic properties of aerogels. These models are often subdivided further into two categories: hard-sphere aggregation models and flexible coarse-grained models. Aggregation models relevant to aerogels are diffusion-limited aggregation (DLA), reaction-limited aggregation (RLA), diffusion-limited cluster aggregation (DLCA) and reaction-limited cluster aggregation (RLCA). In DLA and RLA, clusters grow by the addition of monomers, while in DLCA and RLCA the clusters are themselves mobile and may meet and aggregate.[30] Application of DLCA to aerogels was first reported by Hasmy *et al.*[31,32] They compared the DLCA model of aerogels against short- and long-range results of the small angle neutron scattering (SANS) spectra of colloidal aerogels. The results of DLCA were found to be in good agreement with the SANS data. Later, Ma *et al.* modelled the aerogel structure using DLCA, in which they described the inter-particle bonds as beam elements and studied their linear elastic properties within the framework of finite element analysis.[33–35] Of particular interest was the influence of the dangling bonds on the mechanical response of aerogels. They concluded their reports by identifying the scaling exponent m in the power-law relationship, which is often used for such porous materials. Accordingly,

$$E \propto \rho_{\mathrm{e}}^{\mathrm{m}}$$

in which, E represents the Young's modulus and ρ_{e} denotes the envelope density. Recently, the model proposed by Ma *et al.*[33–35] was extended into a two-level model by Liu *et al.* to simulate the tensile properties of silica aerogels, specifically the Young's modulus and the tensile strength were explored.[36]

Molecular dynamics simulations have gained significant attention towards aerogel modelling in the past decade. Gelb used flexible coarse-graining to model low-density silica aerogels and described their linear elastic properties, primarily the bulk modulus.[37] In his first report, he found the obtained bulk modulus to be lower than the one reported in the corresponding experimental findings. In an extension study, Ferreiro-Rangel and Gelb further explored the bulk modulus using fluctuation analysis and direct compression-expansion simulations.[38] The evaluated power-law exponent was between 3 and 3.15 through simulations, which was in good agreement with the reports from Ma *et al.*[33–35] Recently, they reported the uniaxial tensile-compressive response of silica aerogels of different densities using molecular dynamics simulations and hybrid Monte Carlo methods.[39] The Young's modulus obtained from both methods was in good agreement with the values in the literature, except for very low-density aerogels. The density was found to have a weak dependence on the Poisson ratio. Uniaxial tension and compression of a large magnitude were reported. Low density aerogels were found to possess greater elasticity in comparison to high density ones which failed sooner, suggesting brittle behaviour. This is true, because low density aerogels have higher porosities, giving them larger freedom to move under deformation. The models also displayed auxetic behaviour at larger tensile strains. It is important to note that the aerogel structure, even under extreme tensile stress, showed only very few broken bonds. This can be attributed to the fact that in aggregated clusters and structures, there is one critical path often referred to as the backbone chain that carries most of the load, leaving much of the network not contributing to the mechanical strength of the material.[40–42] Such a silica aerogel backbone was modelled and simulated by Lei *et al.* in order to investigate the Young's modulus of the backbone.[43] Recently, we simulated the tensile-compressive response of silica aerogels under large strains *via* molecular dynamics simulations.[44] Inelastic features such as cyclic stress softening and residual deformation were explored. New phenomenological relations describing these effects were proposed that give an insight into the damage mechanisms in silica aerogels. Although such molecular dynamics simulation models provide a deeper understanding of the nano-structural evolution under deformation of aerogels, they are computationally rather expensive when one thinks of macro- or even sometimes upper micro-scale simulations. For this reason, multiscale simulations incorporating the structural response at all levels: molecular to meso to macro level, is the current need. Mesoscopic to macroscopic material responses can be efficiently studied based on the so called micro-mechanical approaches. Recently, we proposed a micro-mechanical model describing the response of cellulose aerogels prepared from molten salt hydrates.[28] This modelling approach was then adopted to describe the tension and failure in cellulose aerogels.[29] These models will be discussed later in the chapter. Although such mechanical models qualitatively and quantitatively seem to be reliable tools for describing the mechanical response, they need to be motivated from the molecular scale, using for example molecular dynamics simulations to determine nano-scale properties that appear as parameters in mechanical models. Such molecular simulations of polysaccharide and protein based aerogels need to be achieved in order to determine fractal and inter-particle behaviour under deformation.

8.3 Overview of Modelling of Polysaccharides

Molecular modelling of polysaccharides, such as alginate, carrageenan, pectin, galactomannan, and so forth, was first studied in order to determine their three-dimensional structures.[45] Modelling of polysaccharides to obtain their mechanical properties is, however, very limited in the literature. In the next subsections, we shall overview the models, characterizing the mechanical behaviour and properties of different polysaccharides. It should be noted that only the models reporting on the mechanical behaviour of polysaccharides are discussed.

8.3.1 Cellulose

From the modelling aspect, the most well studied polysaccharide is cellulose. The very first study on the modelling of cellulose by using molecular dynamics (MD) was published by Reiling and Brickmann.[46,47] The CHARMM22 program package was applied to perform MD calculations.[48,49] They presented a detailed report on the structural arrangement and properties of cellulose I and II. They also further studied in detail both phases of cellulose I, in example, Iα and Iβ. While the packing differences between the two were reported to be small, the Young's moduli differed significantly. The average values of Young's moduli estimated for cellulose I and II were in excellent agreement with the experimental data. This study was followed by the group of Kroon-Batenburg, who investigated the stability of cellulose structures, in particular Iβ and II.[50,51] In these papers, MD simulations on the GROMOS package were performed to investigate the details of the hydrogen bonding and on the similarities between the two cellulose structures.[52] Cellulose Iβ was described as a nicely ordered structure with parallel orientation of molecules, consisting of hydrogen bonded layers. On the other hand, cellulose II was shown to form a tight 3D hydrogen bonded network. The Young's moduli for cellulose Iβ was reported to be 1.5 times that of cellulose II. The reported values of the elastic moduli were in good agreement with the experiments, and close to the values reported by Reiling and Brickmann. Similar studies on molecular dynamics investigations of amorphous cellulose, Iα and Iβ, in the bulk state were also reported by Mazeau and Heux.[53] They used the second-generation force field PCFF to perform MD calculations.[54–58] Care was taken to make sure that there were no internal stresses, to reach a local minimum of the potential energy surface, and thus ensuring an equilibrium state.

In 2003, Chen, Lickfield, and Yang modelled cellulose based on molecular mechanics. The initial conformation of the cellulose chain in their model was generated using the Monte Carlo method and packed into a 20×20×20 Å^3 cubic cell subjected to periodic boundary conditions.[59] After energy minimization and cell parameters optimization, the relaxed cellulose models were subjected to tension and compression with a strain increment of 0.5% (approximately 0.1 Å) up to a maximum strain of 30%. A linear relationship between stress and strain was observed at low strain values. At higher strains, the material exhibited yielding at about 7–8% strain and an average initial modulus of 10.3 GPa, which was comparable to the corresponding values in the literature. With increasing strains in extension-compression, larger permanent sets (the amount of residual strain after unloading to the stress-free state, is called a permanent set) were observed (see Figure 8.1). The damage was explained by the breakage of hydrogen bonds, as the average elongation of hydrogen bonds decreased close to the yielding point. This is often observed in polymer physics. Chain slippage was observed during tension and was evidenced by the breakage of the original, and formation of new, hydrogen bonds. This study was followed by Chen *et al.*, discussing the effects of crosslinking on the mechanical behaviour of cellulose.[60] Dimethylol dihydroxyl ethylene urea (DMDHEU) and decane were used as rigid and flexible crosslinks, respectively. Molecular model construction for DMDHEU and decane were performed using similar tools as for cellulose in the previous study, for example using Cerius2 software and the molecular mechanics PCFF force

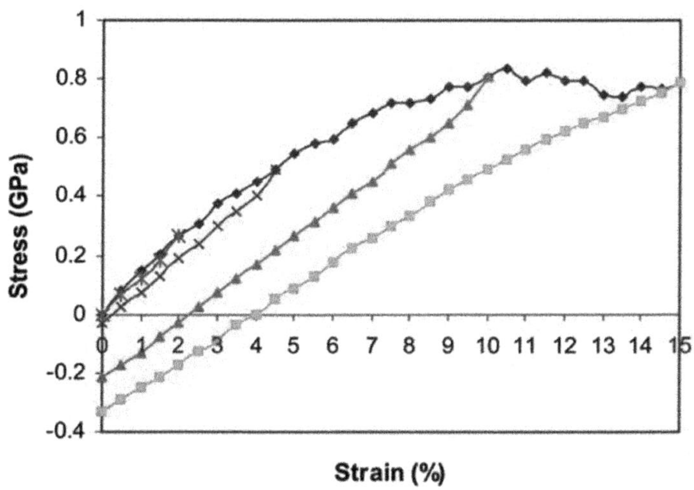

Figure 8.1 Cyclic loading unloading response of amorphous cellulose model subjected to maximum cyclic strains of 2, 4, 10 and 15%, respectively, using molecular mechanics.
Reprinted from *Polymer*, **45**, W. Chen, G. C. Lickfield and C. Q. Yang, Molecular modelling of cellulose in amorphous state. Part 1: model building and plastic deformation study, 1063–1071, Copyright 2004, with permission from Elsevier.[59]

field. Random conformations were created using the Monte Carlo method and these molecules were then relaxed using the NVT (canonical ensemble) molecular dynamics. DMDHEU molecules were combined with a cellulose polymer chain to build cellulose with rigid crosslinks. To this end, 5, 10, and 20% (w/w) DMDHEU was used. Similarly, for building flexible crosslinks, 6, 12 and 24% (w/w) decane was used. The effect of these different levels of DMDHEU, decane, and water concentrations, on the mechanical properties was examined through stress–strain curves. DMDHEU crosslinked cellulose showed an increased initial elastic modulus and less apparent yielding, while decane crosslinked cellulose demonstrated a slightly reduced stiffness, against dry cellulose. A comparison of the constitutive response of DMDHEU, decane and dry cellulose is illustrated in Figure 8.2. The values of 10% DMDHEU and 24% decane were used for comparison, as they both occupied the same number of crosslinking sites in a model cell. DMDHEU crosslinks were observed to resist external stress, while decane crosslinks did not resist such stresses under tension. Upon unloading of samples, the DMDHEU crosslinked cellulose showed excellent recovery against the poor recovery of dry and decane crosslinked cellulose. However, the DMDHEU crosslinked cellulose blocked the chain slippage. In spite of the increased initial elastic modulus and better elastic response, it led to the stress being unevenly distributed among the cellulose chains. Accordingly, the chains in some regions were subjected to higher stress leading to failure. On the other hand, flexible decane crosslinked cellulose demonstrated no improvements with respect to the breaking strain or deformation recovery.

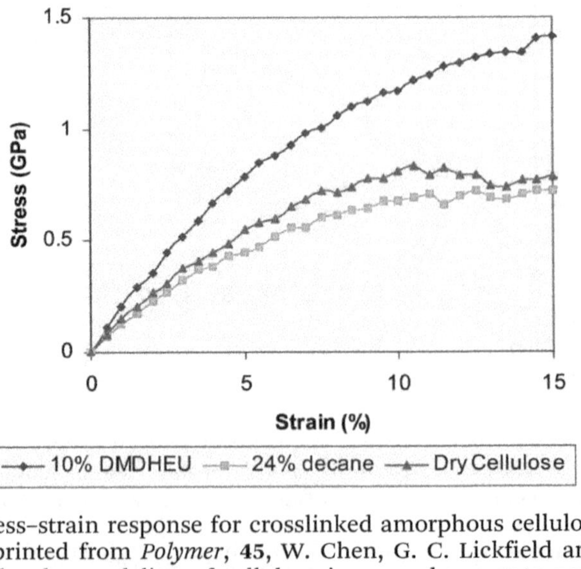

Figure 8.2 Stress–strain response for crosslinked amorphous cellulose models. Reprinted from *Polymer*, **45**, W. Chen, G. C. Lickfield and C. Q. Yang, Molecular modeling of cellulose in amorphous state part II: effects of rigid and flexible crosslinks on cellulose, 7357–7365, Copyright 2004, with permission from Elsevier.[60]

8.3.2 Other Polysaccharides

Modelling reports on other polysaccharides are, however, very limited. Beckham and Crowley[61] examined the decrystallization of the α-chitin by MD simulations using the CHARMM36 carbohydrate force field. Their approach was further expanded by Jin, Feng and Xu[62] who studied the mechanical properties of chitin-protein interfaces by MD simulations on the large-scale atomic/molecular massively parallel simulator (LAMMPS).[63] They investigated the structure of the chitin crystal and simulated the chitin block under tension using MD. The stress–strain relationship was used to determine the tensile modulus which was found to have a value of 88.5 GPa. This was, however, much higher against the corresponding experimental value of 41 GPa,[64] and was attributed to the structural inhomogeneity in the experiments performed. Yang *et al.* recently studied the relationship between the hydrogen bonding energy and the mechanical properties of starch *via* molecular dynamics simulations.[65] To the best of our knowledge, other polysaccharides have not been modelled so far to obtain their mechanical properties and response using molecular modelling. Accordingly, in the next section, we focus on the mechanical modelling of polysaccharide and protein based aerogels.

8.4 Mechanical Modelling of Polysaccharide and Protein Based Aerogels

As discussed previously, aerogels have been primarily of interest to the scientific community for their excellent thermal and acoustic insulation properties. So far, mostly experimental studies were reported. In the past two decades, however, there has been a growing interest in understanding the mechanical and physical properties of aerogels, and it is here that modelling becomes indispensable.

Until recently, there have been almost no contributions on the mechanics of polysaccharide and protein based aerogels under deformation. Neither of the two kinds of aerogels have been modelled using atomistic or multi-scale models. With growing applications that require these aerogels to maintain structural integrity, and bear mechanical loads, an in-depth study of micro-mechanics of such aerogels is becoming increasingly important. A short review on the modelling approaches discussed in this section is focussed on the micro-mechanics of polysaccharidic aerogels. To this end, we first discuss some experimental characterization tools used to identify the structure and properties of polysaccharide based aerogels.

8.4.1 Mechanical Characterisation of Polysaccharide and Protein Based Aerogels

Understanding of the micro-structural morphology is an integral aspect of the micro-mechanical modelling. Subsequently, we first portray these aerogels using scanning electron microscopy (SEM) images. Figures 8.3 and 8.4

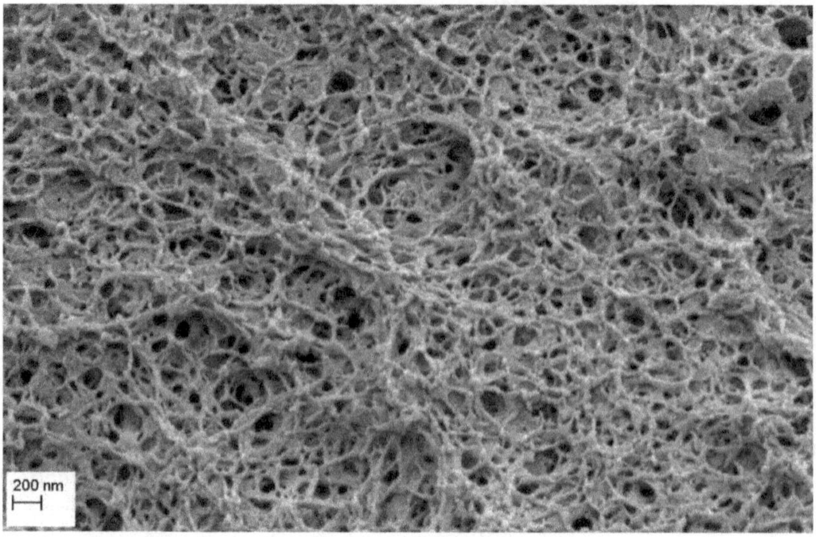

Figure 8.3 SEM Image of $ZnCl_2$ based cellulose aerogel.
Image prepared by Maria Schestakow at the German Aerospace Center, Cologne, Germany.

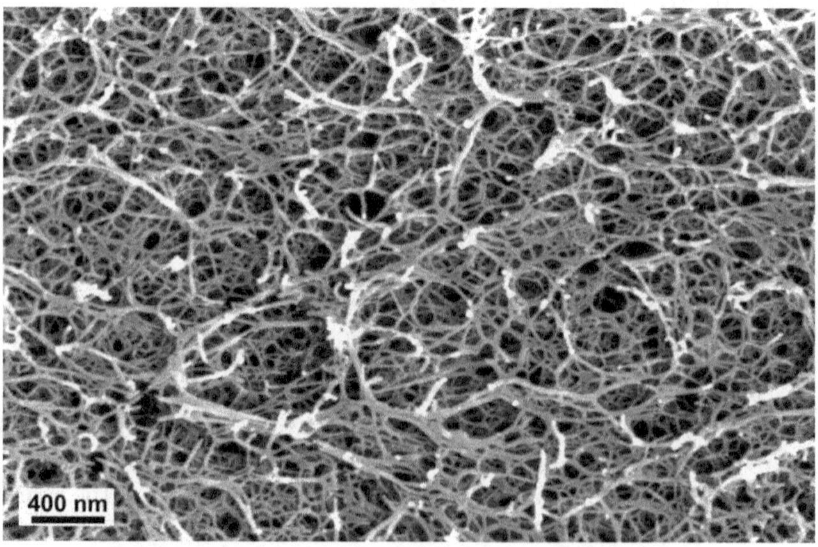

Figure 8.4 SEM image of κ-carrageenan aerogel.
Image prepared by Kathirvel Ganesan at the German Aerospace Center, Cologne, Germany.

show SEM pictures of cellulose and κ-carrageenan aerogels. It can be realized that these aerogels exhibit a somewhat cellular microstructure. Other polysaccharide based aerogels such as pectin, alginate, and starch, along with protein based aerogels show a similar cellular morphology.[17,24,66] The nano-fibrillar structure is chemically bonded (in the form of knots), where two or more fibrils meet forming a cellular appearance. There are two noteworthy critical points. One is that the number of so called microcells within a particular aerogel strongly depends on the chemical contents during the synthesis procedure. Although the number of microcells is not a quantity known to determine the mechanical properties of cellular porous materials, it does have a substantial influence on the pore-space. Secondly, the range of pore-space strongly influences the mechanical behaviour of aerogels. Nitrogen desorption isotherms are often used to determine the pore-size variations using the Barrett–Joyner–Halenda (BJH) model.[67] This model is based on an assumption that the pores are cylindrical in shape, which is far from a reality in aerogels. However, this method has been widely accepted within the aerogel community as a standard procedure for identifying the pore-space in aerogels due to the lack of a more efficient alternative. The other method is mercury porosimetry, which can provide pore-size distributions over a large range. However, it can be applied to only very stiff aerogels, as the more flexible ones are subjected to significant compression during the intrusion experiment. A more recent technique to describe the pore-space is the beam-based positron annihilation and this has been applied to silica aerogels.[68] A more realistic pore-space can also be obtained by X-ray imaging *via* nanotomography. Such a procedure has recently been applied to organic (resorcinol formaldehyde) aerogels, but it still remains rather challenging at the moment.[25] These newer procedures need to be explored in more detail so that if they are proven to provide a better description of the pore-space, they can be used more effectively in future. In our work, the pore-space is described using the standard accepted BJH analysis *via* nitrogen desorption isotherms. Such pore size analysis shows strong variations in the shape and interval of the pore-space, given different chemical quantities of the aerogels. The following quantities show strong influences: (i) types of solvents, (ii) polysaccharide concentration, and (iii) gelation procedure. The apparent average fibril diameter of these aerogels is also influenced, but can be estimated *via* SEM image analysis. Besides image evaluation, the fibril diameters of polysaccharidic and protein aerogels can be calculated more accurately from the Brunauer–Emmett–Teller (BET) specific surface area $S_{\mathrm{m,BET}}$ and the skeletal density ρ_{s} as:[13,16,28,29]

$$d_{\mathrm{f}} = \frac{4}{S_{\mathrm{m,BET}} \cdot \rho_{\mathrm{s}}}.$$

Other factors associated with changes in solvents and polysaccharide/protein concentrations are the density and Young's moduli. These mechanical properties are not discussed here as they have been addressed in the previous chapter. An interesting property of polysaccharide and protein based aerogels is their almost zero Poisson ratio. Under uniaxial tension/compression, these aerogels show almost no dimensional changes in the directions orthogonal to the loading one (a sample picture showing this

Figure 8.5 Illustration of the zero Poisson ratio in polysaccharide based aerogels. The picture shows an undeformed and deformed cellulose aerogel specimen. Reproduced from ref. 28 with permission from the Royal Society of Chemistry.

interesting feature in the case of $ZnCl_2$ based cellulose aerogels is displayed in Figure 8.5)[28,29]

The characteristics of polysaccharide and protein aerogels such as its morphology, pore-space, and certain structural properties form a strong input foundation for proposing mechanical models. This is addressed in the following subsection.

8.4.2 Microcell-based Modelling

Keeping the experimental characterization analysis presented above as a motivation, we recently proposed constitutive models describing the mechanical behaviour of polysaccharide and protein based aerogels, in particular cellulose, under compression[28] and tension.[29] These models are based on the mechanics of cell walls within the cellular network. Although these models have so far only been verified against the experimental data for cellulose aerogels, they work for most of the polysaccharide and protein based aerogels that show a cellular microstructure. The goal of this study was to understand the micro-mechanics of these aerogels under deformation and to be able to tailor their mechanical properties during synthesis based on the model predictions.

As seen earlier, SEM investigations of polysaccharidic aerogels show a characteristic cellular morphology. Cellular materials are often characterized by power-scaling laws.[69] However, these are limited to the linear regime of the mechanical response, and are also based on some phenomenological material constants. These laws describe only the correlation of different mechanical properties like Young's modulus with density, but they cannot construe the mechanical behaviour of materials. The mechanics of such materials are largely dictated by the behaviour of their cell walls.[69] In the microcell-based modelling approach,[28,29] a micromechanical approach is attempted whereby all the material parameters are physically motivated. They can be altered during the synthesis in order to produce aerogels more suited to the desired or a foreseen application. Therefore, to develop the constitutive models, the following assumptions were made:

1. The aerogel network is considered to be made up of idealized square shaped microcells.
2. The microcells have an isotropic spatial distribution within the network.

Under compression, the macroscopic response of porous materials is known to be dictated by the microscopic bending of the clusters or cell walls.[69–72] In our model, this bending is described based on the extended version of the Euler–Bernoulli beam theory that considers geometric nonlinearities.[73,74] Accordingly, the strain energy of a square shaped microcell is formulated as a function of its cell wall length and the applied micro-stretch. An affine deformation concept is used, which means that the microscopic stretches follow the macroscopic deformation gradient. As discussed earlier, the Poisson ratio of cellulose aerogels (in addition to other polysaccharide and protein based aerogels) is almost zero. As a result, there are no displacements in the directions orthogonal to the loading direction. This macroscopic effect is also mimicked at the micro-scale by restricting the microcell deformation in directions orthogonal to the loading one, for example no barrelling movement takes place. This response of one microcell is then used to formulate the total network response as follows. The pore-size variation data obtained using the BJH model is used as an input to account for the distributions of microcell sizes in the model. The one-dimensional response is then generalized to three dimensions using the numerical integration over the unit sphere.[75] Such numerical integration schemes are a helpful and accurate tool as long as there is a continuous distribution of discrete entities within the network (microcells in our case) and their response is not highly nonlinear.[76] The cell walls are shown to be able to withstand the load only so long as their bending stress does not reach a critical value. Beyond this critical value, the cell is considered to collapse, and this serves as a damage criterion. A reader interested in the compression model is referred to Rege *et al.*[28]

On the other hand, under tension these aerogels are very weak and fail at very small strains. In this case, not only the bending of cell walls is important, but also their stretching. A constitutive model is accordingly formulated considering such combined bending and tension energies. The damage criterion is specified based on the breakage of cell walls that occurs upon their reaching a predefined critical normal stress. The model numerically describes the evolution of damage within the network, and accurately predicts the onset of material failure. This failure prediction is accurate within an error of up to 0.2% strain for different specimens studied. A reader interested in the tension model is referred to Rege and Itskov.[29] The number of material parameters stays the same in both compression and tension models discussed above. These include the cell size variations *via* BJH analysis, the cell wall diameter, the Young's modulus of the cell walls, the total number of microcells in the network, and the critical failure stress. Moreover, the input parameters such as the BJH data, the fibril diameter, or the Young's modulus of the cell walls, stay unchanged for a given type of material (for example 3 wt.% cellulose aerogels prepared in $ZnCl_2$) in both models. This correlation between synthesis parameters and the model predictions (for both models) is a result of their true micromechanical background.

Both constitutive models include only physically based variables and no empirical/phenomenological parameters. Such physically motivated models could be helpful in tailoring the mechanical properties of aerogels in the future. However, these models are still far from covering the complete deformation range of aerogels. For example, modelling the densification behaviour in porous cellular materials is still challenging and remains a topic of the contemporary research.

8.4.3 Simulation and Results

Of special interest is the influence of the physical model parameters on the model predictions. The effect of the fibrillar diameter on the constitutive response was studied in Rege *et al.* (see Figure 8.6).[28] As the cell wall or fibril diameter increases, the network response stiffens. The fibril diameter has been shown to be directly related to the chemical content of these polysaccharide and protein-based aerogels and can be controlled during the synthesis procedure. Preparation of materials, to suit the desired requirements, is often benefited by such parameter analysis.

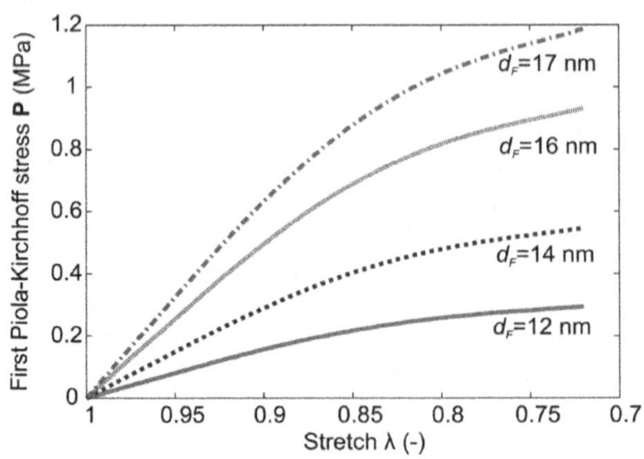

Figure 8.6 Model sensitivity to the change in average fibril diameter on the constitutive response of polysaccharide based aerogels.
Reproduced from ref. 28 with permission from the Royal Society of Chemistry.

The model predictions have now been validated against the experimental data of polysaccharide based aerogels, such as cellulose and κ-carrageenan. As mentioned earlier, there are very few studies on protein based aerogels as yet. Due to the lack of experimental data on their mechanical properties, the model could not be validated against protein aerogels in this contribution. However, their characteristic cellular morphology and material properties suggest that such models can be used as a useful tool to characterise their response. All the experimental data were obtained by conducting uniaxial quasi-static compression tests on these aerogel specimens per ASTM standards (ASTM D695) using a Universal Testing Machine Z100 by Zwick/Roell, Germany. No special care was taken to maintain isothermal conditions in the laboratory and all the tests were conducted at an initial strain rate of 0.1% s^{-1}.[77]

Model predictions of cellulose aerogels have been validated against the experimental data of 3 and 5 wt.% concentrations of both $ZnCl_2$ and $Ca(SCN)_2$ based cellulose aerogels. Both these preparation routes (different solvents) show very different mechanical properties under deformation. The results can be visualized in Figures 8.7 and 8.8. It is seen that as the concentration of cellulose in the aerogel increases, the network response stiffens. This was both qualitatively and quantitatively predicted by our model by varying the number of microcells and the average fibril diameter of cell fibrils in the network. A detailed parameter analysis can be found in a study published by Rege *et al.*[28] κ-carrageenan aerogels with 1 and 3 wt.% concentrations were also tested using the model and show good agreement against corresponding experimental data, as demonstrated in Figure 8.9. Other polysaccharides such as pectin and alginate, along with protein-based aerogels also could be verified using this microcell-based model.

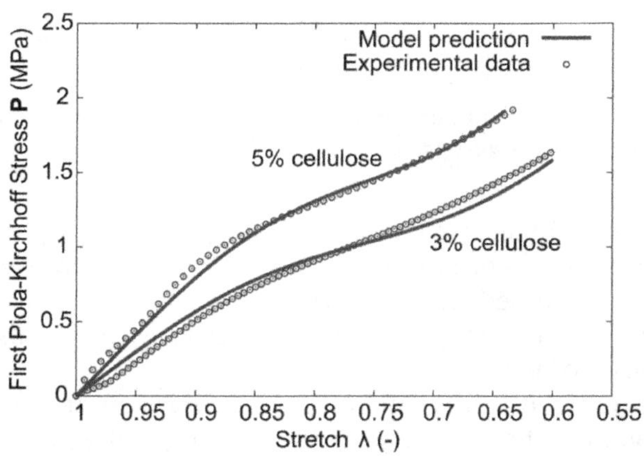

Figure 8.7 Compression model predictions against experimental data for $ZnCl_2$ based cellulose aerogels.
Reproduced from ref. 28 with permission from the Royal Society of Chemistry.

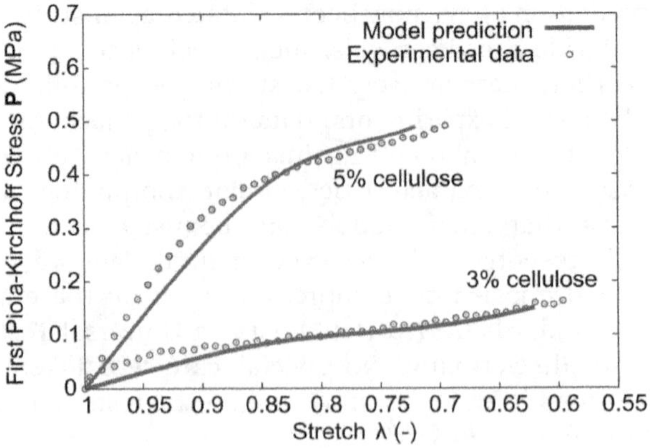

Figure 8.8 Compression model predictions against experimental data for $Ca(SCN)_2$ based cellulose aerogels.
Reproduced from ref. 28 with permission from the Royal Society of Chemistry.

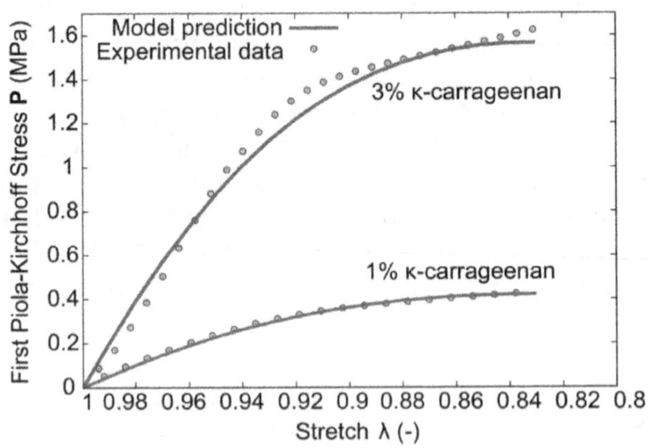

Figure 8.9 Compression model predictions against experimental data for κ-carrageenan aerogels.

The tension model has also been validated against the experimental data of 3 and 7 wt.% concentrations of $ZnCl_2$ based cellulose aerogels (see Figure 8.10). As discussed in the previous subsection, the model not only captures the mechanical response of these aerogels, but it also accurately predicts the onset of failure.

The above presented validations of the compression and tension model predictions against the corresponding experimental data manifest the following:

a. The ability of both models to predict the mechanical behaviour in advance, based on a given set of material parameters.

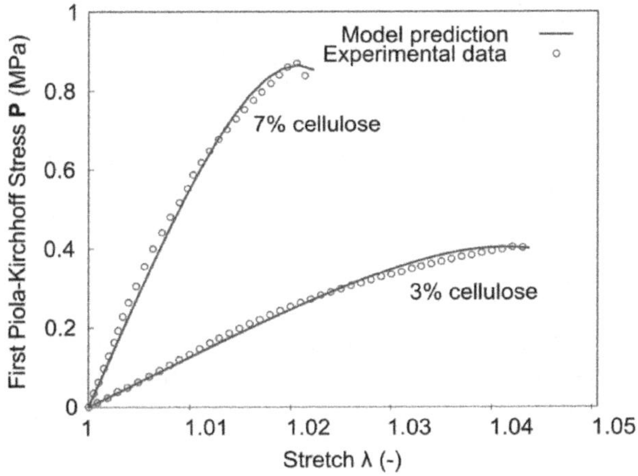

Figure 8.10 Tension model predictions against experimental data for $ZnCl_2$ based cellulose aerogels.

b. If a direct correlation between the physical/structural parameters and the chemical synthesis parameters is established, the models can predict the variation in the mechanical response due to changes in the chemical content of the aerogels.
c. The models fit well for polysaccharide based aerogels as long as they exhibit a cellular morphology.

8.5 Summary and Outlook

Recent developments in the field of aerogel modelling have opened up a pathway to understanding their mechanical properties and response from a quantum mechanical scale to a macroscopic scale. However, there remains many challenges to bridge the inter-scale gap, which need to be addressed. In particular, there is much work to be done on the modelling and simulation of polysaccharide and protein based aerogels. Recent molecular models of silica aerogels have demonstrated their ability to quantify various nano-structural properties. Such models accurately predict the nano-scale properties that can be used as input for mechanical models at the meso-macro scales. Our recent attempts at modelling the mechanical behaviour of polysaccharide-based aerogels within the framework of continuum mechanics show the ability of the models to provide reliable results for the constitutive response and mechanical properties. The models are micro-mechanically motivated, and every model parameter can be altered during the synthesis process in order to tailor the mechanical properties according to the given requirements.

Molecular simulations are also necessary to understand the nano-scale properties which are otherwise difficult to obtain using any state of the art experimental process. Thus, it appears that further improvements in computational modelling of aerogels will result in multiscale models. Only in this way, can one effectively describe the material response at the macro-scale while still being connected to the atomic structure at the nano-scale. The goal of realistic simulations of polysaccharide and protein based aerogels is likely to result from such multiscale simulations and there is a very keen interest in this within the aerogel community.

Acknowledgements

The authors acknowledge Maria Schestakow and Kathirvel Ganesan from the German Aerospace Center, in Cologne, Germany for providing the samples and experimental data on cellulose and κ-carrageenan aerogels.

References

1. J. Fricke and A. Emmerling, Aerogels, *J. Am. Ceram. Soc.*, 1992, **75**(8), 2027–2036.
2. L. W. Hrubesh, Aerogel applications, *J. Non-Cryst. Solids*, 1998, **225**, 335–342.
3. H. Maleki, L. Durães and A. Portugal, An overview on silica aerogels synthesis and different mechanical reinforcing strategies, *J. Non-Cryst. Solids*, 2014, **385**, 55–74.
4. S. S. Kistler, Coherent Expanded Aerogels, *Nature*, 1931, **127**, 741.
5. S. S. Kistler, Coherent Expanded-Aerogels, *J. Phys. Chem.*, 1932, **36**(1), 52–64.
6. S. Zhao, W. J. Malfait, N. Guerrero Alburquerque, M. M. Koebel and G. Nyström, *Angew. Chem.*, 2018, **57**, 7580–7608.
7. C. A. García-González, M. Alnaief and I. Smirnova, Polysaccharide-based aerogels—Promising biodegradable carriers for drug delivery systems, *Carbohydr. Polym.*, 2011, **86**(4), 1425–1438.
8. M. Robitzer, F. D. Renzo and F. Quignard, Natural materials with high surface area. Physisorption methods for the characterization of the texture and surface of polysaccharide aerogels, *Microporous Mesoporous Mater.*, 2011, **140**(1–3), 9–16.
9. C. A. García-González, M. Jin, J. Gerth, C. Alvarez-Lorenzo and I. Smirnova, Polysaccharide-based aerogel microspheres for oral drug delivery, *Carbohydr. Polym.*, 2015, **117**, 797–806.
10. T. Mehling, I. Smirnova, U. Guenther and R. H. H. Neubert, Polysaccharide-based aerogels as drug carriers, *J. Non-Cryst. Solids*, 2009, **355**(50–51), 2472–2479.
11. C. Tan, B. M. Fung, J. K. Newmann and C. Vu, Organic Aerogels with Very High Impact Strength, *Adv. Mater.*, 2001, **13**(9), 644–646.
12. J. Cai, S. Kimura, M. Wada, S. Kuga and L. Zhang, Cellulose aerogels from aqueous alkali hydroxide-urea solution, *ChemSusChem*, 2008, **1**(1–2), 149–154.

13. M. Schestakow, I. Karadagli and L. Ratke, Cellulose aerogels prepared from an aqueous zinc chloride salt hydrate melt, *Carbohydr. Polym.*, 2016, **137**, 642–649.
14. F. Liebner, A. Potthast, T. Rosenau, E. Haimer and M. Wendland, Cellulose aerogels: Highly porous, ultra-lightweight materials, *Holzforschung*, 2008, **62**(2), 129–135.
15. L. Ratke, Monoliths and Fibrous Cellulose Aerogels, in *Aerogels Handbook*, ed. M. Aegerter, N. Leventis, and M. M. Koebel, Springer, New York, 2011, pp. 173–190.
16. K. Ganesan and L. Ratke, Facile preparation of monolithic kappa-carrageenan aerogels, *Soft Matter*, 2014, **10**(18), 3218–3224.
17. P. Gurikov, S. P. Raman, D. Weinrich, M. Fricke and I. Smirnova, A novel approach to alginate aerogels: carbon sioxide induced gelation, *RSC Adv.*, 2015, **5**, 7812–7818.
18. B. Ding, J. Cai, J. Huang, L. Zhang, Y. Chen and X. Shi, *et al.*, Facile preparation of robust and biocompatible chitin aerogels, *J. Mater. Chem.*, 2012, **22**(12), 5801.
19. K. Rinki, P. K. Dutta, A. J. Hunt, D. J. Macquarrie and J. H. Clark, Chitosan Aerogels Exhibiting High Surface Area for Biomedical Application: Preparation, Characterization, and Antibacterial Study, *Int. J. Polym. Mater.*, 2011, **60**(12), 988–999.
20. X. Chang, D. Chen and X. Jiao, Chitosan-Based Aerogels with High Adsorption Performance, *J. Phys. Chem. B*, 2008, **112**, 7721–7725.
21. A. El Kadib and M. Bousmina, Chitosan bio-based organic-inorganic hybrid aerogel microspheres, *Chemistry*, 2012, **18**(27), 8264–8277.
22. R. J. White, V. L. Budarin and J. H. Clark, Pectin-derived porous materials, *Chemistry*, 2010, **16**(4), 1326–1335.
23. A. Veronovski, G. Tkalec, Z. Knez and Z. Novak, Characterisation of biodegradable pectin aerogels and their potential use as drug carriers, *Carbohydr. Polym.*, 2014, **113**, 272–278.
24. I. Selmer, C. Kleemann, U. Kulozik, S. Heinrich and I. Smirnova, Development of egg white protein aerogels as new matrix material for microencapsulation in food, *J. Supercrit. Fluids*, 2015, **106**, 42–49.
25. R. Tannert, M. Schwan, A. Rege, M. Eggeler, J. C. da Silva and M. Bartsch, *et al.*, The three-dimensional structure of flexible resorcinol-formaldehyde aerogels investigated by means of holotomography, *J. Sol-Gel Sci. Technol.*, 2017, **84**(3), 391–399.
26. J. Quintanilla, R. Reidy, B. Gorman and D. Mueller, Gaussian random field models of aerogels, *J. Appl. Phys.*, 2003, **93**, 4584–4589.
27. J. Mikes and K. Dusek, Simulation of polymer network formation by the Monte Carlo method, *Macromolecules*, 1982, **15**, 93–133.
28. A. Rege, M. Schestakow, I. Karadagli, L. Ratke and M. Itskov, Micromechanical modeling of cellulose aerogels from molten salt hydrates, *Soft Matter*, 2016, **12**(34), 7079–7088.
29. A. Rege and M. Itskov, A microcell-based constitutive modeling of cellulose aerogels under tension, *Acta Mech.*, 2018, **229**, 585–593.

30. L. D. Gelb, Simulation and Modeling of Aerogels Using Atomistic and Mesoscale Methods, in *Aerogels Handbook*, ed. M. Aegerter, N. Leventis, M. M. Koebel, Springer, New York, 2011, pp. 565–581.
31. A. Hasmy, M. Foret, J. Pelous and R. Jullien, Small-angle neutron-scattering investigation of short-range correlations in fractal aerogels: Simulations and experiments, *Phys. Rev. B*, 1993, **48**(13), 9345–9354.
32. A. Hasmy, E. Anglaret, M. Foret, J. Pelous and R. Jullien, Small-angle neutron-scattering investigation of long-range correlations in silica aerogels: Simulations and experiments, *Phys. Rev. B*, 1994, **50**(9), 6006–6016.
33. H.-S. Ma, A. P. Roberts, J.-H. Prevost and G. W. Scherer, Mechanical structure – property relationship of aerogels, *J. Non-Cryst. Solids*, 2000, **277**, 127–141.
34. H.-S. Ma and J.-H. Pr, Computer simulation of mechanical structure-±property relationship of aerogels, *J. Non-Cryst. Solids*, 2001, **285**, 216–221.
35. H.-S. Ma, J.-H. Prevost and G. W. Scherer, Elasticity of DLCA model gels with loops, *Int. J. Solids Struct.*, 2002, **39**, 4605/14.
36. Q. Liu, Z. Lu, M. Zhu, Z. Yuan, Z. Yang and Z. Hu, *et al.*, Simulation of the tensile properties of silica aerogels: the effects of cluster structure and primary particle size, *Soft Matter*, 2014, **10**(33), 6266–6277.
37. L. D. Gelb, Simulating silica aerogels with a coarse-grained flexible model and langevin dynamics, *J. Phys. Chem. C*, 2007, **111**(43), 15792–15802.
38. C. A. Ferreiro-Rangel and L. D. Gelb, Investigation of the bulk modulus of silica aerogel using molecular dynamics simulations of a coarse-grained model, *J. Phy. Chem. B*, 2013, **117**(23), 7095–7105.
39. C. A. Ferreiro-Rangel and L. D. Gelb, Computational Study of Uniaxial Deformations in Silica Aerogel Using a Coarse-Grained Model, *J. Phys. Chem. B*, 2015, **119**(27), 8640–8650.
40. W.-H. Shih, W. Y. Shih, S.-I. Kim, J. Liu and I. A. Aksay, Scaling behavior of the elastic properties of colloidal gels, *Phys. Rev. A*, 1990, **42**(8), 4772–4779.
41. A. D. Dinsmore, V. Prasad, I. Y. Wong and D. A. Weitz, Microscopic structure and elasticity of weakly aggregated colloidal gels, *Phys. Rev. Lett.*, 2006, **96**(18), 1–4.
42. R. Dargazany, A. Rege and M. Itskov, Elasticity of colloidal clusters with application to carbon-black aggregates in filled rubbers, *Const. Models Rubber VIII*, 2013, 261–266.
43. J. Lei, Z. Liu, J. Yeo and T. Y. Ng, Determination of the Young's modulus of silica aerogels – an analytical–numerical approach, *Soft Matter*, 2013, **9**(47), 11367.
44. S. P. Patil, A. Rege, Sagardas, M. Itskov and B. Markert, Mechanics of Nanostructured Porous Silica Aerogel Resulting From Molecular Dynamics Simulations, *J. Phys. Chem. B*, 2017, **121**(22), 5660–5668.
45. R. Chandrasekaran, X-ray and Molecular Modeling Studies on the Structure-function Correlations of Polysaccharides, *Macromol. Symp.*, 1999, **140**, 17–29.

46. S. Reiling and J. Brickmann, Theoretical investigations on the structure and physical properties of cellulose, *Macromol. Theory Simul.*, 1995, **4**, 725–743.
47. R. J. Marhöfer, S. Reiling and J. Brickmann, Computer simulations of crystal structures and elastic properties of cellulose, *Ber. Bunsen-Ges. Phys. Chem.*, 1996, **100**(8), 1350–1354.
48. B. R. Brooks, R. E. Bruccoleri, B. D. Olafson, D. J. States, S. Swaminathan and M. Karplus, CHARMM: a program for macromolecular energy, minimization, and dynamics calculations, *J. Comput. Chem.*, 1983, **4**(2), 187–217.
49. L. Nilsson and M. Karplus, Empirical energy functions for energy minimization and dynamics of nucleic acids, *J. Comput. Chem.*, 1986, **7**(5), 591–616.
50. L. M. J. Kroon-Batenburg, B. Bouma and J. Kroon, Stability of Cellulose Structures Studied by MD Simulations. Could Mercerized Cellulose II Be Parallel? *Macromolecules*, 1996, **29**, 5695–5699.
51. L. M. J. Kroon-Batenburg and J. Kroon, The crystal and molecular structures of cellulose I and II, *Glycoconjugate J.*, 1997, **14**, 677–690.
52. W. F. van Gunsteren, H. J. C. Berendsen *GROMOS (Groningen Molecular Simulation Computer Program Package)*, University of Groningen, The Netherlands, 1987.
53. K. Mazeau and L. Heux, Molecular Dynamics Simulations of Bulk Native Crystalline and Amorphous Structures of Cellulose, *J. Phys. Chem. B*, 2003, **107**, 2394–2403.
54. J. R. Maple, U. Dinur and A. T. Hagler, Derivation of force fields for molecular mechanics and dynamics from ab initio energy surfaces, *Proc. Natl. Acad. Sci.*, 1988, **85**(15), 5350–5354.
55. J. Maple, M. J. Hwang, T. Stockfisch and A. Hagler, Derivation of Class II force fields. III. Characterization of a quantum force field for alkanes, *Isr. J. Chem.*, 1994, **34**(2), 195–231.
56. J. R. Maple, M. J. Hwang, T. P. Stockfisch, U. Dinur, M. Waldman and C. S. Ewig, *et al.*, Derivation of class II force fields. I. Methodology and quantum force field for the alkyl functional group and alkane molecules, *J. Comput. Chem.*, 1994, **15**(2), 162–182.
57. H. Sun, S. J. Mumby, J. R. Maple and A. T. Hagler, An ab initio CFF93 all-atom force field for polycarbonates, *J. Am. Chem. Soc.*, 1994, **116**(7), 2978–2987.
58. H. Sun, Ab initio calculations and force field development for computer simulation of polysilanes, *Macromolecules*, 1995, **28**(3), 701–712.
59. W. Chen, G. C. Lickfield and C. Q. Yang, Molecular modeling of cellulose in amorphous state. Part I: model building and plastic deformation study, *Polymer*, 2004, **45**(3), 1063–1071.
60. W. Chen, G. C. Lickfield and C. Q. Yang, Molecular modeling of cellulose in amorphous state part II: effects of rigid and flexible crosslinks on cellulose, *Polymer*, 2004, **45**(21), 7357–7365.

61. G. T. Beckham and M. F. Crowley, Examination of the alpha-chitin structure and decrystallization thermodynamics at the nanoscale, *J. Phys. Chem. B*, 2011, **115**(15), 4516–4522.
62. K. Jin, X. Feng and Z. Xu, Mechanical Properties of Chitin–Protein Interfaces: A Molecular Dynamics Study, *Bionanoscience*, 2013, **3**(3), 312–320.
63. S. Plimpton, Fast Parallel Algorithms for Short-Range Molecular Dynamics, *J. Comput. Phys.*, 1995, **117**, 1–19.
64. T. Nishino, R. Matsui and K. Nakamae, Elastic Modulus of the Crystalline Regions of Chitin and Chitosan, *J. Polym. Sci., Part B: Polym. Phys.*, 1999, **37**(11), 1191–1196.
65. J. Yang, K. Tang, G. Qin, Y. Chen, L. Peng and X. Wan, *et al.*, Hydrogen bonding energy determined by molecular dynamics simulation and correlation to properties of thermoplastic starch films, *Carbohydr. Polym.*, 2017, **166**, 256–263.
66. L. Druel, R. Bardl, W. Vorwerg and T. Budtova, Starch Aerogels: A Member of the Family of Thermal Superinsulating Materials, *Biomacromolecules*, 2017, **18**(12), 4232–4239.
67. E. P. Barrett, L. G. Joyner and P. P. Halenda, The Determination of Pore Volume and Area Distributions in Porous Substances. I. Computations from Nitrogen Isotherms, *J. Am. Chem. Soc.*, 1951, **73**(1), 373–380.
68. M. Petkov, M. Weber, K. Lynn and K. Rodbell, Porosity characterization by beam-based three-photon positrion annihilation spectroscopy, *Appl. Phys. Lett.*, 2001, **79**, 3884.
69. L. Gibson, M. Ashby *Cellular Solids: Structure and Properties*, Cambridge University Press, 1999.
70. L. Weigold and G. Reichenauer, Correlation between mechanical stiffness and thermal transport along the solid framework of a uniaxially compressed polyurea aerogel, *J. Non-Cryst. Solids*, 2014, **406**, 73–78.
71. L. J. Gibson, Biomechanics of cellular solids, *J. Biomech.*, 2005, **38**(3), 377–399.
72. P. R. Onck, T. Koeman, T. van Dillen and E. van der Giessen, Alternative explanation of stiffening in cross-linked semiflexible networks, *Phys. Rev. Lett.*, 2005, **95**(17), 178102.
73. K. E. Bisshopp and D. C. Drucker, Large Deflection of Cantilever Beams, *Q. Appl. Math.*, 1945, **3**(3), 272–275.
74. L. Chen, An integral approach for large deflection cantilever beams, *Int. J. Non-Linear Mech.*, 2010, **45**(3), 301–305.
75. Z. P. Bazant and B. H. Oh, Efficient Numerical Integration on the Surface of a Sphere, *Zeit. Angew. Math. Mech.*, 1986, **66**(1), 37–49.
76. M. Itskov, On the accuracy of numerical integration over the unit sphere applied to full network models, *Comput. Mech.*, 2016, **57**(5), 859–865.
77. R. W. Pekala, C. T. Alviso and J. D. LeMay, Organic aerogels: microstructural dependence of mechanical properties in compression, *J. Non-Cryst. Solids*, 1990, **125**(1–2), 67–75.

CHAPTER 9

Biodegradation of Polysaccharide and Protein Based Aerogels

REKHA ROSE KOSHY*[a,b] AND LALY A. POTHAN[a]

[a] Department of Chemistry, Bishop Moore College, Mavelikara, Kerala, India; [b] Department of Chemistry, CMS College, Kottayam, Kerala, India
*Email: rekhakoshy2@gmail.com

9.1 Introduction

Degradation or biodegradation has become a key word in the development of new technologies, not only for pharmaceutical or medical applications, but also for agricultural and environmental applications. Degradability of aerogels is a relatively new topic and no adequate definitions or test protocols exist. In reality everything degrades, the question is how fast it degrades. The most desirable biodegradable system would be the one that leaves no residual polymer after the degradation.[1]

Since its discovery by Kistler in 1931,[2] different types of aerogels have emerged including inorganic (such as SiO_2 derived from various alkoxysilanes, TiO_2, Al_2O_3, ZrO_2, *etc.*),[3,4] organic (*i.e.* resorcinol–formaldehyde (RF), polyurethane, polyimide, polystyrene, *etc.*)[5] and carbon (*i.e.* carbon, carbon nanotubes, graphene),[6] semiconductor chalcogenide (*i.e.* CdS, CdSe, PbTe),[7] natural based aerogels (*i.e.* cellulose and other polysaccharides and various proteins),[8,9] and more recently SiC-based aerogels.[10] Among these,

Green Chemistry Series No. 58
Biobased Aerogels: Polysaccharide and Protein-based Materials
Edited by Sabu Thomas, Laly A. Pothan and Rubie Mavelil-Sam
© The Royal Society of Chemistry 2018
Published by the Royal Society of Chemistry, www.rsc.org

biodegradable and bio-based polymers are of increasing interest, because of the limits in fossil fuel resources and their excellent biodegradability.[11] Also, for pharmaceutical, medical and food applications, the biodegradability and biocompatibility of aerogels is a limiting factor.[12] Hence, biodegradability of bio-based aerogels such as polysaccharide aerogels (*i.e.* cellulose, chitosan, alginate, starch, pectin, agar, *etc.*) and protein-based aerogels such as gelatin, whey protein, soy, egg white or silk fibroin need to be discussed in conjunction with their applications. Such studies will help reduce landfill shortage problems initiated by the abundant waste of non-biodegradable conventional plastics.

9.2 Definition of Biodegradation

Earlier, biodegradation was defined as a decomposition of substances by the action of microorganisms. This action leads to the recycling of carbon, the mineralisation (CO_2, H_2O and salts) of organic compounds and the generation of new biomass. For biodegradable polymers, biodegradation may mean fragmentation, loss of mechanical properties, or sometimes degradation through the action of microorganisms such as bacteria, fungi, and algae.[13] The biodegradation of polymeric materials includes several steps and the process can stop at each stage.[1] Various steps involved in the biodegradation process are:

i. Biodeterioration
   In this step, the combined action of microbial communities, other decomposer organisms and/or abiotic factors fragment the biodegradable materials into tiny fractions.
ii. Depolymerization
   In this process microorganisms secrete catalytic agents (*i.e.* enzymes and free radicals) that are able to cleave polymeric molecules, progressively reducing their molecular weight. This process generates oligomers, dimers and monomers.
   Some molecules are recognized by receptors of microbial cells and can go across the plasmic membrane. The other molecules stay in the extracellular surroundings and can be the object of different modifications.
iii. Assimilation
   In the cytoplasm, transported molecules integrate the microbial metabolism to produce energy, new biomass, storage vesicles and numerous primary and secondary metabolites.
iv. Mineralization
   Concomitantly, some simple and complex metabolites may be excreted and reach the extracellular surroundings (*e.g.* organic acids, aldehydes, terpens, and antibiotics, *etc.*). Simple molecules such as CO_2, N_2, CH_4, H_2O and different salts from intracellular metabolites that are completely oxidised are released into the environment.

The term "biodegradation" indicates the predominance of biological activity in the degradation of organic matter. However, in nature, biotic and abiotic factors act synergistically to decompose organic matter. Several studies on the biodegradation of some polymers show that the abiotic degradation precedes microbial assimilation. Consequently, the abiotic degradation must not be neglected.

Bacteria important in the biodegradation process include, among others, Bacillus (capable of producing thick-walled endospores that are resistant to heat, radiation and chemical disinfection), Pseudomonas, Klebsiella, Actinomycetes, Nocardia, Streptomyces, Thermoactinomycetes, Micromonospora, Mycobacterium, Rhodococcus, Flavobacterium, Comamonas, Escherichia, Azotobacter and Alcaligenes (some of them can accumulate polymers up to 90% of their dry mass). Temperature is one of the most important factors affecting microorganism growth. Also of importance are the sources of carbon and nitrogen, and pH. Fungi active in the biodegradation process are Sporotrichum, Talaromyces, Phanerochaete, Ganoderma, Thermoascus, Thielavia, Paecilomyces, Thermomyces, Geotrichum, Cladosporium, Phlebia, Trametes, Candida, Penicillium, Chaetomium, and Aerobasidium.[14] The biodegradation process can be divided into (9.1) aerobic and (9.2) anaerobic degradation.

$$\text{Aerobic biodegradation: Polymer} + O_2 \rightarrow CO_2 + H_2O + \text{biomass} + \text{residue(s)} \tag{9.1}$$

$$\text{Anaerobic biodegradation: Polymer} \rightarrow CO_2 + CH_4 + H_2O + \text{biomass} + \text{residue(s)} \tag{9.2}$$

If oxygen is present, aerobic biodegradation occurs and carbon dioxide is produced. If there is no oxygen, anaerobic degradation occurs and methane is produced instead of carbon dioxide. When conversion of biodegradable materials or biomass to gases (such as carbon dioxide, methane, and nitrogen compounds), water, salts, minerals and residual biomass occurs, this process is called mineralization.[15] Mineralization is complete when all the biodegradable materials or biomass are consumed and all the carbon is converted to carbon dioxide.[16]

Biodegradable materials have the proven capability to decompose in the most common environments where the material is disposed, within one year, through natural biological processes into non-toxic carbonaceous soil, water or carbon dioxide. The chemical structure (responsible for functional group stability, reactivity, hydrophilicity and swelling behaviour) is the most important factor affecting the biodegradability of polymeric materials. Other important factors are, amongst others, physical and physico-mechanical properties (*e.g.*, molecular weight, porosity, elasticity and morphology).

9.3 Biodegradability of Polysaccharide and Protein Based Aerogels

Polysaccharide-based aerogels accomplish the biodegradability that is lacked in silica aerogels and hence have received considerable attention from both academic and industrial researchers over the past decade. Among those aerogels based on polysaccharides, cellulose fiber has a higher crystallinity than the others that exhibit similar molecular structures (starch, CMC, agar, *etc.*). Therefore, their degradable behaviour should be different. On the other hand, these polymeric aerogels have a high porosity and may show a distinct biodegradability in regard to their corresponding solid materials. The decomposition rates of solid films are lower than their aerogel counterparts. This is due to the higher specific surface area of aerogels. The porous aerogel structure is beneficial to organism cell adhesion, growth and proliferation. As a result, a faster consumption of the aerogel matrix occurs in comparison to films, which will in turn be depicted by an increased CO_2 released rate.

Studies published by Chen *et al.* showed that polysaccharide-based pectin aerogels degraded very fast in a compost media within a duration of only 10 days, and its degree of biodegradability was as high as 50%.[11] The degree of biodegradation (expressed as %) of pectin aerogels was evaluated by monitoring the CO_2 yield, with the help of a Micro-Oxymax Respirometer System using the equation:

$$D(\%) = \frac{V(CO_2)_t - V(CO_2)_b}{V(CO_2)_{th}} \times 100$$

In which $V(CO_2)_t$ and $V(CO_2)_b$ represents the amounts of CO_2 released within the assay and the blank reactor, respectively, and $V(CO_2)_{th}$ is the theoretical amount of CO_2 available from the samples.

It was also found that addition of clay and Al^{3+} resulted in a higher biodegradation rate. Figure 9.1 shows the degree of biodegradation of pectin and pectin/clay aerogels in compost media.

Wang studied the degradation behavior of an aerogel made from starch, recycled cellulose fibres (RCF)/sodium carboxylmethylcellulose (CMC) and polyvinylchloride (PVC) using a micro respirometer system, and found that both the cellulose and starch based aerogels exhibited a higher decomposition rate and released more CO_2 than the one based on polyvinyl alcohol (PVOH).[17] This indicated that the natural cellulose and starch have a better biodegradability than synthesized PVOH. On the other hand, the pure starch aerogel decomposed faster than the CMC/RCF hybrid aerogel. This is due to the higher degree of crystallinity of RCF in comparison to starch and CMC. To analyse the effect of porosity, the selected aerogels samples were compressed to make films and their biodegradability was compared with the corresponding aerogels. The decomposition rates of solid

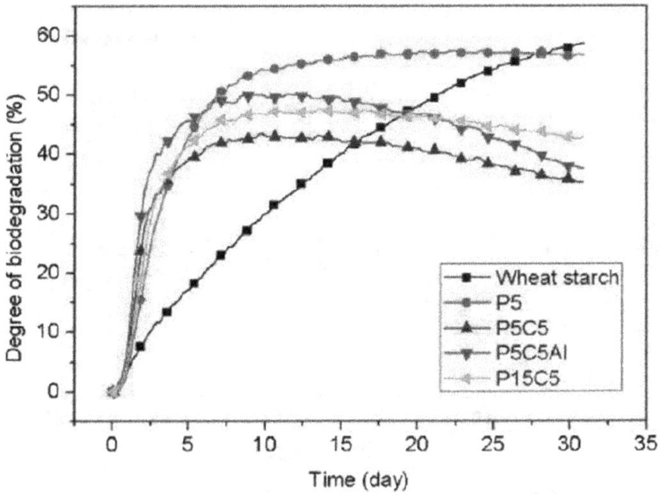

Figure 9.1 Degree of biodegradation of the pectin and pectin/clay aerogels in compost media.
Reprinted with permission from H. Chen, B. Chiou, Y. Wang and D. A. Schiraldi, *Appl. Mater. Interfaces*, 2013, **5**, 1715 – 1721, Copyright 2013 American Chemical Society.

films were lower than their aerogel counterparts. This is due to the higher specific surface area of aerogels.

Unlike polysaccharide aerogels, protein-based aerogels composites exhibited relatively low degradation rates compared with their solid counterparts. The existence of the layered "house of cards" architecture in the case of proteins probably increased the time for degradation to occur because the interior region was more difficult to access for the microorganisms, or the aerogels presented a tortuous path through which degradation microorganisms must diffuse. These findings were proved beyond doubt by Pojanavaraphan *et al.* who investigated the biodegradability of low density bio-based casein/clay aerogel composites for the first time using a compost media under controlled conditions.[18] It was observed that the casein aerogels and casein/clay aerogel composites exhibited relatively low degradation rates compared to wheat starch aerogels. Their biodegradability reached only 24–30%, even after 45 days of immersion in compost media. However, in the presence of a chemical cross-linking agent, DL-glyceraldehyde (GC), the degradation rate was found to increase. This implies that the addition of GC resulted in a different mode of attack and disruption on the casein component of the tested samples. The microorganisms in the soil directly attacked the GC component to achieve metabolism first, leading to a faster biodegradation compared to that in a neat casein aerogel. The degradation rate of the casein/clay aerogel was also higher than the pure casein aerogel (Figure 9.2) which showed that the existence of the clay aerogel may change the integrity of the casein matrix, thus promoting the biodegradation process.

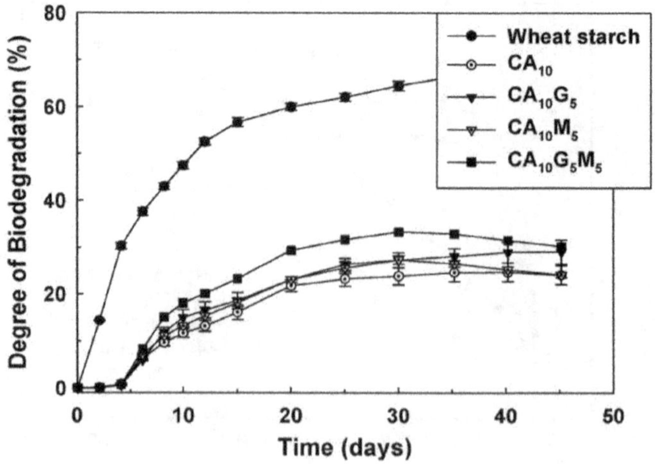

Figure 9.2 Degree of biodegradation (%) of the casein aerogels and casein/clay aerogel composites under compost at 23 ± 2 °C.
Reprinted with permission from T. Pojanavaraphan, R. Magaraphan, B. Sen Chiou and D. A. Schiraldi, *Biomacromolecules*, 2010, **11**, 2640–2646, Copyright 2010 American Chemical Society.

9.4 Conclusion

In the present scenario of environmental pollution caused by the disposal of non-biodegradable polymer waste into the environment, fabrication of biodegradable materials is gaining rapid interest and aerogels are no exception. The use of biopolymers can to some extent reduce the effect of environmental pollution, but these aerogels lack the mechanical strength and thermal stability that normal silica based aerogels possess. Nature-like experiments are difficult to realise in the laboratory due to the great number of parameters occurring during biogeochemical recycling. In actual fact, all these parameters cannot be entirely reproduced and controlled *in vitro*. In particular, the diversity and efficiency of microbial communities and catalytic abilities to use and to transform a variety of nutrients cannot be anticipated. Hence, a detailed study of the biodegradation process in aerogels needs to be performed with a special focus on biodeterioration, biofragmentation and assimilation. One solution consists of the labelling of the initial polymer with a fluorochrome, a radioelement or a stable isotope. However, these methods are expensive due to the need for specific chemicals, analytical equipment and qualified technicians.

References

1. N. Lucas, C. Bienaime, C. Belloy, M. Queneudec, F. Silvestre and J. E. Nava-Saucedo, *Chemosphere*, 2008, 73, 429–442.
2. S. S. Kistler, *Nature*, 1931, **127**, 741.
3. C. M. Anderson, *Aerogel Handb.*, 2011, 47–77.

4. M. Moner-Girona, A. Roig, E. Molins, I. De Ci and C. Met, *J. Sol-Gel Sci. Technol.*, 2003, **26**, 645–649.
5. Y. Tao, A. Hirofumi Kanoh and K. Kaneko, *Langmuir*, 2004, **21**, 504–507.
6. T. Horikawa, J. Hayashi and K. Muroyama, *Carbon N. Y.*, 2004, **42**, 169–175.
7. H. Maleki, L. Duraes, C. A. Garcia-Gonzalez, P. del Gaudio, A. Portugal and M. Mahmoudi, *Adv. Colloid Interface Sci.*, 2016, **236**, 1–27.
8. C. A. Garcia-Gonzalez, M. Alnaief and I. Smirnova, *Carbohydr. Polym.*, 2011, **86**, 1425–1438.
9. M. Ahmadi, A. Madadlou and A. A. Saboury, *Food Chem.*, 2016, **196**, 1016–1022.
10. E. Zera, R. Campostrini, P. R. Aravind, Y. Blum and G. D. Soraru, *Adv. Eng. Mater.*, 2014, **16**, 814–819.
11. H. Chen, B. Chiou, Y. Wang and D. A. Schiraldi, *ACS Appl. Mater. Interfaces*, 2013, **5**, 1715–1721.
12. M. Alnaief, M. A. Alzaitoun, C. A. García-González and I. Smirnova, *Carbohydr. Polym.*, 2011, **84**, 1011–1018.
13. J.-W. Rhim, H.-M. Park and C.-S. Ha, *Prog. Polym. Sci.*, 2016, **38**, 1629–1652.
14. K. Leja and G. Lewandowicz, *Pol. J. Environ. Stud.*, 2010, **19**, 255–266.
15. J. P. Maran, V. Sivakumar, K. Thirugnanasambandham and R. Sridhar, *Carbohydr. Polym.*, 2014, **101**, 20–28.
16. I. Kyrikou and D. Briassoulis, *J. Polym. Environ.*, 2007, **15**, 125–150.
17. L. Wang, Universitat Politècnica de Catalunya, 2015.
18. T. Pojanavaraphan, R. Magaraphan, B. Sen Chiou and D. A. Schiraldi, *Biomacromolecules*, 2010, **11**, 2640–2646.

CHAPTER 10

Thermal, Electrical, Insulation and Fire Resistance Properties of Polysaccharide and Protein-based Aerogels

MIGUEL SÁNCHEZ-SOTO,*[a] LIANG WANG,[b] TOBIAS ABT,[a] LUCÍA G. DE LA CRUZ[a] AND DAVID A. SCHIRALDI[c]

[a] Department of Materials Science, Centre Català del Plàstic, Universitat Politècnica de Catalunya, BarcelonaTech (UPC), C/Colom 114, E-08222 Terrassa, Spain; [b] School of Textiles, Tianjin Polytechnic University, Binshui West Road 399, 300387, Tianjin, China; [c] Department of Macromolecular Science & Engineering, Case Western Reserve University, Cleveland, OH 44106-7202, USA
*Email: m.sanchez-soto@upc.edu

10.1 Thermal and Insulation Properties

The need for efficient insulating materials has drastically increased in recent years due to worldwide efforts to decrease energy consumption. Around 25–40% of this energy is consumed in dwelling houses, of which 50–60% is required for heating and cooling of the buildings.[1,2] Therefore, new efficient thermal insulation materials such as vacuum insulation panels and aerogels can drastically reduce this energy consumption. The outstanding thermal insulation properties of these materials are due to their reduced gas phase conductivity, which is dependent on vacuum and on pore sizes smaller than 100 nm.[1]

Green Chemistry Series No. 58
Biobased Aerogels: Polysaccharide and Protein-based Materials
Edited by Sabu Thomas, Laly A. Pothan and Rubie Mavelil-Sam
© The Royal Society of Chemistry 2018
Published by the Royal Society of Chemistry, www.rsc.org

10.1.1 Thermal Conductivity Mechanisms

The heat transport in porous materials is governed by the thermal conductivity λ and the thermal diffusivity (D), which may be calculated from each other if the specific heat capacity c_p and the foam density ρ are known.[1–4]

$$\lambda = D \cdot c_p \cdot \rho \tag{10.1}$$

The thermal conductivity of porous materials can be approximated as the sum of four separate components

$$\lambda = \lambda_s + \lambda_g + \lambda_c + \lambda_r \tag{10.2}$$

In which λ_s is the heat conductivity through the solid phase, λ_g is the heat conductivity through the gas phase, λ_c is the heat convection in the gas phase and λ_r is the radiation through the foam cells. The total heat conductivity is schematically represented in Figure 10.1 and is mainly governed by the conductivity through the solid phase. Therefore, insulation materials should preferably be highly porous, having only a small solid fraction. It was found for SiO_2- and carbon-based aerogels that the solid thermal conductivity scales with the density, ρ:[4]

$$\lambda_s = C \cdot \rho^{\alpha} \tag{10.3}$$

In which C is a pre-factor that depends on the interconnectivity of the particles in the aerogels. The exponent α was found to be around 1.5 for both SiO_2- and carbon-based aerogels.

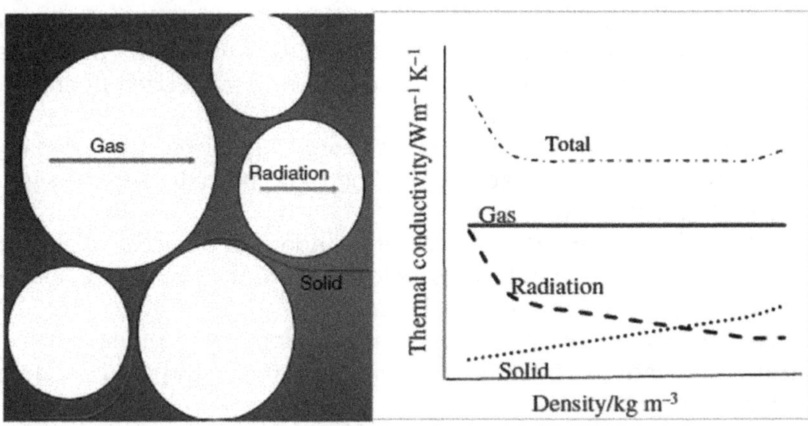

Figure 10.1 Thermal conductivity in porous materials as a result of conduction through porous solids, conduction through gas contained in the pores and radiation.
Reprinted from *Journal of Thermal Analysis and Calorimetry*, Thermal properties of polysaccharide aerogels, **127**, 2017, 363–370, G. Horvat, T. Fajfar A. P. Uzunalić, Ž. Knez, Z. Novak, © Akadémiai Kiadó, Budapest, Hungary 2016.[7] With permission of Springer.

The conductivity of the gas phase can be minimized by reducing the cell size to the nanometer range in order to exploit the Knudsen effect. Knudsen diffusion occurs when the pore size is similar or smaller than the mean free path of gas molecules, which is around 70 nm in air at standard conditions.[1,2,4,5] In this case, the gas molecules collide more often with molecules forming the solid cell walls than with each other and thus the energy transfer through the gas phase is reduced. Then λ_g in cellular structures can be described by the Knudsen equation:

$$\lambda_g = \frac{\lambda_{g0}}{1 + \beta \cdot Kn} \tag{10.4}$$

In which λ_{g0} is the thermal conductivity of the gas at standard conditions ($\lambda_{air} = 26 \text{ mW m K}^{-1}$), β is a factor for the energy transfer between gas molecules and the surrounding cell walls (≈ 2 for air) and Kn is the Knudsen number. The latter is a characteristic quantity for the thermal conductivity of the gas in a porous material and is defined as follows:

$$Kn = \frac{l_g}{\phi} \tag{10.5}$$

where l_g is the mean free path of the gas molecules and ϕ is the pore diameter.

The cell size of polysaccharide and protein-based aerogels depends on several factors, including the nature of the raw material, molecular weight, concentration of precursor solution or drying method, to mention a few. For instance, cellulose-based aerogels with pore sizes ranging from several tens of nm up to several hundreds of nm were obtained as a function of the concentration and route followed to dissolve cellulose I.[6] In any case, the aerogels that address insulation purposes are characterized by small cell sizes (typically in the range of 1 to 100 nm).

As the mean free path of air molecules is around 70 nm, it is comparable to the average cell size. λ_g in aerogels is therefore partially suppressed and can be even lower than that of still air.[4,7]

It is widely accepted that the convection component λ_c can be neglected in closed cell materials with cell sizes below 4 mm, as well as in open cell materials with cell sizes lower than 2 mm.[5]

The radiative conductivity λ_r in aerogels is a local phenomenon because the mean free path of infrared photons is small compared to aerogel dimensions and can be described by:[4]

$$\lambda_r = \frac{(16/3) \cdot n^2 \cdot \sigma \cdot T^3}{[e(T) \cdot \rho]} \tag{10.6}$$

In which n is the mean index of refraction of the specimen, σ is the Stefan–Boltzman constant, T is the absolute temperature, $e(T)$ is the mass specific extinction coefficient and ρ is the density of the aerogel.

The radiative conductivity λ_r can be reduced by drastically increasing the mass specific extinction coefficient. This can be achieved by integrating carbonaceous materials that are known to be strong infrared absorbers, such as carbon black.[8]

10.1.2 Influence of Aerogel Structure on Insulation Properties

Aerogel insulation properties profoundly depend on aerogel structure. As discussed in Section 10.1.1, reducing the pore size from the micrometer to the nanometer range greatly reduces heat conductivity due to the Knudsen effect. Moreover, pore size reduction and pore density increase, which also lead to a reduction of the pore wall thickness. This reduction of the pore size to the nanometer range leads to a confinement of both the solid and the gaseous phase and increases the tortuosity. The latter is defined as the ratio between the distance of any real path and the shortest possible distance between two points, as shown in Figure 10.2. The confinement of the polymer matrix and the increase in tortuosity hinder the heat transfer through the aerogel because a longer distance to transmit the heat results in increased phonon scattering.[2]

10.1.3 Insulation with Polysaccharide and Protein-based Aerogels

Thermal insulation materials with a thermal conductivity below that of still air (*i.e.* <26 mW m K^{-1}) are desired. However, air-based insulation materials cannot fall below the thermal conductivity of still air. Therefore, so-called super-insulation materials with nanometer-sized pores and greatly reduced

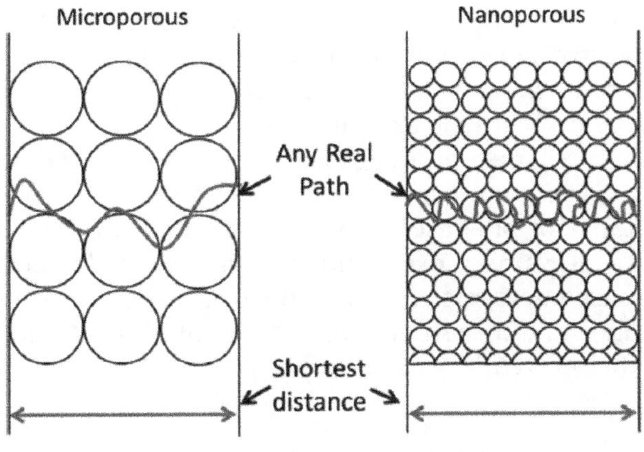

Figure 10.2 Scheme of the tortuous path through the solid phase inside a microporous (left) and a nanoporous (right) material.
Reprinted from *Progress in Materials Science*, **78**, B. Notario, J. Pinto, M. A. Rodriguez-Perez, Nanoporous polymeric materials: A new class of materials with enhanced properties, 93–139, Copyright 2016, with permission from Elsevier.[2]

Table 10.1 Thermal conductivities of different bio-based aerogels.

Material	Porosity (%)	Pore size ϕ (nm)	Thermal conductivity λ (mW m K^{-1})	References
Still air	—	—	26	2
Wheat gluten aerogel	83–85	35–60 μm	39–83	3
Cellulose–SiO_2 aerogel	98–99	40–70	18–28	54
Pectin aerogel	97	17–19	21–23	7
Xanthan aerogel	95	20	27	7
Guar aerogel	87	14	89	7
Alginate aerogel	95	15	42–81	7
Lignin aerogel	75–88	12–27	50	55

heat conductivity are required for applications in which minimizing the heat losses is crucial.[9] The insulating performance of bio-based materials such as wood chips, recycled paper or cork is relatively poor with thermal conductivities of 40–50 mW m K^{-1} as compared to conventional thermal insulators such as extruded polystyrene (XPS)/expanded polystyrene (EPS), polyurethane (PUR) foams or mineral wools (20–40 mW m K^{-1}).[1,2,7] On the other hand, silica-based aerogels exhibit super-insulation properties with a thermal conductivity of ~17–21 mW m K^{-1}, but they are brittle and difficult to prepare in large sizes.[10] Nevertheless, the main disadvantage of aerogels as thermal insulators is their high cost, which is around 10 times higher as compared to traditional insulation materials having the same thermal performance.[9] An excerpt of the thermal performance of some selected polysaccharide-based aerogels is shown in Table 10.1. Only a few reports on the insulation properties of protein-based aerogels are available. Nevertheless, Blomfeldt *et al.* prepared wheat gluten-based aerogels, both unplasticized and plasticized with glycerol.[3] The wheat gluten aerogels had densities of 0.13–0.18 g cm^{-3}, porosities of around 85% and thermal conductivities as low as 39 mW m K^{-1}.

Regarding polysaccharide-based aerogels, thermal super-insulating pectin aerogels prepared *via* dissolution–gelation–coagulation and subsequent drying with supercritical CO_2 exhibit a thermal conductivity ranging from 16–22 mW m K^{-1}.[11] Some nanocellulose-based aerogels were shown to have thermal conductivities even lower than 16 mW m K^{-1}.[12] To overcome the poor fire resistance of bio-based insulation materials, hybrid aerogels have been investigated. Graphene oxide and sepiolite nanorods were incorporated into cellulose-based aerogels. The hybrid aerogels exhibited excellent combustion resistance and a thermal conductivity of 15 mW m K^{-1}, which was about half of that of XPS.[13]

10.2 Electrical and Magnetic Properties

10.2.1 Bio-based Aerogels and Electromagnetic Mechanisms

The use of bio-based aerogels to impart electrical and magnetic characteristics is a relatively recent development, and is based on the concept of using the three dimensional (3-D) aerogel network as a tool for moulding, templating and structuring of a second phase that introduces the desired properties. Usually, this phase is formed by the incorporation of magnetic or electrical substances. The magnetic components include Fe, Co, Ni and their corresponding compounds, whereas the electrical conductive components may be metal nanoparticles (NPs), conductive polymers (CPs) or carbon nanomaterials (graphene, carbon nanotubes, *etc.*). CPs have π-conjugated double bonds along the backbone of the polymer. π-bonded electrons can migrate freely if counterions (anion for *n*-type or cation for *p*-type) exist in the polymer chains, leading to their excellent electrical conductivities. In the case of inorganic semiconductors such us graphene, electrons or holes are assigned as charge carriers and the charge transport is described by a band model, in which the electrical properties are dominated by the width of the energy gap. The latter is defined as a difference in energy between the valence band and conducting band (Figure 10.3). For example, a single graphene carbon layer shows a much higher electrical conductivity than multi-layered graphite due to a near zero band gap.[14] When the conductive component loadings are above the percolation threshold at which the dispersed conductive phase becomes continuous, the conductivity of aerogels starts to increase in parallel to its amount.[15] In addition, bio-based polymers, such as polysaccharides and proteins contain carbon in the backbone. Therefore, they can be transformed into electrically conductive porous carbon materials after a carbonization process.

10.2.2 Synthesis Methods

Conductive components can be introduced into bio-based aerogels by three methods: (1) Incorporating conductive fillers into the precursors of aerogels;

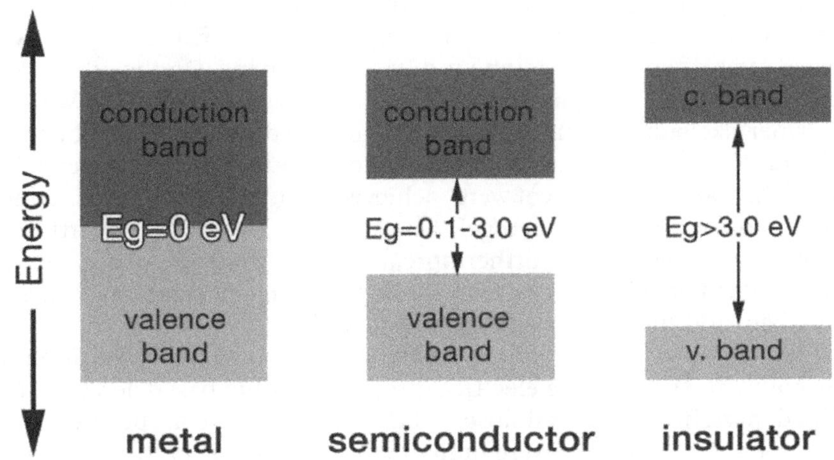

Figure 10.3 Energy band diagram demonstrating different band gap energies. Reproduced from ref. 52 with permission from The Royal Society of Chemistry.

(2) Creating a conductive 3-D network over the skeleton of aerogels by dipping into conductive polymer solutions; (3) *In situ* polymerization of conductive polymers into the aerogels' network.

The porous aerogels can play several roles: (1) endowing the conductive components with the mechanical properties and polymer processability; (2) as a template to control size, shape and organization; and (3) as a dispersing agent to prevent conductive NPs from aggregating in the aqueous precursor.[16]

To prepare conductive carbon porous materials, a carbonization process under an inert atmosphere with temperatures above 600 °C is necessary to transform bio-aerogels into their electrically conductive carbon counterparts.

There are very few publications on magnetic, bio-based aerogels, although some work has been done. An important class to consider is the magnetic, cellulose-$CoFe_2O_4$ aerogels, which can be actuated by a small household magnet. Furthermore, the foam is able to absorb water, and release it upon compression.[17] Another class that has been investigated is γ-Fe_2O_3 NPs-loaded pectin aerogels, which have been synthesized as monoliths and as microspheres.[18] The superparamagnetic properties of the γ-Fe_2O_3 NPs were preserved, giving the material its magnetic properties.

10.2.3 Bio-based Aerogels Modified with Conductive Materials

In general, two different strategies have been followed to obtain bio-based aerogels with electrical conductivity, namely the use of conductive polymers and the addition of carbon-based fillers.

The first CPs-modified bio-based aerogels were reported by Ikkala *et al.*[19] Nanocellulose aerogel networks were modified by dipping the aerogels in an electrically conductive polyaniline (PANI)/surfactant solution. Flexible aerogels with relatively high conductivity of approximate 1 $S\,m^{-1}$ were obtained after rinsing off the unbound PANI and drying. On the other hand, aniline was recently polymerized *in situ* to PANI, which interacted with the biopolymer skeletons, such as in pectin bio-based aerogels, as illustrated in Figure 10.4.[20] Favorable electrical conductivities as well as excellent electrochemical performance were achieved due to the interpenetrated PANI networks within the aerogel. The detailed electrical properties are displayed in Table 10.2. Furthermore, Ag NPs were deposited on the obtained PANI/cellulose aerogels, leading to an increase in electrical conductivity and highest specific capacitance.[21]

Polypyrrole (PPy) is another type of CP that has gained increased attention for its ability to enhance the electrical properties of bio-based aerogels. PPy-modified nanofibrillated cellulose (NFC) aerogels with a specific charge capacity of about 220 $C\,g^{-1}$ were prepared by Stromme *et al.* through *in situ* polymerization of a pyrrole monomer onto the cellulose-based materials.[22] Nano-scale cellulose served as a template to control the growth and distribution of PPy. Reports on aerogels modified with other CPs are rare, mainly

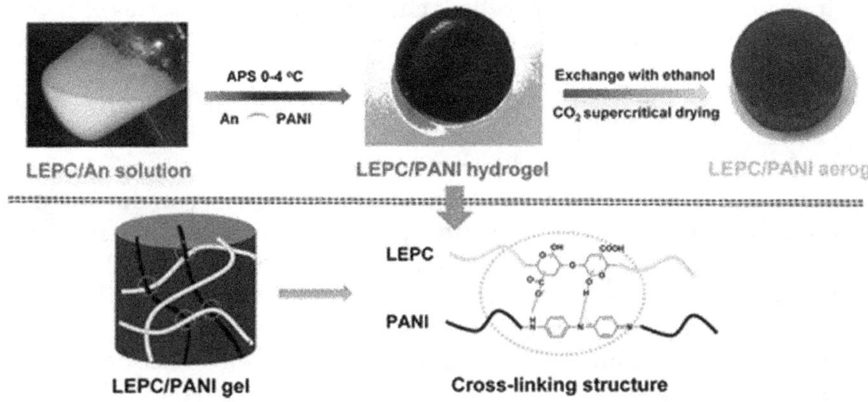

Figure 10.4 Schematic illustration of pectin/PANI aerogel formation. Reprinted with permission from H.-B. Zhao, L. Yuan, Z.-B. Fu, C.-Y. Wang, X. Yang, J.-Y. Zhu, J. Qu, H.-B. Chen and D. A. Schiraldi, *ACS Applied Materials & Interfaces*, 2016, **8**, 9917–9924, Copyright 2016 American Chemical Society.[20]

Table 10.2 Comparison of electrical properties of modified bio-based aerogels.[a]

Polymer matrix	Conductive components	Electrical conductivity ($S\,m^{-1}$)	Highest specific capacitance ($F\,g^{-1}$)	Electrochemical stability	References
Pectin	PANI	2×10^{-3}–0.1	184 at 0.5 $A\,g^{-1}$	74% CR after 1000 CDCs	20
Cotton cellulose	PANI	3.45×10^{-2}	145 at 0.1 $A\,g^{-1}$		21
Cotton cellulose	PANI/Ag NPs	4.59×10^{-2}–0.94	217 at 0.1 $A\,g^{-1}$	83% CR after 1000 CDCs	21
Chitosan	F-MWCNT	up to 0.5			23
Chitosan	MWCNT	0.32–1.7			24
Cellulose	MWCNT	up to 1.8			25
NFC	MWCNT/PANI		791 at 0.2 $A\,g^{-1}$	82.2% CR after 3000 CDCs	27
NFC	MWCNT		178 at 5 $mV\,s^{-1}$	99.9% CR after 1000 CDCs	29
NFC	Graphene		207 at 5 $mV\,s^{-1}$	99.1% CR after 5000 CDCs	28
NFC	Graphene/MWCNT	up to 12	252 at 0.5 $A\,g^{-1}$	99.5% CR after 1000 CDCs	56

[a]CDCs: charge-discharge cycles; CR: capacitance retention; F-MWCNT: functionalized multi-wall carbon nanotube; NFC: nanofibrillated cellulose; MWCNT: multi-wall carbon nanotube; PANI: polyaniline.

due to the lack of suitable sol–gel chemistry for the corresponding conductive polymers.

Carbon nanofillers such as graphene and carbon nanotubes (CNT) are considered to be the best second phase to improve electrical properties due to their good conductivity and excellent chemical stability. A series of hybrid biopolymer/carbon nanofiller composite aerogels have been created for different applications.

For example, Yan *et al.* used carbon nanotubes to prepare chitosan/CNT aerogels which demonstrated an electrical conductivity that was 8 orders of magnitude greater than that of pure chitosan aerogels. This marked improvement was made possible by the presence of a quasi-two-dimensional pathway that allows charges to be transported through the aerogel.[23] Del Monte *et al.* found that the length of multi-wall CNTs (MWCNTs) had a positive effect on the electrical conductivity of chitosan aerogels.[24] Li *et al.* demonstrated that the electrical conductivity of cellulose/CNT aerogels increased with increasing CNT content, agreeing with the percolation threshold theory described previously.[25] Mäder *et al.* investigated electrically conductive CNT/cellulose aerogels as vapor sensors. The electrical resistance of composite aerogels changed upon exposure to several volatile organic compound vapors, as seen in Figure 10.5.[26] This phenomenon is attributed to the 3-D porous structure of aerogels that provides the capacity to contact more directly and efficiently with vapors. Li *et al.* prepared a NFC/MWCNT/PANI composite aerogel that had a low charge-transfer resistance (1.24 Ω). The combination with conductive PANI further improved the electrical properties of the aerogels.[27]

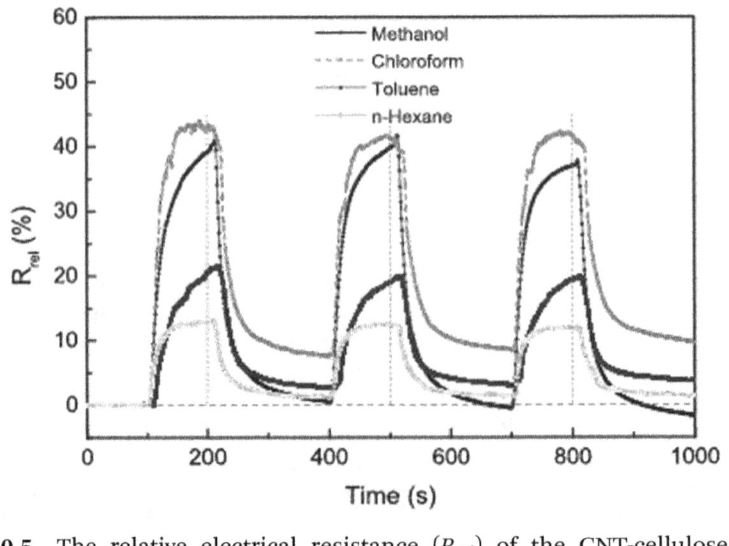

Figure 10.5 The relative electrical resistance (R_{rel}) of the CNT-cellulose aerogel (3 wt% CNT) to different saturated vapors at 25 °C *versus* time during three vapor/air exposure cycles.
Reproduced from *Sensors and Actuators B: Chemical*, **213**, H. Qi, J. Liu, J. Pionteck, P. Pötschke, E. Mäder, Carbon nanotube–cellulose composite aerogels for vapour sensing, 20–26, Copyright 2015, with permission from Elsevier.

Graphene is also used to improve the electrical properties of bio-based aerogels. However, given their strong π–π stacking and van der Waals interactions, graphene sheets tend to form irreversible agglomerates or even restack to form graphite, leading to a great decrease in electrical properties. Therefore, graphene oxide (GO) with functional groups was usually used as a replacement for graphene, forming a uniform and stable precursor for preparing aerogels. GO is subsequently reduced to graphene (RGO) which acts as the conductive phase in aerogel composites. Several bio-based polymer/graphene hybrid aerogels have been prepared in accordance with this method.

Gao *et al.* compared the electrical properties of NFC/MWCNT and NFC/RGO aerogels prepared *via* the same strategy.[28,29] The NFC/RGO aerogels showed better electrochemical performance than the NFC/MWCNT ones under the same testing conditions. Gong *et al.* prepared a NFC/RGO/MWCNT ternary aerogel with better electrochemical properties than that containing a single carbon component,[30] as seen in Table 10.2.

10.2.4 Bio-based Aerogel-templated Conductive Carbon Materials

Panayiotou and co-workers were the first to prepare porous carbon by carbonizing a freeze-dried chitin aerogel.[31] Most recently, several works have been reported on electrically conductive porous carbon materials using polysaccharide aerogels as precursor templates. Robust aerogel-like carbon was derived from bacterial cellulose (BC) aerogels, displaying an electrical conductivity of 20.6 $S\,m^{-1}$.[32] Another type of graphitized aerogel was synthesized based on BC/lignin biomass, which had a specific capacitance of 124 $F\,g^{-1}$ at 0.5 $A\,g^{-1}$, as well as outstanding cycle stability (98% capacity retention after 10 000 charge–discharge cycles).[33] Carbon nanofillers were also reported as additives to improve the electrical properties of pyrolysed carbon aerogels. Shao *et al.* prepared micro/meso-porous carbon materials by carbonizing graphene/silk fibroin aerogels.[34] The obtained materials showed enhanced electrochemical properties (256 $F\,g^{-1}$ of capacitance at 0.5 $A\,g^{-1}$ and 73.4% capacitance retention after 10 000 charge–discharge cycles).

In summary, bio-based aerogels with excellent electrical properties can be prepared by creating a continuous conductive phase in their network skeletons. These functional aerogels could be used in multiple applications such as electrodes, sensors or supercapacitors to mention a few.

10.3 Fire Resistance Properties

Every year, much effort is directed at reducing the impact that fire can have on daily life. As such, finding materials that are able to prevent ignition, inhibit combustion, and reduce flame propagation has become a matter of great interest in materials science. Polysaccharides are made of long organic carbohydrate chains and are generally highly combustible materials which need to be modified by the addition of flame retardant additives.[35] However, some proteins, such as casein and whey behave differently due to their nitrogen, sulphur and phosphate content, and have been recently used as natural fire retardant agents for textiles as well as for bulk and polymer foams.[36]

Three main characterization techniques are used to analyze and compare the fire resistance properties of materials: the Limited Oxygen Index (LOI), the Standard for Safety of Flammability of Plastic Materials for Parts in Devices and Appliances testing (UL94) and cone calorimetry. On the one hand, LOI corresponds to the minimal oxygen concentration needed to allow combustion to continue or to completely burn a sample in a controlled [O_2/N_2] atmosphere.[35–38] As air contains 21% oxygen, materials with an LOI below 21 are recognized as "flammable", whereas those with an LOI above 21 are catalogued as "non-flammable", as their combustion cannot be maintained without an external energy source at room temperature.[36] Consequently, the higher the LOI of a material is, the more flame retardant it is.

On the other hand, UL94 tests are used to measure the flammability of polymer-based materials. The most common test is UL94 V, which can be used to determine the ignitability of the sample, as well as how readily the flame spreads throughout the material. The test consists of burning a vertically oriented sample over a cotton layer for 10 seconds, where the flame ought to propagate from the bottom of the sample to the top. The materials are classified as V-0, V-1 or V-2 determined by three factors: burning time after the application of a test flame, total burning duration and presence or absence of drips during the burning process.[36,38]

Finally, the most effective test is cone calorimetry which evaluates the flammability by the mass loss during the experiment.[35] In this test, the sample is placed on a load cell before being uniformly irradiated from above by a conical electrical heater. Subsequently, the combustion is generated by an electric spark and the gas products drift through the heating cone, where they are collected and measured in the exhaust duct, along with smoke density and the concentrations of O_2, CO and CO_2.[38]

The measurements of the gas flow and oxygen concentration are critical to calculate the heat release rate (HRR), which expresses the quantity of heat released per unit of surface area and time reflecting ($kW\,m^{-2}$). The peak of heat release (pHRR) is the maximum of the HRR curve and it is dependent on both the test conditions and the intrinsic fire properties of the material. Intuitively, the time to the peak of heat release (TTpHRR) is indicative of flame propagation, although the fire growth index (FIGRA), defined as the ratio of pHRR to TTpHRR, is more commonly used to evaluate the flame propagation rate. The cone calorimeter test also provides the integration of the HRR *versus* time curve to obtain the total heat released (THR), and allows the user to characterize other properties including time to ignition (TTI), time of combustion (TOF), mass loss during combustion, total smoke released (TSR), opacity and the CO and CO_2 quantities, the former being related to the toxicity of the gases released in a fire.[39]

10.3.1 Flame Retardancy Mechanisms

Flame retardants are used to stop or delay fire. Depending on their nature and properties, they are able to interfere with the various processes involved in polymer combustion (see Figure 10.6).[37] They are classified into two groups: those that block the spread of fire by physical interactions and those that utilize chemical reactions to halt the combustion.

There are multiple physical processes that help inhibit or slow combustion such as cooling, dilution or by forming a protective layer.[37] The cooling is the result of an endothermic reaction in the presence of fire flame-retardants where a diminution of temperature occurs by heat consumption. As a result of this consumption, the reaction medium reaches a lower temperature than is required for the polymer to undergo combustion. Due to the formation of inert gases such as H_2O, CO_2 and NH_3, the combustion process becomes restricted by the reduction of the concentration of reagents *via* dilution.[38]

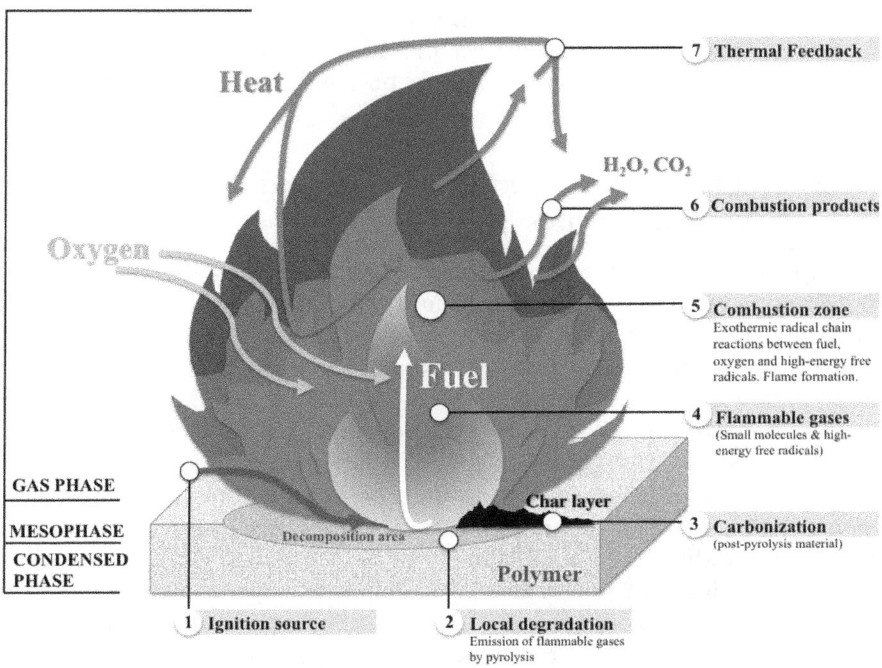

Figure 10.6 Combustion process.
Adapted from ref. 53. Chemtura, About Flame Retardants, available online: www.chemturaflameretardants.com/flameRetardantsInfo.html. Accessed Nov. 28, 2016.

The last mechanism occurs when a protective layer is formed between two phases: the gaseous one, where combustion takes place and the solid phase where thermal degradation occurs. This layer can be either solid or gaseous, but must be able to restrict the transfer of combustible volatile gases and oxygen, which leads to a decrease in the production of decomposition gases.

Two different chemical reactions have to be distinguished depending on the medium in which they are placed: in the gas or the solid phase. In the gas phase, the combustion gases released from the material are mixed with the oxygen from the air and free radicals, which leads to different chemical reactions resulting in the appearance of visible flames.[37] In this situation the fire can be stopped by incorporating additives that are able to inhibit highly reactive species (OH– or H–), reducing the fuel source.[35] This leads to a decrease in the exothermicity of the reaction, followed by reduction in temperature, a decrease in the quantity of fuel and the end of the combustion process.

Alternatively, in the solid phase, the flame retardants can react in two possible ways: *via* char-forming or intumescence. The char-forming process consists of the generation of a carbonated barrier between the surface of the polymer and the flame. This char acts as a physical insulator between the gas and the solid phase. As fewer hot gases are able to continue combustion, pyrolysis decreases and further reduces the production of additional gases. In contrast, the intumescence effect provides insulation through the formation of a protective barrier caused by the swelling of blowing agents.[37]

10.3.2 Combustion Properties of Bio-based Aerogels and Their Composites

The majority of foams used in applications such as packaging or insulation are derived from petroleum feedstock and are non-biodegradable, causing a significant environmental problem after end-use. The use of industrial crop or food crop by-products as raw materials in those applications is an interesting and cost-effective alternative to traditional petroleum-based materials. Low-density bio-based aerogels present a powerful and sustainable solution, provided that good thermal insulation and fire-resistance properties are achieved. However, due to their biodegradability, bio-based aerogels are mostly utilized in short-term applications, and therefore only a few combustion studies have been carried out on these materials.

Among the different polysaccharides, cellulose is probably the most frequently used for making aerogels[19,40] and different approaches have been followed to increase its fire resistance. To this end, the combined effect of montmorillonite clay (MMT) and ammonium polyphosphate (APP) were found to have a positive effect on the thermal stability and fire resistance properties of aerogels made from recycled cellulose fibers and carboxymethyl cellulose.[41] It was observed that both clay nanoplatelets and APP contribute to a high decrease of the HRR, pHRR and fire growth rate due to both the barrier against heat and volatile gases created by clay and the thermal decomposition of APP which yields ammonia and polyphosporic acid. This acid reacts with the hydroxyl groups of cellulose causing dehydration and formation of char. In the same way, thermal and gasification experiments performed on nanofibrillated cellulose-MMT aerogels with an equiaxed and honeycomb-like pore structure[13] showed the strong interaction between cellulose and MMT, which helped to maintain the aerogel stiffness at temperatures up to 300 °C. Graphene oxide and cellulose nanofibers (CNF) were combined with sepiolite in a colloidal suspension to form aerogels by freeze-drying.[8] The presence of well-distributed nanoparticles imparted a significant resistance to fire, as the samples showed no self-propagation under vertical burning (UL94). Furthermore, the LOI was 34, much higher than the typical LOI of fire-retardant polymer foams, which lay in the range between 22 and 25. Moreover, with the proper combination of nanofillers, the foams display a 25% decrease in pHRR when compared with cellulose nanofiber aerogels (Table 10.3). By combining CNF and similar nanofillers, a very low value of the pHRR (20 $kW\,m^{-2}$) was obtained in cellulose aerogels when boric acid anions were present.[42] The value of the pHRR was even lower in the presence of the borate anions, which slow down char oxidation. The formation of a CNF-borate hybrid promoted charring and graphitization of the polysaccharide structure, explaining the high flame retardant behavior of the aerogels.

Using the 3-D cellulose hydrogel as a template, magnesium hydroxide nanoparticles (MHn) were synthetized *in situ* by dipping in magnesium chloride and sodium hydroxide solutions.[43] In comparison to pure cellulose, the resultant aerogels showed delayed combustion velocity and were self-extinguishing when the MHn content reached 82 phr. In a similar manner, aluminum nanoparticles were incorporated *in situ* to obtain transparent cellulose aerogels that exhibited a 12-fold lower pHRR than the untreated system (Table 10.3).[44]

Alginate is an anionic polysaccharide derived from seaweed that has been reported to be inherently flame-retardant, in the form of fibers showing a LOI value of 48 and a pHRR of 4.99 $kW\,m^{-2}$.[45] Taking advantage of this behavior and its ability to gel in water, low-flammability aerogels based on ammonium alginate and MMT clay were prepared.[46] Under high heat flux conditions (50 $kW\,m^{-2}$) pure alginate foams burnt slowly with very low heat released. The pHRR was 64 $kW\,m^{-2}$ for low-density alginate aerogels and 99 $kW\,m^{-2}$ for the denser ones. The addition of clay resulted in a further reduction of the flammability showing no open fire during the test. Moreover, another important issue to consider is the shape of the specimen after burning. The relative high amount of clay (50% of total solids) acted as a barrier layer, limiting heat and mass transport and decreasing the rate of burning. The lamellar structure created due to the rearrangement of clay nanoplatelets during lyophilization was an important factor that contributed to the inhibition of flame propagation.

Table 10.3 Cone calorimeter combustion parameters of different bio-based aerogels and foams (heat flux of 50 $kW\,m^{-2}$).[a,b]

Sample	TTI (s)	TTpHRR (s)	pHRR ($kW\,m^{-2}$)	THR ($sm^2\,kW^{-1}$)	FIGRA ($kW^{-1}\,s^{-1}\,m^{-2}$)	Residue (%)	References
EPS	9	45	256.0	9.1	5.7	8.6	46
Cellulose[c]	n/a	351	294.0	11.3	n/a	14.0	44
Cellulose/AlH[c]	n/a	247	24.0	1.7	n/a	60.0	44
CNF-GO-SEP[d]	8	22	47.0	n/a	n/a	n/a	8
Alginate	96	38	64.0	13.2	2.6	3.3	46
A5C5	No flame	45	32.0	12.0	0.7	53.3	46
A5ALH5	No flame	110	20.4	2.0	0.2	52.0	47
GA	10	36	232.9	19.0	6.5	2.9	49
GA5C5	15	24	56.4	4.9	2.4	45.4	49
Xanthan Gum	16	48	202.3	7.8	4.5	5.1	48
Wheat Gluten[d]	18	n/a	325	n/a	n/a	12	3

[a]EPS: expanded polystyrene; CNF: cellulose nano-fibers; GO: graphene oxide; Sep: sepiolite; A: alginate; C: clay; AlH: aluminum hydroxide; GA: gum arabic; TTI: time to ignition; TTpHRR: time to the peak of heat release; pHRR: peak of heat release; THR: total heat released; FIGRA: fire growth index.
[b]A5C5 means aerogel was prepared using 5 gr of alginate and 5 gr of clay in 100 ml water.
[c]Measured in microcone calorimeter.
[d]Heat flux 35 $kW\,m^{-2}$.

In further research, inorganic particles such us magnesium or aluminum hydroxides were introduced into alginate aerogels.[47] The alginate/flame retardant aerogel composites showed no visible flame and had pHRR values lower than 25 ($kW m^{-2}$). In addition, the key characteristic parameters such us TTI and FIGRA were found to decrease with the introduction of these additives, of which aluminum hydroxide was the most effective. After burning, the structure of the MMT network remained unchanged, only exhibiting an increase in the pore size.

Xanthan gum,[48] agar and gum Arabic[49] are other polysaccharide-based materials that have been successfully used to create aerogels following an environmentally friendly process that uses water as the only solvent. The fire response of these aerogels is similar to what is expected of pure polysaccharides, as the short TTI and large pHRR values indicated a high burn rate and a propensity to flash over. Again, the combination with clay (50% w/w) led to a four-fold decrease in the pHRR values due to the clay enrichment of the sample surfaces during burning, reducing the decomposition rate of the underlying polymer fraction.

Proteins such us wheat gluten or casein are other promising bio-based materials for creating non-flammable aerogels that also exhibit very low thermal conductivities. Casein contains a large amount of phosphorous compounds, which makes this by-product a very promising material for future fire-resistant foams.[50] On the other hand, wheat gluten foams exhibit low thermal conductivities ($\lambda = 40$ $mW m K^{-1}$) similar to that of glass wool, although the water uptake and poor flammability are still limiting its use as insulation material. Cone calorimetry studies carried out on wheat gluten foams revealed vigorous combustion with high pHRR, though combustion was slow to initiate (Table 10.3).[3] To improve the fire retardancy, wheat gluten was combined with *in situ* formed silica using tetraethyl orthosilicate (TEOS) as a precursor.[51] The hybrid gluten-based aerogels containing 30–50% of TEOS were difficult to ignite and self-extinguished after removal of the external flame in a UL94 test. The material demonstrated no dripping or open flame and was thus classified as V-0. The reason behind this behavior was attributed to the formation of a silica phase, the dissipation of water and the relative containment of fuel products.

10.4 Conclusions

Polysaccharide and proteins are abundant renewable natural polymers that can be easily transformed into highly porous aerogels due to their gelation capabilities. The performance of the formed aerogels depends on the selected matrix and the obtained microstructure. However, incorporation of functional fillers can modify or impart additional properties to the aerogels without significant changes to the porous nature. The reduction in the pore size to the nanoscale is a key factor to improve the mechanical behavior and specially to decrease the thermal conductivity of aerogels. In addition, different polysaccharides (*i.e.* cellulose or alginate) and proteins (*i.e* gluten or casein) have been successfully combined with fillers to obtain fire-resistant aerogels which opens up new opportunities for the development of light, renewable and cost-effective materials for commercial applications. Many other possibilities and properties such as magnetic or electric conductivity have been obtained by taking advantage of the 3-D network structure of bio-based aerogels as a tool for moulding, templating and structuring a second phase that introduces the desired properties. In general, these composite aerogels are very promising materials for advanced engineering applications where multifunctional requirements are needed.

Acknowledgements

The authors acknowledge the financial support received from the Spanish Ministry of Economy and Competitiveness through the Project MAT2013-40730-P.

References

1. S. Liu, J. Duvigneau and G. J. Vancso, *Eur. Polym. J.*, 2015, **65**, 33–45.
2. B. Notario, J. Pinto and M. A. Rodriguez-Perez, *Prog. Mater. Sci.*, 2016, **78–79**, 93–139.
3. T. O. J. Blomfeldt, F. Nilsson, T. Holgate, J. Xu, E. Johansson and M. S. Hedenqvist, *ACS Appl. Mater. Interfaces*, 2012, **4**, 1629–1635.
4. X. Lu, R. Caps, J. Fricke, C. T. Alviso and R. W. Pekala, *J. Non-Cryst. Solids*, 1995, **188**, 226–234.
5. B. Notario, J. Pinto, E. Solorzano, J. A. de Saja, M. Dumon and M. A. Rodríguez-Pérez, *Polymer*, 2015, **56**, 57–67.
6. A. Rege, M. Schestakow, I. Karadagli, L. Ratke and M. Itskov, *Soft Matter*, 2016, **12**, 7079–7088.
7. G. Horvat, T. Fajfar, A. Perva Uzunalić, Ž. Knez and Z. Novak, *J. Therm. Anal. Calorim.*, 2016, 1–8.
8. B. Wicklein, A. Kocjan, G. Salazar-Alvarez, F. Carosio, G. Camino, M. Antonietti and L. Bergström, *Nat. Nanotechnol.*, 2015, **10**, 277–283.
9. E. Cuce, P. M. Cuce, C. J. Wood and S. B. Riffat, *Renewable Sustainable Energy Rev.*, 2014, **34**, 273–299.
10. L. Zuo, Y. Zhang, L. Zhang, Y.-E. Miao, W. Fan and T. Liu, *Materials*, 2015, **8**, 5343.
11. C. Rudaz, R. Courson, L. Bonnet, S. Calas-Etienne, H. Sallée and T. Budtova, *Biomacromolecules*, 2014, **15**, 2188–2195.
12. W. Chen, Q. Li, Y. Wang, X. Yi, J. Zeng, H. Yu, Y. Liu and J. Li, *ChemSusChem*, 2014, 7, 154–161.
13. A. E. Donius, A. Liu, L. A. Berglund and U. G. K. Wegst, *J. Mech. Behav. Biomed. Mater.*, 2014, **37**, 88–99.
14. Y. Zhu, S. Murali, W. Cai, X. Li, J. W. Suk, J. R. Potts and R. S. Ruoff, *Adv. Mater.*, 2010, **22**, 3906–3924.
15. H.-X. Tan and X.-C. Xu, *Compos. Sci. Technol.*, 2016, **128**, 155–160.

16. A. Millan, F. Palacio, E. Snoeck, V. Serin and P. Lecente, in *Polymer Nanocomposites*, ed. Y. W. Mai and Z. Z. Yu, Woodhead Publishing, Cambridge, 2006, pp. 440–484.
17. R. T. Olsson, M. A. Samir, G. Salazar-Alvarez, L. Belova, V. Ström, L. A. Berglund, O. Ikkala, J. Nogues and U. W. Gedde, *Nat. Nanotechnol.*, 2010, **5**, 584–588.
18. C. A. García-González, E. Carenza, M. Zeng, I. Smirnova and A. Roig, *RSC Adv.*, 2012, **2**, 9816–9823.
19. M. Pääkkö, J. Vapaavuori, R. Silvennoinen, H. Kosonen, M. Ankerfors, T. Lindström, L. A. Berglund and O. Ikkala, *Soft Matter*, 2008, **4**, 2492–2499.
20. H.-B. Zhao, L. Yuan, Z.-B. Fu, C.-Y. Wang, X. Yang, J.-Y. Zhu, J. Qu, H.-B. Chen and D. A. Schiraldi, *ACS Appl. Mater. Interfaces*, 2016, **8**, 9917–9924.
21. J. Tian, D. Peng, X. Wu, W. Li, H. Deng and S. Liu, *Carbohydr. Polym.*, 2017, **156**, 19–25.
22. D. O. Carlsson, G. Nystrom, Q. Zhou, L. A. Berglund, L. Nyholm and M. Stromme, *J. Mater. Chem.*, 2012, **22**, 19014–19024.
23. J. Yan, H. Wang, T. Wu, X. Li and Z. Ding, *Composites, Part A*, 2014, **67**, 1–7.
24. S. Nardecchia, M. C. Serrano, M. C. Gutierrez, M. L. Ferrer and F. d. Monte, *J. Mater. Chem. B*, 2013, **1**, 3064–3072.
25. H.-D. Huang, C.-Y. Liu, D. Zhou, X. Jiang, G.-J. Zhong, D.-X. Yan and Z.-M. Li, *J. Mater. Chem. A*, 2015, **3**, 4983–4991.
26. H. Qi, J. Liu, J. Pionteck, P. Pötschke and E. Mäder, *Sens. Actuators, B*, 2015, **213**, 20–26.
27. C. Yang, C. Chen, Y. Pan, S. Li, F. Wang, J. Li, N. Li, X. Li, Y. Zhang and D. Li, *Electrochim. Acta*, 2015, **182**, 264–271.
28. K. Gao, Z. Shao, J. Li, X. Wang, X. Peng, W. Wang and F. Wang, *J. Mater. Chem. A*, 2013, **1**, 63–67.
29. K. Gao, Z. Shao, X. Wang, Y. Zhang, W. Wang and F. Wang, *RSC Adv.*, 2013, **3**, 15058–15064.
30. Q. Zheng, Z. Cai, Z. Ma and S. Gong, *ACS Appl. Mater. Interfaces*, 2015, **7**, 3263–3271.
31. C. Tsioptsias, C. Michailof, G. Stauropoulos and C. Panayiotou, *Carbohydr. Polym.*, 2009, **76**, 535–540.
32. Z.-Y. Wu, C. Li, H.-W. Liang, J.-F. Chen and S.-H. Yu, *Angew. Chem.*, 2013, **125**, 2997–3001.
33. X. Xu, J. Zhou, D. H. Nagaraju, L. Jiang, V. R. Marinov, G. Lubineau, H. N. Alshareef and M. Oh, *Adv. Funct. Mater.*, 2015, **25**, 3193–3202.
34. Y. Wang, Y. Song, Y. Wang, X. Chen, Y. Xia and Z. Shao, *J. Mater. Chem. A*, 2015, **3**, 773–781.
35. G. Malucelli, F. Bosco, J. Alongi, F. Carosio, A. Di Blasio, C. Mollea, F. Cuttica and A. Casale, *RSC Adv.*, 2014, **4**, 46024–46039.
36. P. Kiliaris and C. D. Papaspyrides, *Prog. Polym. Sci.*, 2010, **35**, 902–958.

37. T. R. Hull and A. A. Stec, in *Fire Retardancy of Polymers: New Strategies and Mechanisms*, ed. T. R. Hull and B. K. Kandola, Royal Society of Chemistry, 1st edn, 2009, pp. 6–11.
38. F. Laoutid, L. Bonnaud, M. Alexandre, J. M. Lopez-Cuesta and P. Dubois, *Mater. Sci. Eng., R*, 2009, **63**, 100–125.
39. B. Schartel and T. R. Hull, *Fire Mater.*, 2007, **31**, 327–354.
40. R. Gavillon and T. Budtova, *Biomacromolecules*, 2008, **9**, 269–277.
41. L. Wang and M. Sanchez-Soto, *RSC Adv.*, 2015, **5**, 31384–31391.
42. B. Wicklein, D. Kocjan, F. Carosio, G. Camino and L. Bergström, *Chem. Mater.*, 2016, **28**, 1985–1989.
43. Y. Han, X. Zhang, X. Wu and C. Lu, *ACS Sustainable Chem. Eng.*, 2015, **3**, 1853–1859.
44. B. Yuan, J. Zhang, J. Yu, R. Song, Q. Mi, J. He and J. Zhang, *Sci. China: Chem.*, 2016, **59**, 1335–1341.
45. J. Zhang, Q. Ji, X. Shen, Y. Xia, L. Tan and Q. Kong, *Polym. Degrad. Stab.*, 2011, **96**, 936–942.
46. H.-B. Chen, Y.-Z. Wang, M. Sánchez-Soto and D. A. Schiraldi, *Polymer*, 2012, **53**, 5825–5831.
47. K. Shang, W. Liao, J. Wang, Y.-T. Wang, Y.-Z. Wang and D. A. Schiraldi, *ACS Appl. Mater. Interfaces*, 2016, **8**, 643–650.
48. L. Wang, D. A. Schiraldi and M. Sánchez-Soto, *Ind. Eng. Chem. Res.*, 2014, **53**, 7680–7687.
49. L. Wang, M. Sánchez-Soto and T. Abt, *Ind. Crops Prod.*, 2016, **91**, 15–21.
50. D. A. Schiraldi, in *Green Polymer Chemistry: Biobased Materials and Biocatalysis*, American Chemical Society, 2015, vol. 1192, pp. 471–482.
51. Q. Wu, R. L. Andersson, T. Holgate, E. Johansson, U. W. Gedde, R. T. Olsson and M. S. Hedenqvist, *J. Mater. Chem. A*, 2014, **2**, 20996–21009.
52. M. E. Abdelhamid, A. P. O'Mullane and G. A. Snook, *RSC Adv.*, 2015, **5**, 11611–11626.
53. Chemtura, About Flame Retardants, available online: www.chemtura-flameretardants.com/flameRetardantsInfo.html. Accessed Nov. 28, 2016.
54. C. Jiménez-Saelices, B. Seantier, B. Cathala and Y. Grohens, *Carbohydr. Polym.*, 2017, **157**, 105–113.
55. L. Perez-Cantu, F. Liebner and I. Smirnova, *Microporous Mesoporous Mater.*, 2014, **195**, 303–310.
56. A. Abarrategi, M. C. Gutiérrez, C. Moreno-Vicente, M. J. Hortigüela, V. Ramos, J. L. López-Lacomba, M. L. Ferrer and F. del Monte, *Biomaterials*, 2008, **29**, 94–102.

CHAPTER 11

Mechanical, Rheological and Viscoelastic Properties of Polysaccharide and Protein Based Aerogels

MPHO C. LEFATLE[a,b] AND MAYA J. JOHN*[a,b]

[a] CSIR Materials Science and Manufacturing, Polymers and Composites Department, P.O Box 1124, Port Elizabeth 6000, Republic of South Africa; [b] Department of Chemistry, Faculty of Science, Nelson Mandela Metropolitan University, P.O Box 1600, Port Elizabeth 6000, Republic of South Africa
*Email: MJohn@csir.co.za

11.1 Introduction

Aerogels are exceptional materials that are becoming increasingly popular, finding application both technically and commercially due to their versatile properties.[1] These materials were first created in the early 1930s by Kistler. He replaced liquids in the gels with gas without destroying the gel solid network. This supercritical drying method prohibited liquid–vapor meniscuses from forming at the exit of gel pores, responsible for the mechanical tension in the liquid and pressure at the pore walls which resulted in gel shrinkage.[2] Aerogels are now synthesized through the use of low temperature sol–gel chemistry followed by supercritical drying (Figure 11.1).[3,8]

Aerogels have interesting properties, for example in the solid state they are able to maintain a fixed volume and shape. Their densities can vary from

Green Chemistry Series No. 58
Biobased Aerogels: Polysaccharide and Protein-based Materials
Edited by Sabu Thomas, Laly A. Pothan and Rubie Mavelil-Sam
© The Royal Society of Chemistry 2018
Published by the Royal Society of Chemistry, www.rsc.org

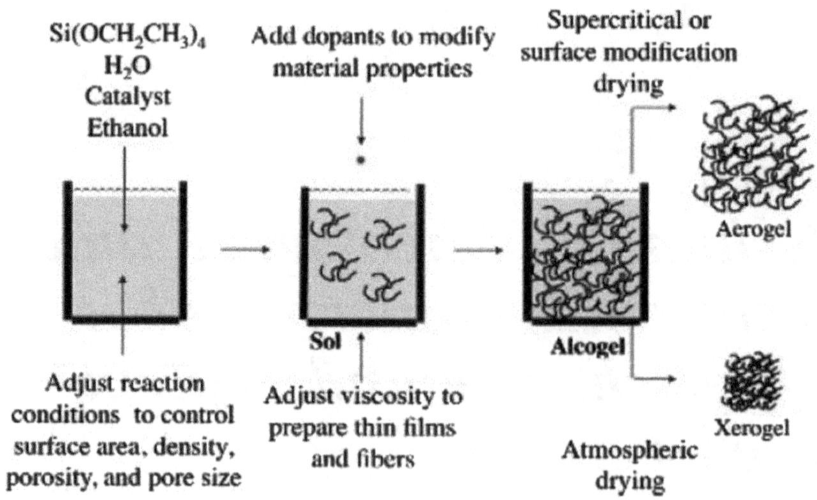

Figure 11.1 Preparation of an aerogel.[8]

1000 $kg\,m^{-3}$ (in the solid state) to approximately 1 $kg\,m^{-3}$ (less than the density of air).[4] They can be produced to have a low thermal conductivity (<0.02 $W\,mK^{-1}$), to be translucent and to have a high surface area (>1000 $m^2\,g^{-1}$). The combination of structural, thermal and optical properties make them applicable for various applications, such as platforms for chemical sensors and thermal conductivity (Table 11.1).[5] They can be further used for water treatment, as aerocapacitors, and molds for casting of aluminum metal and for impregnation of flavor in food and beverages.[6,7]

Various types of aerogels have been prepared and these include oxide aerogels (consisting of silica and non-silica), carbon aerogels (consisting of carbonized plastic, CNT and graphene) and organic aerogels (these are resin based and cellulose based). However, silica based aerogels are more common as their preparation is reliable and easier.[4,8] Silica-based aerogels are hygroscopic in nature and possess poor mechanical properties, and the use of silica-based aerogels in industrial application is limited. As cellulose-based, polyimide-based (PI-based) and resin-based aerogels possess diverse properties and structures, they are usually used in applications related to biomedicine.[9] This chapter will focus on polysaccharide and protein aerogels, specifically on the factors affecting the mechanical, rheological and viscoelastic properties of these aerogels.

11.2 Bio-based Aerogels

11.2.1 Polysaccharide-based and Protein-based Aerogels

Bio-based aerogels are of extreme interest as they are being developed as a replacement for petroleum based foams. These materials exhibit typical

Table 11.1 Properties and applications of aerogels. Reprinted from *Journal of Non-Crystalline Solids*, **225**, L. W. Hrubesh, Aerogel applications, 335–342, Copyright 1998, with permission from Elsevier.

Property	Features	Applications
• Thermal conductivity	• Best insulating solid • Transparent • High temperature • Lightweight	• Architectural and appliance insulation, portable coolers, transport vehicles, pipes, cryogenic skylights • Space vehicles and probes, casting molds
• Density/ porosity	• Lightest synthetic solid • Homogeneous • High specific surface area • Multiple compositions	• Catalysts, Sorbers, Sensors, Fuel Storage, Ion Exchange • Targets for ICF, X-ray lasers
• Optical	• Low refractive index solid • Transparent • Multiple compositions	• Cherenkov detectors, lightweight optics, lightguides, special effect optics
• Acoustic	• Lowest sound speed	• Impedance matchers for transducers, range finders, speakers
• Mechanical	• Elastic • Lightweight	• Energy absorber, hypervelocity particle trap
• Electrical	• Lowest dielectric constant • High dielectric strength • High surface area	• Dielectrics for ICS, spacers for vacuum electrodes, vacuum display spacers, capacitors

elastic–plastic behavior and low thermal conductivity. The raw material for the aerogels is obtained from forestry or agricultural feedstock, mainly based on polysaccharides and proteins.

Polysaccharides aerogels are gaining interest among researchers because they are renewable, biodegradable and their raw material is abundant.[10] What makes them attractive is that their production is not expensive and they can be obtained from a variety of raw materials, such as paper waste and rice straws.[9] These type of aerogels have been used as diagnostic agents and artificial tissues, and drug delivery for medical application due to their biocompatibility (soluble in the body) and biodegradability, making them suitable for enzymatic decomposition in the body.[11,12] Polysaccharide aerogels can be produced from cellulose, starch, alginate, chitosan and pectin.[13,14] Protein aerogels based on soy, egg white, silk fibroin and whey protein have also been investigated and show promising opportunities. Just like polysaccharide aerogels, protein aerogels are biodegradable and biocompatible making them ideal for applications in life sciences and biomedicine.[14] A range of polysaccharides and proteins that have been used to develop aerogels are listed in Table 11.2.

Table 11.2 Examples of polysaccharides and proteins used to prepare aerogels. Reprinted with permission from Q. Mi, S. Ma, J. Yu, J. He and J. Zhang, *ACS Sustainable Chem. Eng.*, 2016, **4**, 656, Copyright 2016 American Chemical Society.

Polysaccharide	References	Proteins	References
Cellulose	77	Silk fibroin	78
Pectin	79	Whey protein	31
Carrageenan	80	Egg white protein	34
Starch	81	Soy protein	36
Alginate	81		
Chitosan	82		

11.2.2 Mechanical Properties

Low economic costs and superior mechanical properties make most materials desirable. Hence, it is important to have knowledge and understanding about the mechanical behavior and how the mechanical properties can be altered to fit the desired use of application.[15] In order for an aerogel to qualify for load bearing applications, several extensive mechanical experiments that are identical or similar to the actual loading conditions are required, so as to determine the mechanical behavior under real time conditions. As actual service conditions for aerogels can be diverse, many loading conditions have to be carried out so as to determine the mechanical response. These include compression, bending, torsion, tension, and multiaxial stress states under quasistatic, dynamic and fatigue loading conditions.[16]

The biggest challenge with aerogels is that they have poor mechanical strength.[17] Their mechanical strength is reported to be similar to that of cotton fiber balls, meaning that they can easily be crushed and/or under low stress levels be irreversibly damaged.[18] However, methods such as cross-linking and filler reinforcement have been shown to improve mechanical properties with a reduced amount of matter (Table 11.3).[19,20]

Salama *et al.* investigated the effects of cross-linking on the strength of the hemicellulose citrate–chitosan aerogel foams. It was discovered that hemicellulose citrate and chitosan do not form foams after freeze drying. The freeze drying resulted in the formation of a brittle hard glass condensed pellet. However, once the two materials were cross-linked, a flexible aerogel foam that can be deformed over a large strain and still regain its shape once the load had been removed was formed.[21]

The mechanical behavior of alginate aerogels have been studied by Shang *et al.*, in which the samples were prepared with different nanofillers (aluminium hydroxide, sodium montmorillonite, magnesium hydroxide, layered double hydroxide and kaoline) using the freeze drying method. The mechanical strength of the alginate aerogels were improved by the post cross-linking processes (Table 11.4).[19] When alginate was crosslinked with calcium ions, the resulting aerogels had similar mechanical properties as

Table 11.3 Mechanical properties of polysaccharide aerogels. Reprinted from *Trends in Food Science & Technology*, **34**, K. S. Mikkonen, K. Parikka, A. Ghafar and M. Tenkanen, Prospects of polysaccharide aerogels as modern advanced food materials, 124–136, Copyright 2013, with permission from Elsevier.[a]

Type of polysaccharide aerogel	Mechanical properties	References
Calcium–Alginate monolith	Compressive strength ~20–40 kPa	83
Calcium–Alginate monolith reinforced with carboxymethylcellulose	Compressive strength ~20–60 kPa	83
Calcium–Alginate monolith reinforced with *N,N*-methylenebisacrylamide	Compressive strength ~20–70 kPa	83
NFC cellulose microfibril foam by sulfuric acid hydrolysis method	Compressive strength 30.7 kPa	84
NFC dispersions based on TEMPO-oxidation pretreatment	Compressive strength 3–240 kPa	85
NFC dispersions based on TEMPO-oxidation pretreatment	Compressive strength 35–2800 kPa	85
NFC dissolved in hydrated calciumthiocyanate melt	Bending strength 2 MPa	86
Starch reinforced with NFC	Young's modulus 1.7–7.0 MPa	87
Glucuronoarabinoxylan	Compressive modulus 57 kPa	88
Glucuronoarabinoxylan and CNC	Compressive modulus 158 kPa	88
Glucuronoxylan–chitosan	Tensile strength 0.54 N mm^{-2}	21
Glucuronoxylan citrate–chitosan	Tensile strength 1.61 N mm^{-2}	21
Starch–chitosan	Tensile strength 1.0 N mm^{-2}	21
Starch citrate–chitosan	Tensile strength 1.8 N mm^{-2}	21
Commercial cellulose foam	Tensile strength 23.8 N mm^{-2}	21

[a]NFC: nanofibrillar cellulose; CNC: Cellulose nanocrystals; TEMPO: 2,2,6,6-tetramethyl-piperidine-1-oxyl.

rigid PU foams and moduli similar to that of balsa. The compressive modulus also improved with an increase in solid content from 1 to 97 MPa (Table 11.5) as the microstructures were altered from a layered structure to a network structure.[22] Starch biofoams that have been freeze dried have shown reduced compression strength and this is due to the moisture absorption that reduces cell wall properties. The moisture plasticized the cell wall resulting in the reduction of the Young's modulus (from approximately 1.9 to 1.0 GPa) and the glass-transition temperature (from 230 °C to room temperature) for 15% glycerol plasticized starch.[23] This in turn meant that the cell structure of starch foams is not always suitable for maximum mechanical performance. However, to combat this, the hygroscopic amylopectin starch matrix was reinforced with microfibrillated cellulose (MFC) and this resulted in an improved yield strength (from

Table 11.4 Mechanical properties of aerogels prepared from alginate with different fillers. Reprinted with permission from K. Shang, W. Liao, J. Wang, Y. Wang, Y. Wang and D. Schiraldi, *ACS Appl. Mater. Interfaces*, 2016, **8**, 643, Copyright 2016 American Chemical Society.[a]

Sample	Modulus (MPa)	Specific modulus ($m^2 s^{-2}$)
A5	0.94 ± 0.12	14.2 ± 2.3
A5Ca	3.45 ± 0.53	50.0 ± 6.5
A5MH5	4.92 ± 0.75	47.3 ± 5.7
A5MH5Ca	7.07 ± 1.12	64.2 ± 8.7
A5ATH5Ca	8.63 ± 0.97	75.1 ± 6.3
A5LDH5Ca	8.42 ± 1.23	82.5 ± 7.8
A5MMT5Ca	10.28 ± 1.12	110.4 ± 9.8
A5Kaolin5Ca	10.04 ± 1.54	92.1 ± 8.6

[a]A: alginate; 5: percentage of materials in water; Ca: post cross-linking process; MH: magnesium hydroxide; ATH: aluminum hydroxide; LDH: layered double hydroxides; MMT: montmorillonite.

130 to 480 kPa) and modulus (from 3.4 to 20.3 MPa) for up to 40% MCF content and this was attributed to the MFC stabilizing the moisture of the plasticized starch.[24]

Composite aerogels made from cellulose and $CoFe_2O_4$ nanoparticles by freeze drying have been shown to have great flexibility, however, they tend to be brittle. These composite aerogels can be easily deformed to a large extent before any fracture appears. When the cellulose composite aerogels were compared to the cellulose aerogel, the composites had a higher compressibility and increased instantaneous modulus (85 MPa for the cellulose aerogel compared to 133 and 180 MPa for composites with 0.01 and 0.1 M of $CoCl_2$, respectively). Nanocellulose aerogels made from native cellulose I crystals and dried using supercritical CO_2 also had a high modulus and strength (Table 11.6). They were less brittle, flexible and could be compressed to a strain of more than 50% without disintegration (Figure 11.2).[25,26]

Rudaz *et al.* illustrated that super insulating pectin aerogels obtained by CO_2 supercritical drying and evaluated by uniaxial compression tests were less fragile as compared to silica aerogels, and were plastically deformed to a maximum strain of 50–60%. Also, the Young's modulus varied from 4 to 18 MPa with aerogels prepared from 2–6% gelled solution.[27]

Gluten foams with different mechanical properties can be produced through the variation of gluten concentration and using fibers or plasticizer.[28] Gluten protein from wheat is stiff and brittle in its dry unplasticized form. However, once plasticized (with glycerol), its mechanical properties were similar to plasticized polyvinyl chloride (PVC) and low density polyethylene (LDPE). The improved mechanical properties of gluten, such as surface tension, surface compressibility, rigidity, and surface viscosity made the gluten more compressible and stable than other proteins. The authors reported that gluten monolayers did not collapse up to 32 dyne cm^{-1}

Table 11.5 Mechanical properties (modulus, density and specific modulus) of alginate-MMT aerogels. Reprinted from *Polymer*, **53**, H.-B. Chen, Y.-Z. Wang, M. Sanchez-Soto and D. A. Schiraldi, Low flammability, foam-like materials based on ammonium alginate and sodium montmorillonite clay, 5825–5831, Copyright 2012, with permission from Elsevier.

Alginate	Property	A5	A7.5	A10	A12.5	A15
No clay	Modulus (MPa)	0.99 ± 0.06	3.2 ± 0.2	9.4 ± 1.4	20 ± 2	46 ± 9
	Density ($g\,cm^{-3}$)	$0.047 \pm <0.001$	$0.066 \pm <0.001$	$0.085 \pm <0.001$	$0.108 \pm <0.001$	0.131 ± 0.002
	Specific modulus ($MPa\,cm^3\,g^{-1}$)	21 ± 2	49 ± 3	110 ± 17	185 ± 16	350 ± 64
5% Clay	Modulus (MPa)	5.8 ± 0.7	21 ± 3	42 ± 5	70 ± 7	97 ± 11
	Density ($g\,cm^{-3}$)	$0.085 \pm <0.001$	0.108 ± 0.002	0.130 ± 0.002	0.152 ± 0.002	0.174 ± 0.002
	Specific modulus ($MPa\,cm^3\,g^{-1}$)	68 ± 8	196 ± 27	324 ± 36	458 ± 48	557 ± 65

Table 11.6 Mechanical properties of cellulose aerogels prepared from (a) deionized water (Aero-0), (b) 20% AMIMCl aqueous solution (Aero-20), (c) 40% AMIMCl aqueous solution (Aero-40) and (d) 60% AMIMCl aqueous solution (Aero-60). Reprinted with permission from Q. Mi, S. Ma, J. Yu, J. He and J. Zhang, *ACS Sustainable Chem. Eng.*, 2016, **4**, 656, Copyright 2016 American Chemical Society.

Sample	Young's modulus (MPa)	Compressive stress at 50% strain (MPa)
Aero-0	15.2	1.28
Aero-20	23.4	1.40
Aero-40	23.7	1.58
Aero-60	27.3	1.99

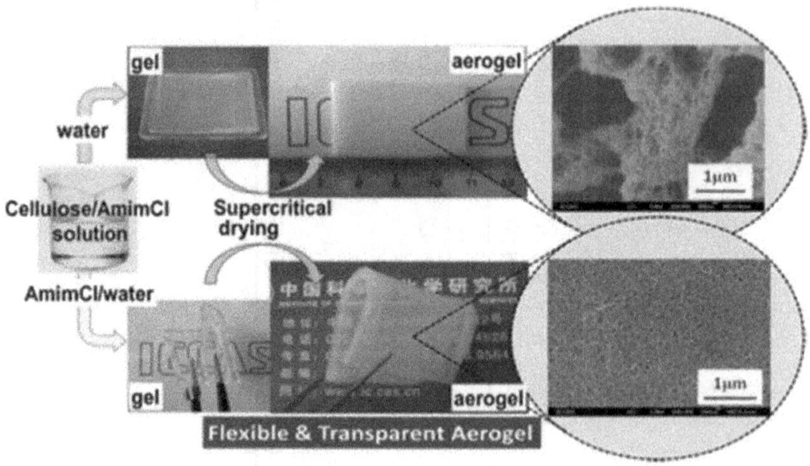

Figure 11.2 Flexible cellulose aerogel.

pressure as compared to other protein monolayers that collapsed at approximately 20 dyne cm^{-1} pressure.[29] The improved properties of gluten in its plasticized form could possibly be attributed to the high molecular weight glutenin molecules.[30] Blomfeldt *et al.* reported that gluten foams showed a quick increase in the modulus with increasing density for gluten foams with no plasticizer (approximately 0.4 to 3 MPa) or reinforced with fibers (approximately 0.4 to 3.1 MPa). However, addition of glycerol resulted in foams that have a low modulus (about 0.3 MPa) and a high strain recovery (approximately 95%).[28]

Blending whey protein with nanocrystalline cellulose (NCC) followed by freeze drying resulted in an increased Young's modulus from approximately 490–550 kPa for the aerogels and NCC-blended aerogel, respectively. This was due to extensive texture tightening during drying. During drying, the NCC rods positioned themselves towards the positively charged domains of the whey proteins, therefore operating as cross-connectors in the final structure of the aerogel.[31] Mechanical properties of whey protein aerogels are influenced by the pH at which the hydrogel is gelled.[32] Low pH, or a pH below the isoelectric point (p*I*) (pH at which the net charge of protein molecule or amino acid is zero)[33] of native proteins (pH 1.5 and 3) resulted in brittle and cracked aerogels. The resulting poor mechanical stability from aerogels obtained from acidic hydrogels was due to the partial hydrogel disintegration that already occurs during the solvent exchange step. However, aerogels obtained from hydrogels that were gelled at neutral (pH 7) and basic (pH 8 and 10) environment exhibited a compact and hard consistency.[32]

A similar trend was observed for aerogels formed from white egg proteins by CO_2 supercritical drying. The compression tests revealed that the stiffness and the primary breakage point of the aerogels were pH dependent. A pH close to the isoelectric point of the main protein ovalbumin resulted in aerogels that were very soft and this was attributed to less connected aggregate structures. While a pH below and higher than the p*I* of protein ovalbumin gave relatively stiff aerogels due to dense and well connected structures with small pores.[34] The utilization of aerogels as effective stem cells scaffolds relies on careful tuning of the scaffold mechanical properties. Silk fibroin (SF) aerogels obtained by supercritical CO_2 drying have shown potential to be used as scaffolding material for tissue engineering. The increase in SF concentration from 2 to 6 wt%, results in the increase in the compressive moduli of the aerogel from 19 to 174 kPa, respectively. This can likely be ascribed to the creation of denser gel networks at higher SF concentrations.[35]

The development of solid materials through the use of soy protein (SP) can be a daunting task because SP tends to be brittle and lack mechanical strength. However, once the soy protein was reinforced with a low concentration of nanofibrillar cellulose (NFC), followed by freeze drying, the resulting composite aerogel exhibited high compression strength compared to aerogels with only SP. The improved stiffness of the composite aerogels compared to the SP aerogels as can be seen in the stress–strain curve (Figure 11.3) is attributed to the high stiffness of NFC and the difference between the two microstructures. Furthermore, the composite aerogels were less prone to structural damage upon contact with non-polar and polar solvents.[36] The results indicate that the more expensive fibrillar cellulose can be replaced by SP and still maintain the high compression modulus similar to that of pure NFC (Figure 11.3).

The mechanical properties of gum arabic (GA)–clay aerogels have also been investigated. GA is composed of a mixture of arabinogalactan (which is the dominant component) and hydroxyproline protein.[37,38] GA aerogels obtained *via* the freeze drying method exhibited a brittle behavior and this was attributed to the lack of polymer connections between layers, however, addition of clay to these aerogels resulted in the improvement of the compressive mechanical properties (Table 11.7).[20]

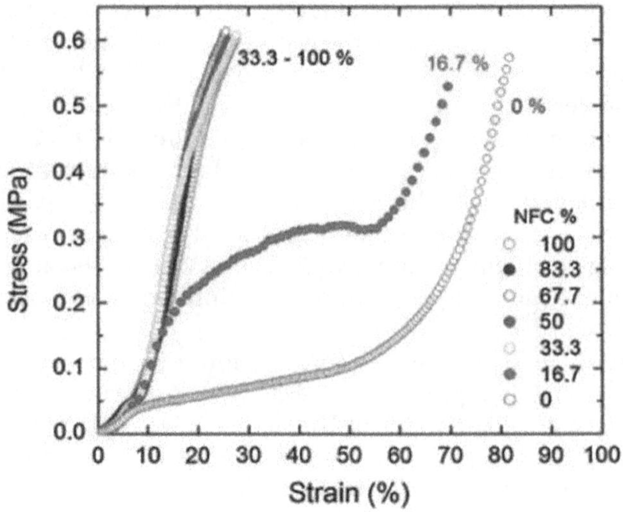

Figure 11.3 Compression stress–strain curve for SP aerogels with different NFC concentrations with measured stress at 18% strain.
Reproduced from *Cellulose*, Soy protein-nano-cellulose composite aerogels, **20**, 2013, 2417–2426, J. C. Arboleda, © Springer Science+Business Media Dordrecht 2013. With permission of Springer.

Table 11.7 Compressive mechanical properties of GA aerogel blended with clay. Reprinted from *Industrial Crops and Products*, **91**, L. Wang, M. Sanchez-Soto and Tobias Abt, Properties of bio-based gum Arabic/clay aerogels, 15–21, Copyright 2016, with permission from Elsevier.[a]

Sample	E (MPa)	E_s (MPa g cm^{-3})	σ_{max} (MPa)	E_a (kJ cm^{-3})
GA7.5	0.6 ± 0.1	7.5 ± 2.1	—	12 ± 2
GA10	1.4 ± 0.3	10.6 ± 2.1	0.25 ± 0.1	49 ± 12
GA15	2.1 ± 0.7	10.9 ± 2.9	1.10 ± 0.1	240 ± 45
GA5C5	1.2 ± 0.2	13.9 ± 2.0	0.35 ± 0.1	44 ± 3
GA7.5C5	2.3 ± 0.1	19.4 ± 1.1	0.40 ± 0.1	62 ± 15
GA10C5	4.0 ± 0.6	30.0 ± 3.4	0.60 ± 0.2	70 ± 1
GA15C5	25.8 ± 4.2	143 ± 23	1.30 ± 0.1	315 ± 20

[a] E: compressive modulus; E_s: specific compressive modulus; σ_{max}: compressive stress at 70% strain; E_a absorbed energy. Concentrations used for GA and clay were (7.5, 10, 15 wt%) and 5 wt%, respectively.

11.2.3 Rheological and Viscoelastic Properties

Rheology is related to the flow and deformation of matter and plays a vital role in various applications.[39,40] For example, additives are employed to impart required flow behavior. These properties give a mathematical description of the viscoelasticity behavior of matter.[41] Factors such as stress and strain play a key role in rheological evaluation.[42] The properties are influenced by the type of structure (*i.e.* entanglement, association, or cross-links) present in the system.[43] Rheological properties can differ from viscous fluids to elastic solids and illustrate the spectrum of possible responses of materials to stress, strain and shear rate.[44] The properties are determined by assessing the force and deformation with respect to time.[42] Rheological techniques measure the force, torque or pressure rate, output, frequency or rotational speed; and temperature.[45] During a dynamic rheological test, two main parameters are determined, namely storage or elastic modulus (G') which indicates the amount of energy stored elastically in the structure and the loss or viscous modulus (G'') which indicates the amount of energy loss or the viscous response.[46] To the best of our knowledge, rheological studies on bio-based aerogels are limited in number. Instead, rheological properties are studied using hydrogels and these hydrogels are then converted to aerogels. According to Bao *et al.*, gels can be categorized as either weak or strong gels based on the G' and G'' ratio. A true gel has a G'/G'' ratio higher than 3, this is where the intermolecular junctions have a higher binding energy. A viscous gel behavior is achieved when the slope of the double logarithmic plot of storage modulus against frequency equals 1.[47] However, when the slope tends towards zero, the gel exhibits an elastic property. Natural polymers that typically form hydrogels include polysaccharides such as starch and alginate and proteins such as collagen and gelatin.[48]

External factors such as the concentration, temperature, and presence of salts or sugars influence the rheological behavior of polysaccharides.[49] Rheological studies of cellulose–NaOH solutions have shown that gelation of cellulose occurs with time and that gelation time is reduced when the concentration of cellulose and temperature increases.[50,51] G' and G'' as a function of time can be used to estimate the gelation temperature (T_{gel}) of cellulose–NaOH–water. Although this method is not precise, it provides a quick comparison for different sample behaviors under the same conditions.[51]

For example, a study by Liu *et al.* illustrated that the rheological method is a reliable and direct way of investigating sol–gel transition behavior of cellulose–NaOH–urea aqueous solution and 1, 4-butanediol diglycidyl ether (BDE), a diepoxide-based bifunctional linker. When G' and G'' were measured at a constant frequency of 1 Hz the shear as a function of temperature (from 20–90 °C) for cellulose–NaOH–urea aqueous solution (from 20 to 63 °C), it was found that the G' was lower than the G'' indicating the common viscoelastic behavior of a liquid. This was attributed to the cellulose chains being attracted by the NaOH hydrate through the formation of new hydrogen bonded networks, while the urea hydrates self-assembled at the surface of the NaOH hydrogen bonded cellulose, forming an inclusion complex (IC). Above 63 °C, the G' increased more than G'', showing the formation of an elastic gel network. This was due to the destruction of the IC resulting in self-association entanglements and molecular interactions on the cellulose backbone as the temperature increased. From this it was clear that 63 °C was the gel point where G' is equal to G''. However, for BDE–cellulose–NaOH–urea solution, the G' was always higher than G'' implying the formation of an elastic gel network (Figure 11.4).[52]

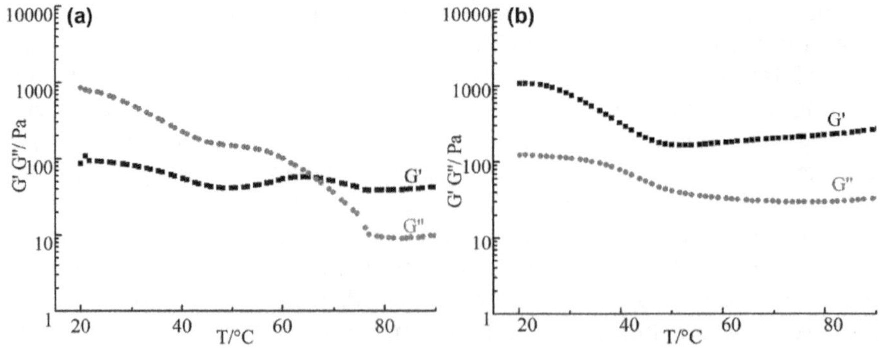

Figure 11.4 Storage and loss modulus of (a) cellulose–NaOH–urea aqueous solution and (b) BDE–cellulose–NaOH–urea solution as a function of temperature.
Reproduced from ref. 52 with permission from the Royal Society of Chemistry.

Rheological properties of starch hydrogels are affected by crystallinity, such as the amylose/amylopectin ratio, granule size and the distribution and quality of smaller components (such as enzyme, phosphorus, proteins and lipids). During the heating and cooling process of starch dispersions, the storage modulus increased with the increase in starch concentration. While at 30% starch concentration, the loss modulus varied dramatically, however it was insignificant at 20% concentration.[53]

The addition of carrageenans results in improved texture with kappa-carrageenans, giving a firm and brittle texture, while with iota-carrageenans (ι-CAR) they give a soft and elastic texture. The difference between the two hydrocolloids is the number of sulphate groups per disaccharide, with kappa-carrageenans and iota-carrageenans having one and two, respectively. A study on the effect of iota-carrageenans on rheological properties of starches was conducted by Tecante and Doublier. The variation of G' and G'' as a function of frequency under 1% deformation at 25 °C for corn starch (CS) and ι-CAR–CS mixture can be seen from Figure 11.5. It can be observed that G' was at least three times larger than G'' for all examined frequencies. This is typical of a gel behavior and indicates a dependence of frequency above 1 Hz. The presence of ι-CAR changed the gel structure of CS, resulting in a viscoelastic solution (Figure 11.5B) as indicated by the crossover of G' and G'' at 0.2 Hz.[54] Tecante and Doublier showed that the storage modulus is strongly dependent on the kappa-carrageenan concentration when mixed with amylose.[55]

Polysaccharides from *Auricularia auricular-judae* (AP), a popular mushroom in China, have also been investigated as hydrogels and they exhibited a shear-thinning flow behavior. Its rheological properties were significantly affected by pH and the presence of salts, and concentration. Furthermore, when the dependence of G' and G'' on frequency were investigated at 25 °C

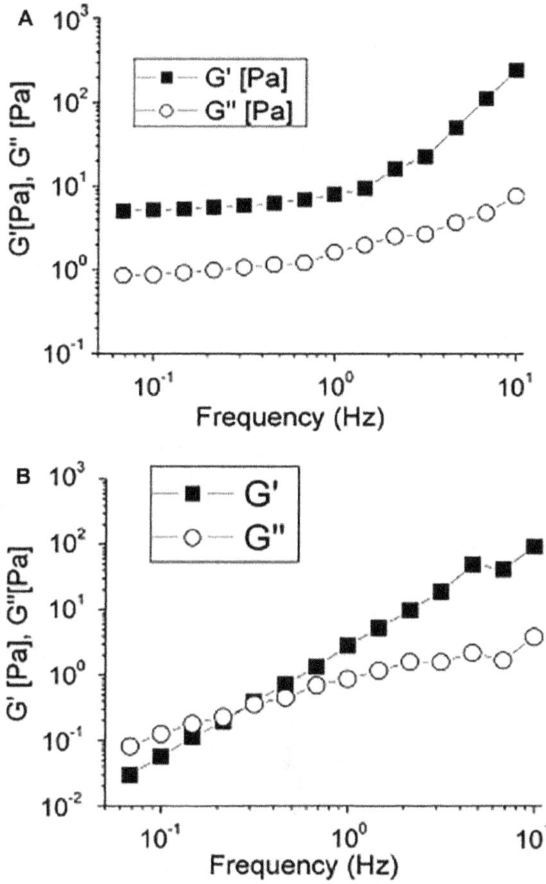

Figure 11.5 Storage and loss modulus as a function of frequency at 1% deformation for (A) corn starch and (B) ι-CAR–CS mixture.
Reprinted from *Carbohydrate Polymers*, **65**, P. C. Sousa, F. Tischer, M. D. Noseda, R. A. de Freitas, M. R. Sierakowski, M. Eugenia and R. Duarte, Effects of iota-carrageenan on the rheological properties of starches, 49–57, Copyright 2006, with permission from Elsevier.

for 0.5, 1 and 2% (w/v), G' was higher than G'' showing little dependence on frequency, which is typical of a gel-like system. An elastic strong gel was formed at 2% AP dispersion, while at concentrations lower than 1% weak viscous gels are formed.[47]

In a study dealing with the rheological properties of five polysaccharides (low-methoxyl pectin, high-methoxyl pectin, alginate, guar gum and xathan), the authors found that low-methoxyl pectin gave the highest difference between G' and G''. The results showed that G'' was around 30% of G', which indicated a fully elastic system. Similar results were observed for the other four polysaccharide (high-methoxyl pectin, alginate, guar gum and xathan) gels ($G' > G''$). However, the difference between the two was lower. An

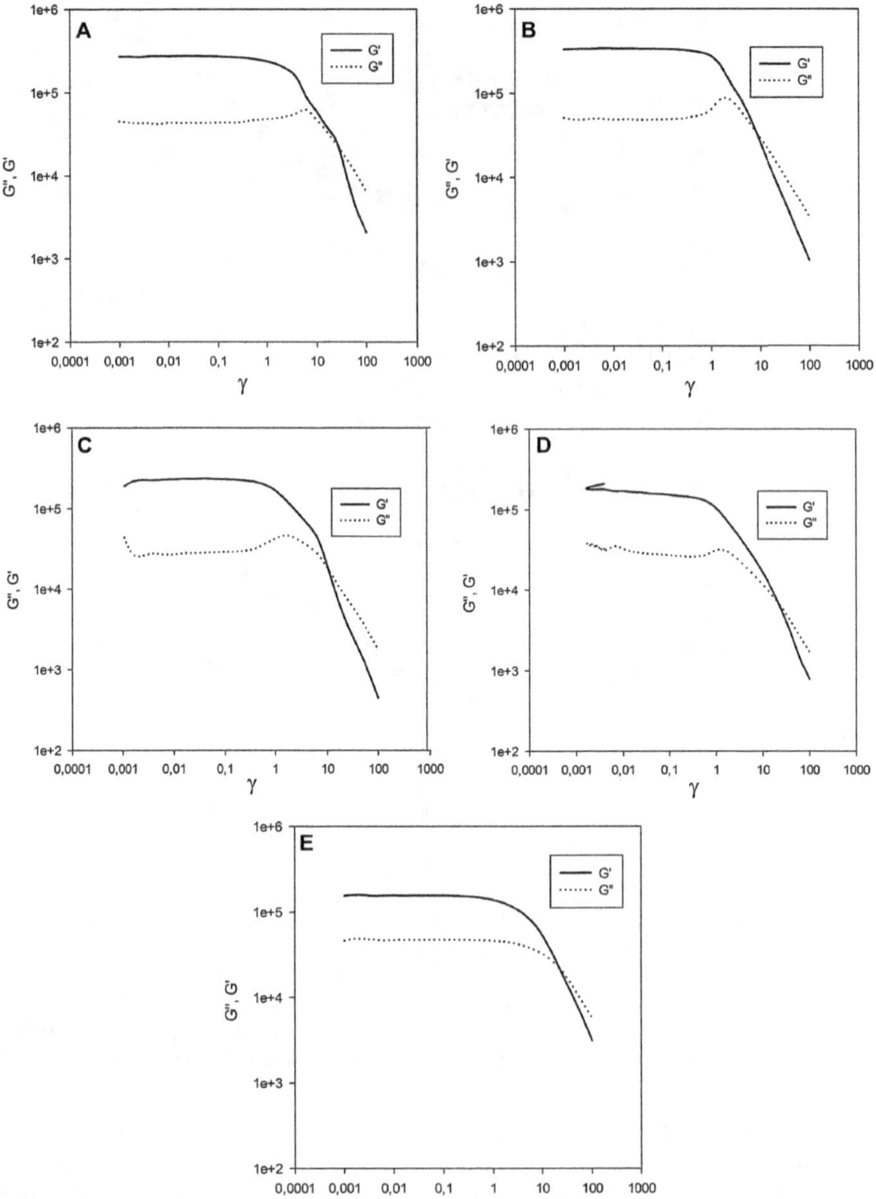

Figure 11.6 Strain sweep measurements of polysaccharide gels at constant frequency (A) high-methoxyl pectin, (B) low-methoxyl pectin, (C) xanthan, (D) alginate and (E) guar gum.
Reproduced from ref. 56 with permission from the Royal Society of Chemistry.

increase in strain resulted in the break-down of the elastic structure, leading to a decrease of G'. At the point where G'' is greater than G' (crossover strain), the elastic structure was found to be broken. G'' was similar for all four polysaccharide gels (high-methoxyl pectin, alginate, guar gum and xathan) as can be seen from Figure 11.6.[56]

Protein may form various types of fluid gels during heating and at different pHs, namely fine-stranded, branched or particulate gels.[57] Particulate gels are formed near the p*I*, while stranded or branched gels are formed away from the p*I*.[58] Several factors affect the flow curve and formation of protein gels and these include protein concentration, molecular weight of the protein, pH and solubility.[59] At rest, protein fluid gels behave like solids, but above the critical value of applied stress exhibit flow behavior. The rheological properties of proteins may be manipulated to meet the requirements for desired applications.[60] Gelatin consists of hydroxyl, carboxyl and amino groups that make it easy to dissolve in water and consequently form a physical thermal reversible gel at reasonably low temperature.[61] Temperature and the concentration of gelatin influences the rheological properties of thermo-reversible gelatin gels.[46]

Rheological properties of gelatin solution and gelatin mixed with formaldehyde at low temperatures (10 °C) and a time span of 2 h for dynamic time sweep tests have been investigated by Wang *et al.* The storage modulus of the gelatin solution monotonically increased and reached a plateau value at 5.0 Pa, while the loss modulus showed a similar trend, however, with a smaller plateau value of 0.5 Pa. From this, it can be postulated that the gelatin formed a physical hydrogel in which the molecule chains recover the collagen triple-helix. For the gelatin-formaldehyde mixture, the G' started at a lower value and this was due to the slowing down of the triple-helix structure formation. The increase in G' resulted in a higher value than that of gelatin, but gradually reached a plateau value of 6.8 Pa, while G'' was 0.37 Pa. Once the crosslinking was achieved (indicated by a plateau being reached), both moduli were greater than those of the gelatin alone. Dynamic oscillatory tests in frequency of the gels indicated that both gels had a soft hydrogel behavior. The increase in G' suggested additional cross-links were introduced by the formaldehyde and these hydrogels could be converted to aerogels *via* freeze drying (Figure 11.7).[61]

Gelling and melting temperatures of chicken skin gelatin and bovine gelatin gels were compared at 6.67% (w/v) concentration. Sarbon *et al.* reported a significant difference between the melting temperature of chicken gelatin and bovine gelatin. The melting temperature was higher for the chicken gelatin compared to the bovine gelatin, and no significant difference was observed for the gelling temperature. Furthermore, the maximum values of G' and G'' were higher for the chicken gelatin than that for bovine gelatin (Table 11.8).[46]

Whey protein concentrates exhibit gel characteristics at moderate concentration (9% w/w). Modification of whey protein is required as increased protein content might result in a product with high firmness, giving a

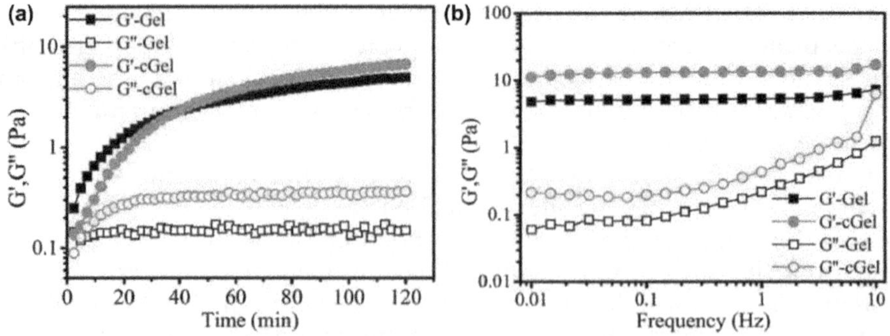

Figure 11.7 Illustrations of (a) rheological time sweep and (b) frequency sweep measurements for gelatin and gelatin–formaldehyde mixture. Reproduced from ref. 61 with permission from the Royal Society of Chemistry.

Table 11.8 Rheological properties of chicken and bovine gelatin with gelling temperature, melting temperature and G′ and G″ modulus after heating to 40 °C and cooling to 10 °C. Reprinted from *Food Hydrocolloids*, **30**, N. M. Sarbon, F. Badii and N. K. Howell, Preparation and characterization of chicken skin gelatin as an alternative to mammalian gelatin, 143–151, Copyright 2013, with permission from Elsevier.

			Maximum value after cooling	
Gelatin	Gelling temp. (°C)	Melting temp. (°C)	G' (Pa)	G'' (Pa)
Chicken	24.88 ± 0.27	33.57 ± 0.52	8273 ± 1016	6639 ± 1192
Bovine	24.43 ± 0.91	31.55 ± 0.04	4330 ± 31	4121 ± 59

rubbery mouth-feel and undesirable texture change with time (hardening).[59] When the rheological properties of transglutaminase cross-linked cold-set whey protein gels were investigated, the elastic modulus and yield/fracture stress and strain were found to be higher than those of non-cross-linked whey protein gels (Figure 11.8).[62]

Furthermore, the effect of pressure application (high pressure processing (HPP)) on the rheological properties of gelatin-whey protein mixtures have been explored. Application of 600 MPa at 30 °C for 15 min resulted in lower G' values for mixed gels compared to G' values of thermally treated (at 80 °C) mixed gels. The mixed system in the solution state had a higher G' compared to the mixed system in the gel form.[63] Valuable information about the protein-polysaccharide texture characteristics and its stability can be obtained from rheological tests. The effects of flaxseed gum (FG) on the rheological properties of peanut protein isolate (PPI) were determined using steady shear and small amplitude oscillatory measurements. The results indicated that the viscosity increased with the increase in FG concentration. FG addition further decreased the gelation time and increased the strength of the FG-PPI gel from 2000 to more than 3000 Pa. Storage modulus

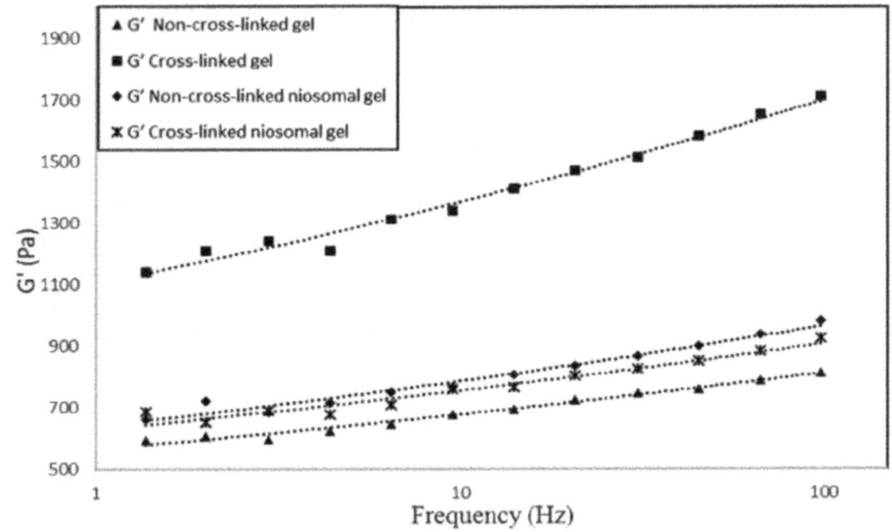

Figure 11.8 Illustration of the frequency sweep test of the hydrogels. Reprinted from *Food Chemistry*, **196**, A. Abaee and A. Madadlou, Niosome-loaded cold-set whey protein hydrogels, 106–113, Copyright 2016, with permission from Elsevier.

was much higher than loss modulus indicating that its behavior was of a physical gel and that the storage modulus was only slightly dependent on frequency.[64]

In reality, materials are not entirely elastic. Under certain conditions they exhibit non-elastic properties. This strongly applies to polymers, which may possess non-elastic deformation when subjected to certain conditions, while metals may be considered purely elastic.[65] Polymers can become complex as they are both viscous and elastic.[40] A viscoelastic material possesses both viscous and elastic properties.[66] This indicates that viscoelasticity deals with numerous different phenomena and examples include retardation of volume, stress relaxation, creep and dynamic mechanical behavior (all in shear and tensile deformation).[65] To our knowledge, only Wang *et al.* have investigated viscoelastic properties on aerogels. Most of the viscoelastic properties have been carried out on hydrogels and the hydrogels are then converted into aerogels.

Polysaccharides are materials that are viscoelastic, meaning that they possess both solid and liquid characteristics simultaneously.[49] Polysaccharides are hydrophilic in nature and it is this property that renders them suitable precursors for hydrogels that can further be changed into aerogels.[67] In a study by Dong *et al.* NFC functionalized with silver nanoparticles (NFC-Ag) was prepared *via* the freeze drying method. The dynamic viscoelastic properties of NFC and NFC-Ag gels were measured and the results showed that the addition of silver nitrate ($AgNO_3$) resulted in rapid gelation. The strong and colorless gel formed could not be deformed by shaking. The storage modulus was greater than the loss modulus, indicating that the gel behaved as an elastic solid. The G' leveled off and a crossover between G' and G'' was not observed at the investigated frequencies (0.1–100 rad s^{-1}) (Figure 11.9).[68]

Aerogels produced from hemicellulose have limited application due to the chemical heterogeneity of hemicellulose and low molecular weight. Hence, chemical modification, reinforcement with other biopolymers or enzymatic treatment is required to remedy these limitations.[21,69,70] Spruce galactoglucomannas (GGM) are a group of hemicelluloses that can be converted to aerogels. However, because of its relative low molecular weight, the native GGM has a relatively low viscosity in aqueous media and even when the concentration is increased they still exhibit a poor ability to form gels.[71]

A study by Alakalhunmaa *et al.* investigated formation of wood based aerogels *via* freeze drying of cross-linked (with ammonium zirconium carbonate-AZC) aided hydrogels from GGM reinforced with cellulose nanofibrils (CNF). From the results, it is observed that pure native GGM resulted in the formation of a non-viscous solution with no gelling ability. However, crosslinking GGM with AZC improved the viscosity, but did not induce the formation of elastic hydrogels, while GGM with CNF alone resulted in weak hydrogels. The combination of AZC and CNF with GGM gave noticeable gel formation, due to the increase in viscosity after treatment with heat. The elasticity of the hydrogel was observed to be dependent on the CNF content.[67]

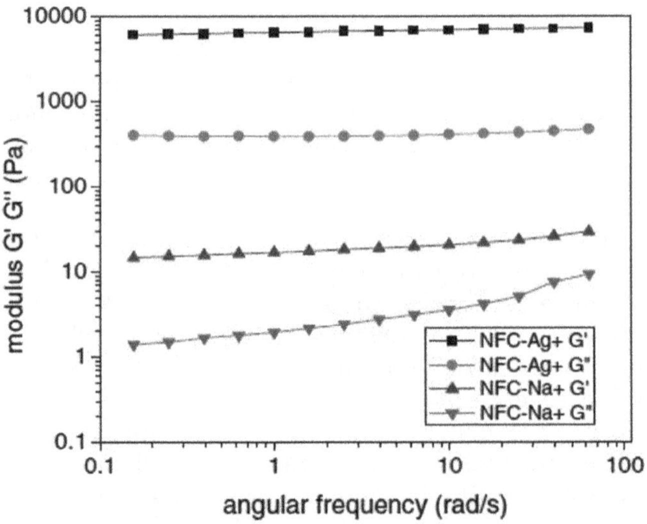

Figure 11.9 Frequency sweep of 1.27% NFC gelled with 50 mM of $AgNO_3$ and $NaNO_3$ at 25 °C.
Reprinted from *Carbohydrate Polymers*, **95**, H. Dong, J. F. Snyder, D. T. Tran and J. L. Leadore, Hyrdrogel, aerogel and film of cellulose nanofibrils functionalized with silver nanoparticles, 760–767, Copyright 2013, with permission from Elsevier.

Starch displays a unique viscosity behavior with change in concentration, temperature and shear rate. Maximum viscosity at a certain concentration reveals the ability of the particles to freely swell before their physical breakdown. During the cooling period, the increase in viscosity is indicative of the inverse relationship between viscosity and temperature.[72] Studies on the viscoelastic behavior of crosslinked (with potassium chloride-KCl) starch with κ-carrageenan gels have been investigated by Tecante and Doublier. From the results, at 60 °C the gels displayed a solid like behavior which was attenuated at lower starch concentration, and when KCl and κ-carrageenan were added. At 25 °C a similar trend was observed, however, the G' was slightly higher and an increased elastic character at higher starch concentrations was observed.[73]

In a study by Wanga *et al.* chitosan aerogels doped with graphene oxide (GO) prepared by lyophilization process were tested for the effective removal of methyl orange and amido black 10B. The results revealed that the G' was greater than G'' for both chitosan and the chitosan-GO (CS-GO) aerogels for the studied frequencies (0.1–15 rad s^{-1}), indicating that the elastic response was dominating, hence the aerogels have a permanent network. The addition of 1% GO resulted in an increase in G' (2.2–3.1 to 3.3–4.0 MPa) while a decrease (0.31–0.66 to 0.1–0.4 MPa) in G'' was observed. This indicates that GO has the ability to improve the elastic response of the aerogels.[74]

Whey proteins have the ability to form heat induced viscoelastic gels under favourable conditions.[59] Different network structures are formed depending on the pH. For example, aggregated particulate networks are formed at an intermediate pH (4–6), while at high or low pH away from the p*I* fine stranded networks are formed. In turn these structural variations affect the viscoelastic properties.[75] Another study by Zheng *et al.* showed that mixing of cellulose (stiff chains) and soy protein isolate (hydrophilic groups) crosslinked with epichlorohydrin resulted in a porous hydrogel, and the hydrogels were further freeze dried to obtain aerogels. The results showed that at the beginning, the solutions behaved as viscous liquids ($G' < G''$), with time G' increased and the gelling point was reached ($G' = G''$) and it was at this point that the solutions became hydrogels. This was attributed to the molecular weight of the system increasing quickly close to the gel point, and hence leading to an increase in viscosity. Also, the crosslinking networks contribute to the conversion of liquid to gel. Furthermore the results indicated that cellulose content played a significant role in the formation of the hydrogel (a decrease in the cellulose content resulted in an increase in gel points of the hydrogels).[76]

When chicken skin gelatin gels were compared to bovine gelatin during cooling (from 40 to 10 °C), the chicken gelatin had a higher G' at low temperatures and this was indicative of its ability to refold into a triple helix (Figure 11.10). The increased G' of chicken gelatin gel indicated that a higher thermal transition is needed compared to bovine gelatin. This implies that bovine gelatin is more heat stable. Mammalian gelatin tends to have increased G' values and thermo-stability and this is attributed to

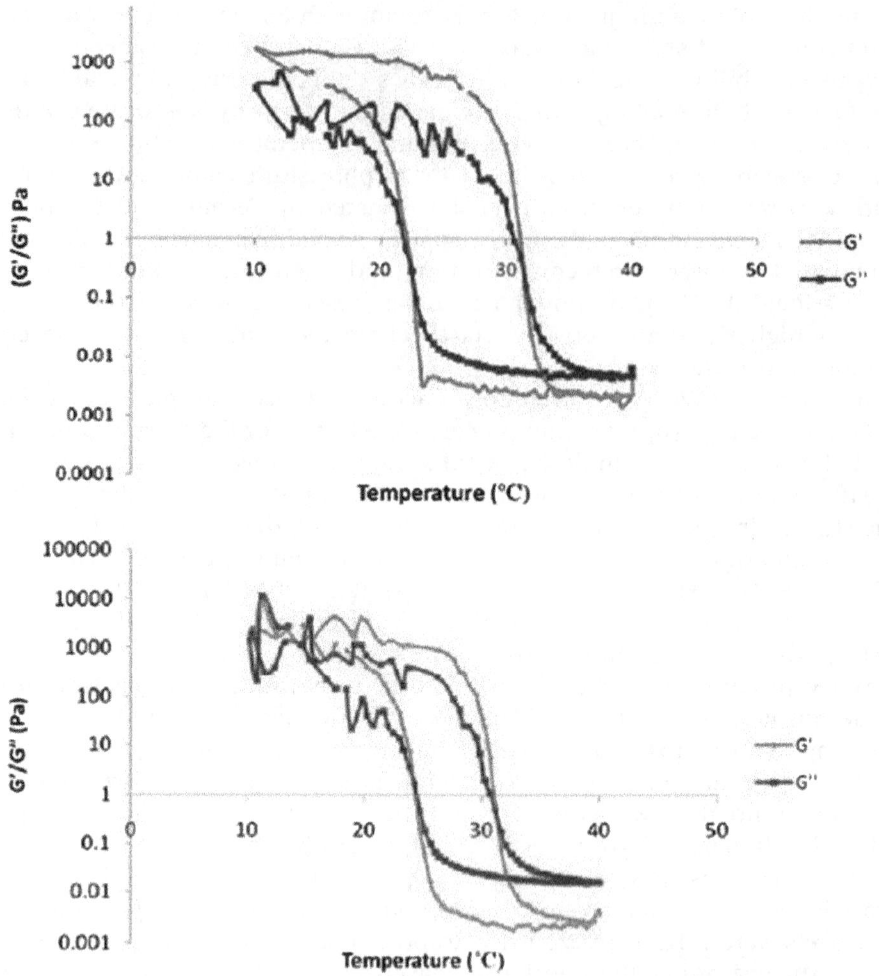

Figure 11.10 Viscoelastic properties of chicken skin gelatin (top) and bovine gelatin (bottom) at 6.67% concentration upon cooling from 40 °C to 10 °C and heating from 10 °C to 40 °C.
Reprinted from *Food Hydrocolloids*, **30**, N. M. Sarbon, F. Badii and N. K. Howell, Preparation and characterization of chicken skin gelatin as an alternative to mammalian gelatin, 143–151, Copyright 2013, with permission from Elsevier.

the composition of imino acid with hydroxyproline stabilizing the triple helix.[46]

11.3 Conclusions

Aerogels are a class of porous material that can be prepared from various materials. However, increasing environmental concern, limitation of the employment of petroleum based synthetic polymers and aerogels that can fulfill requirements for food application has led researchers to find new precursors that are more environmentally and ecofriendly. Bio-based aerogels are a viable alternative to petroleum as they are mostly derived from polysaccharides and protein feedstocks. These type of aerogels (cellulose, starch, alginate, whey protein, silk fibrion, *etc.*) exhibit exceptional properties such as low density, high porosity, biodegradability and biocompatibility. However, they are mechanically delicate materials.

The mechanical properties of these bio-based aerogels can be improved by cross-linking or with reinforcement with fillers. The cross-linking results in aerogel that are flexible, but the rheological properties are affected by the type of structure (*i.e.* cross-links and entanglement). Rheological properties can be used to determine the flow behavior of the bio-based hydrogels. For instance, the relationship between storage modulus and loss modulus reveals whether the hydrogel will exhibit an elastic or viscous behavior. When $G' > G''$ the hydrogel used to prepare aerogels exhibits elastic properties and when $G'' > G'$ the hydrogel exhibits viscous properties. It is generally observed that an increase in concentration results in an increase in mechanical properties of proteins, while a decrease in pH generally results in a decrease in the mechanical properties. For polysaccharides like cellulose the gelation time decreases when the concentration of cellulose and temperature increase. Therefore, mechanical, rheological and viscoelastic properties of these bio-based aerogels are influenced by concentration, pH and temperature.

References

1. S. Riffat and G. Qiu, *Int. J. Low-Carbon Technol.*, 2013, **8**, 1.
2. A. Pierre, in *Aerogels Handbook*, ed. M. Aegerter, N. Leventis and M. Koebel, Springer Science and Business Media, New York, 2011, pp. 3–18.
3. A. Pierre and G. Pajonk, *Chem. Rev.*, 2002, **102**, 4243.
4. A. Du, B. Zhou, Z. Zhang and J. Shen, *Materials*, 2013, **6**, 941.
5. A. Anderson and M. Carroll, in *Aerogels*, ed. M. Aegerter, N. Leventis and M. Koebel, Springer-Verlag, New York, 2011, pp. 47–77.
6. L. Hrubesh, *J. Non-Cryst. Solids*, 1998, **225**, 335.
7. C. Yan, P. Given, G. Huvard, R. Mallepally and M. McHugh, *US Pat.*, EP3185699A2, 2016.
8. A. Acharya, D. Joshi and V. Gokhale, *Int. J. Chem. Process. Eng. Res.*, 2013, **9**, 1.
9. L. Zuo, Y. Zhang, L. Zhang, Y. Miao, W. Fan and T. Liu, *Materials*, 2015, **8**, 6806.
10. M. Alnaief, M. Alzaitoun, C. Garcia-Gonzalez and I. Smirnova, *Carbohydr. Polym.*, 2011, **84**, 1011.
11. Z. Ulker and C. Erkey, *J. Controlled Release*, 2014, **177**, 51.
12. M. Pantić, Ž. Knez and Z. Novak, *J. Non-Cryst. Solids*, 2016, **432**, 519.

13. K. Mikkonen, K. Parikka, A. Ghafar and M. Tenkanen, *Trends Food Sci. Technol.*, 2013, **34**, 124.
14. H. Maleki, L. Duraes, C. Garcia-Gonzalez, P. Gaudio, A. Portugal and M. Mahmoudi, *Adv. Colloid Interface Sci.*, 2016, **236**, 1.
15. L. Neilsen and R. Landel, in *Mechanical Properties of Polymers and Composites*, Marcel Dekker Inc, New York, 1994, pp. 1–28.
16. H. Lu, H. Luo and N. Leventis, in *Aerogels Handbook*, ed. M. Aegerter, N. Leventis and M. Koebel, Springer, New York, 2011, p. 499.
17. H. Maleki, *Chem. Eng. J.*, 2016, **300**, 98.
18. K. Finlay, M. Gawryla and D. Schiraldi, *Ind. Eng. Chem. Res.*, 2008, **47**, 615.
19. K. Shang, W. Liao, J. Wang, Y. Wang, Y. Wang and D. Schiraldi, *ACS Appl. Mater. Interfaces*, 2016, **8**, 643.
20. L. Wang, M. Sanchez-Soto and T. Abt, *Ind. Crops Prod.*, 2016, **91**, 15.
21. A. Salama, R. Vendittia, J. Pawlaka and K. El-Tahlawyb, *Carbohydr. Polym.*, 2011, **84**, 1221.
22. H. Chen, Y. Wang, M. Sánchez-Soto and D. Schiraldi, *Polymer*, 2012, **53**, 5825.
23. A. Dufresne, D. Dupeyre and M. Vignon, *J. Appl. Polym. Sci.*, 2000, **76**, 2080.
24. A. Svagan, L. Berglund and P. Jensen, *ACS Appl. Mater. Interfaces*, 2011, **3**, 1411.
25. S. Liu, Q. Yan, D. Tao, T. Yu and X. Liu, *Carbohydr. Polym.*, 2012, **89**, 551.
26. Q. Mi, S. Ma, J. Yu, J. He and J. Zhang, *ACS Sustain. Chem. Eng.*, 2016, **4**, 656.
27. C. Rudaz, R. Courson, L. Bonnet, S. Calas-Etienne, H. Sallée and T. Budtova, *Biomacromolecules*, 2014, **15**, 2188.
28. T. Blomfeldt, R. Kuktaite, E. Johansson and M. Hedenqvist, *Biomacromolecules*, 2011, **12**, 1707.
29. T. Mita, K. Nikai, T. Hiraoka, S. Matsuo and H. Matsumoto, *J. Colloid Interface Sci.*, 1977, **59**, 172.
30. T. Blomfeldt, R. Olsson, M. Menon, D. Plackett, E. Johansson and M. Hedenqvist, *Macromol. Mater. Eng.*, 2010, **295**, 796.
31. M. Ahmadi, A. Madadlou and A. Saboury, *Food Chem.*, 2016, **196**, 1016.
32. C. Garcia-Gonzalez, R. Subrahmanyam, I. Smirnova and U. Kulozik, *J. Supercrit. Fluids*, 2012, **72**, 111.
33. W. Bryan, *Biochem. Educ.*, 2010, **6**, 14.
34. I. Selmer, C. Kleemann, U. Kulozik, T. Heinrich and I. Smirnova, *J. Supercrit. Fluids*, 2015, **106**, 42.
35. R. Mallepally, M. Marin, V. Surampudi, B. Subia, R. Rao, S. Kundu and M. McHugh, *Biomed. Mater.*, 2015, **10**, 1.
36. J. Arboleda, M. Hughes, L. Lucia, J. Laine, K. Ekman and O. Rojas, *Cellulose*, 2013, **20**, 2417.
37. Y. Akiyama, S. Eda and K. Katō, *Agric. Biol. Chem.*, 1984, **48**, 235.
38. F. Weinbreck, R. Tromp and C. D. Kruif, *Biomacromolecules*, 2004, **5**, 1437.

39. H. Barnes, J. Hutton and K. Walters, in *An Introduction to Rheology*, ed. K. Walters, Elsevier Science Amsterdam, 1989, pp. 1–10.
40. H. Münstedt, *Soft Matter*, 2011, 7, 2273.
41. O. Sadiku-Agboola, E. Sadiku, A. Adegbola and O. Biotidara, *Mater. Sci. Appl.*, 2011, **2**, 30.
42. G. Tabilo-Munizagaa and G. Barbosa-Cánovas, *J. Food Eng.*, 2005, **67**, 147.
43. S. Gulrez, S. Al-Assaf and G. Phillips, in *Progress in Molecular and Environmental and Bioengineering – From Analysis and Modeling to Technology Application*, ed. A. Carpi, InTech, Rijeka, 2011, p. 117.
44. P. Fischer, M. Pollard, P. Erni, I. Marti and S. Padar, *C. R. Phys.*, 2009, **10**, 740.
45. L. Leblanc, *Prog. Polym. Sci.*, 2002, **27**, 627.
46. N. Sarbon, F. Badii and N. Howell, *Food Hydrocolloids*, 2013, **30**, 143.
47. H. Bao, S. You, L. Cao, R. Zhou, Q. Wang and S. Cui, *Food Hydrocolloids*, 2016, **57**, 30.
48. E. Ahmed, *J. Adv. Res.*, 2015, **6**, 105.
49. J. Shi, J. Zhang, Y. Sun, Q. Xu, L. Li, C. Prasad and Z. Wei, *Int. J. Biol. Macromol.*, 2016, **91**, 760.
50. A. Demilecamps, G. Reichenauer, A. Rigacci and T. Budtova, *Cellulose*, 2014, **21**, 2625.
51. R. Gavillon and T. Budtova, *Biomacromolecules*, 2008, **9**, 269.
52. H. Liu, A. Wang, X. Xu, M. Wang, S. Shang, S. Liu and J. Song, *RSC Adv.*, 2016, **6**, 42854.
53. S. Lu, J. Chen, Y. Chen, C. Lii, P. Lai and H. Chen, *J. Cereal Sci.*, 2011, **53**, 31.
54. P. Tischer, M. Noseda, R. deFreitas, M. Sierakowski and M. Duarte, *Carbohydr. Polym.*, 2006, **65**, 49.
55. A. Tecante and J. Doublier, *Carbohydr. Polym.*, 2002, **49**, 177.
56. G. Tkalec, Z. Knez and Z. Novak, *RSC Adv.*, 2015, **5**, 77362.
57. H. Hudson, C. Daubert and E. Foegeding, *J. Agric. Food Chem.*, 2000, **48**, 3112.
58. T. Nicolai and D. Durand, *Curr. Opin. Colloid Interface Sci.*, 2013, **18**, 249.
59. C. Sun, R. Liu, K. Ni, T. Wu, X. Luo, B. Liang and M. Zhang, *J. Food Eng.*, 2016, **186**, 69.
60. Y. Wang, Y. Chang, Y. Xue, Z. Li, Y. Wang and C. Xue, *Food Hydrocolloids*, 2016, **52**, 510.
61. J. Wang, D. Zhao, K. Shang, Y. Wang, D. Ye, H. Kang, W. Liao and Y. Wang, *J. Mater. Chem. A*, 2016, **4**, 9381.
62. A. Abaee and A. Madadlou, *Food Chem.*, 2016, **196**, 106.
63. A. Devi, R. Buckow, Y. Hemar and S. Kasapis, *Food Chem.*, 2014, **156**, 243.
64. C. Chen, X. Huang, L. Wang, D. Li and B. Adhikari, *LWT-Food Sci. Technol.*, 2016, **74**, 528.
65. D. vanKrevelen and K. teNijenhuis, in *Properties of Polymers: Their Correlation with Chemical Structure; Their Numerical Estimation and*

Prediction from Additive Group Contributions, Elsevier, Amsterdam, 2009, pp. 383–500.
66. C. Verdier, J. Etienne, A. Duperray and L. Preziosi, *C. R. Phys.*, 2009, **10**, 790.
67. S. Alakalhunmaa, K. Parikka, P. Penttilä, T. Cuberes, S. Willför, L. Salmén and K. Mikkonen, *Cellulose*, 2016, **23**, 3221.
68. H. Dong, J. Snyder, D. Tran and J. Leadore, *Carbohydr. Polym.*, 2013, **95**, 760.
69. T. Köhnke, T. Elder, H. Theliander and A. Ragauskas, *Carbohydr. Polym.*, 2014, **100**, 24.
70. K. Mikkonen, K. Parikka, J. Suuronen, A. Ghafar, R. Serimaa and M. Tenkanen, *RSC Adv.*, 2014, **4**, 11884.
71. C. Xu, S. Willför, P. Holmlund and B. Holmbom, *Carbohydr. Polym.*, 2009, **75**, 498.
72. J. Singh, L. Kaur and O. McCarthy, *Food Hydrocolloids*, 2007, **21**, 1.
73. A. Tecante and J. Doublier, *Carbohydr. Polym.*, 1999, **10**, 221.
74. Y. Wanga, G. Xia, C. Wua, J. Sun, R. Song and W. Huang, *Carbohydr. Polym.*, 2015, **115**, 686.
75. J. Shim and S. Mulvaney, *J. Sci. Food Agric.*, 2001, **81**, 709.
76. K. Zheng, J. Zhang and J. Chen, *J. Appl. Polym. Sci.*, 2016, **133**, 43853.
77. I. Karadagli, M. Schestakow, B. Milow, T. Gries and L. Ratke, *J. Supercrit. Fluids*, 2015, **106**, 105.
78. M. Marin, R. Mallepally and M. McHugh, *J. Supercrit. Fluids*, 2014, **91**, 84.
79. C. Garcia-Gonzalez, E. Carenza, M. Zeng, I. Smirnova and A. Roig, *RSC Adv.*, 2012, **2**, 9816.
80. K. Ganesan and L. Ratke, *Soft Matter*, 2014, **10**, 3218.
81. C. Garcia-Gonzalez, M. Jin, J. Gerth, A. Alvarez-Lorenzo and I. Smirnova, *Carbohydr. Polym.*, 2015, **117**, 797.
82. K. Rinki, P. Dutta, A. Hunt and J. Clark, *Int. J. Polym. Mater*, 2011, **60**, 988.
83. Y. Cheng, L. Lu, W. Zhang, J. Shi and Y. Cao, *Carbohydr. Polym.*, 2012, **88**, 1093.
84. J. Lee and Y. Deng, *Soft Matter*, 2011, 7, 6034.
85. H. Sehaqui, Q. Zhou and L. Berglund, *Compos. Sci. Technol.*, 2011, **71**, 1593.
86. S. Hoepfner, L. Ratke and B. Milow, *Cellulose*, 2008, **15**, 121.
87. A. Svagan, A. Samir and L. Berglung, *Adv. Mater.*, 2008, **20**, 1263.
88. T. Köhnke, A. Lin, T. Elder, H. Theliander and A. Ragauskas, *Green Chem.*, 2012, **14**, 1864.

CHAPTER 12

Tuning Microscopic and Mechanical Properties of Bio-based Aerogels

JUN LIU,[a,b] STEFAN WILLFÖR[a] AND CHUNLIN XU*[a]

[a] Åbo Akademi University, Faculty of Science and Engineering, Laboratory of Wood and Paper Chemistry, Turku 20500, Finland; [b] Uppsala University, Department of Engineering Sciences, Nanotechnology and Functional Materials, Uppsala 75121, Sweden
*Email: cxu@abo.fi

12.1 Introduction

The majority of the chemicals and materials used today for constructing structural materials such as aerogels are synthetic products from oil or natural gas. These products have encountered problems due to the shortage of supply and of being detrimental to the environment. Therefore, it has become crucial to find natural alternatives instead of synthetic ones. Natural resources are available all over the world in such forms as animals, plants, bacteria, and other organisms, and can be used as building blocks for the design of biorenewable and biodegradable materials. In the design and construction of aerogels, bio-based polymers such as proteins and polysaccharides, which are usually extracted from natural resources, and synthetic polymers rooted from bioderived sources such as polylactic acid, are two major groups that can be classified as being bio-based or bio-derived.[1–4]

Green Chemistry Series No. 58
Biobased Aerogels: Polysaccharide and Protein-based Materials
Edited by Sabu Thomas, Laly A. Pothan and Rubie Mavelil-Sam
© The Royal Society of Chemistry 2018
Published by the Royal Society of Chemistry, www.rsc.org

Taking into account of the applications in biomedical treatments, to date a number of polymers have been utilized to construct scaffolds that can mimic extracellular matrices (ECM). Synthetic materials are strong, cheap, and reliable in terms of reproducibility, however, they do not share the same biochemical signatures expressed in the native tissues.[5] Polymers of natural origin are particularly attractive options due to their similarities with the ECM, as well as chemical versatility and biological performance.[6] The widely considered natural polymers include collagen, gelatin, alginate, chitosan, and cellulose. Many biopolymers are in fact components of the ECM and can regulate division, adhesion, differentiation, and migration of cells more favourably than synthetic polymers.[7] There are several reports that suggest the beneficial cellular responses of biopolymers and their performance over synthetic polymers for tissue regeneration.[8–10] Compared to synthetic polymers, bio-based ones have more functional groups and thus offer diverse and highly selective coupling chemistry. Traditional scaffolds from natural substances, such as collagen, fibronectin, and gelatinous protein mixtures, such as Matrigel, are widely used in tumor biology and therapeutic studies because they normally provide better support for cell functions than those made from synthetic materials.

Recently, cellulosic materials as a new group of bio-based sources have shown promising properties for constructing medical scaffolds.[11–13] In particular, microbial cellulose, due to its good biocompatibility and high mechanical strength, has been extensively investigated for biomedical applications on its own or in composites.[11,14,15] Recently, application of wood biopolymers for biomedical scaffolds has also been explored. Ferraz's group has studied the characteristics of nanocellulose films on the behaviour of *in vitro* cultured monocytes/macrophages. The same group also incorporated algae nanocellulose into wood nanocellulose to investigate the effect of structure of the composite films on their cytocompatibility.[16] Lou *et al.* successfully cultured human pluripotent stem cells in 3D nanocellulose hydrogels.[17] Nanocellulose was tailored to aerogels with different physical properties and showed that they support crucial cellular processes during cell growth and proliferation.[18,19] Thus, in this section, we will focus more on the cellulose-based aerogels.

To satisfy the desired applications in biomedical treatment, the bio-based scaffolds have to be able to act like a natural ECM. They need to possess the desired microscopic properties such as suitable pore sizes allowing interactions with cells of different sizes.[20,21] The scaffolds act as a carrier for the biologically active substances activated surface. Moreover, mechanical characteristics such as ECM rigidity and alignment or organization are believed to play essential roles in biological processes, such as cell differentiation, migration, and wound healing.[22] Therefore, choice of fabrication approaches and adoption of the strategies to tune the microscopic and mechanical properties of the scaffolds is essential and will be elaborated in the following discussion. Moreover, the approaches may have found a use for the extended range of bio-based polymers.

12.2 Tuning of Microscopic Properties

12.2.1 Control of the Inherent Properties of the Raw Materials

A selection of different raw materials for nanocellulose preparation and control of their surface chemistry has provided approaches to tune the microscopic properties of nanocellulose aerogels from sources.[23]

The surface chemistry of nanocelluloses directly affects their dispersion behavior and the microscopic structure of the resultant aerogels. The introduction of positively or negatively charged functional groups, for example carboxyl groups by TEMPO-mediated oxidation or periodate oxidation,[24,25] carboxymethyl groups by carboxymethylation,[26] sulfate groups by sulfation,[27] and trimethylammonium groups by quaternization[28] onto the cellulosic fibers not only facilitates the defibrillation during the nanocellulose preparation, but also endows the cellulose nanofibrils (CNF) stable dispersibility due to the electrostatic repulsion. Therefore, control of the surface charge properties of nanocellulose is a convenient approach to tune the microscopic and mechanical properties of aerogels.

Chen *et al.* prepared four types of nanocellulose with different surface charge properties and compared the morphology and microstructure of the resultant aerogels.[29] Nanocellulose prepared from a hardwood poplar using high-intensity ultrasonication without or with TEMPO-medicated oxidation (which introduces carboxyl groups) can be freeze-dried into an aerogel with 3D-web-like structures at <0.2 wt% concentration (Figure 12.1a and e). For comparison, the nanocellulose prepared from cotton fibers using HCl and H_2SO_4 hydrolysis can only be freeze-dried into an aerogel with a 2D-sheet-like structure regardless of the suspension concentration (Figure 12.1c and g). Similarly, Silva *et al.* reported the effect of surface charge density (degree of oxidation) of nanocellulose on the morphology of the resultant aerogels (Figure 12.2).[30] They reported that the higher surface charge density results in a higher porosity and pore size, and the higher the surface charge density of the aerogels, the more homogeneous the morphology.

12.2.2 Control of the Aerogel Processing Parameters

Formation of gels and removal of solvents from gels are key steps during aerogel production. Therefore, the control of parameters in these steps provides opportunities to tune the microscopic properties of the resultant aerogels.

Cellulose concentration, regeneration solvent systems, and drying methods play key roles in tuning the structure and pore size of cellulose aerogels. Dissolving or dissolution of cellulose in a proper solvent, followed by regeneration of the cellulose hydrogel and freeze-drying or supercritical CO_2 drying is one simple way to prepare a cellulose aerogel. Cai *et al.* prepared highly porous cellulose aerogels with large surface areas (400–500 $m^2\,g^{-1}$)

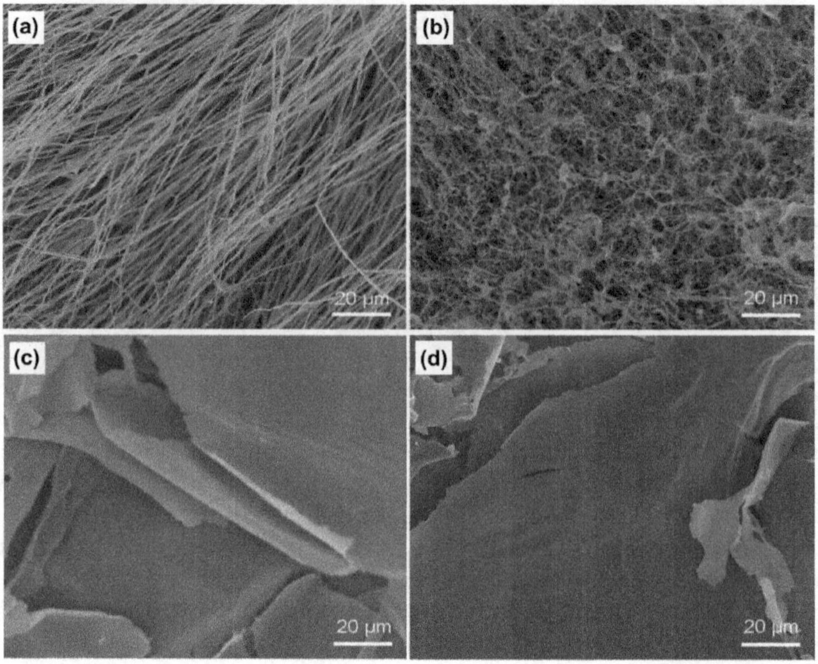

Figure 12.1 SEM images of aerogels obtained from 0.2 wt% nanocellulose fiber suspensions prepared from: (a) high-intensity ultrasonication treatment, (b) (TEMPO)-mediated oxidation, (c) HCl hydrolysis, and (d) H_2SO_4 hydrolysis.
Adapted from ref. 29 with permission from John Wiley and Sons, Copyright ©2014 WILEY-VCH Verlag GmbH & Co. KGaA, Weinheim.

from aqueous alkali hydroxide/urea solutions.[31] Both NaOH/urea and LiOH/urea aqueous solutions can dissolve different celluloses and yield transparent cellulose hydrogels and finally be dried to aerogels. However, the use of LiOH instead of NaOH allowed the formation of more transparent cellulose hydrogels and resulted in aerogels with homogeneous fine fibrils and a highly porous network structure. When all other conditions were the same (cellulose type and concentration, regeneration bath composition), the cellulose aerogel prepared from the LiOH/urea system showed higher specific surface areas.

Ionic liquids, for example 1-butyl-3-methylimidazolium chloride, are another group of cellulose solvents and have also been used to dissolve cellulose for aerogel preparation.[32,33]

Shrinkage and hornification commonly occur during drying of cellulose hydrogels due to the strong hydrogen bonding and high surface tension effects (capillary action), which result in a lower specific surface area and the collapse of the pore structure. Therefore, ethanol and tert-butanol are widely used to exchange with water in a gradient to minimize such shrinkage due to their lower surface tension. Generally, the solvent exchange should be

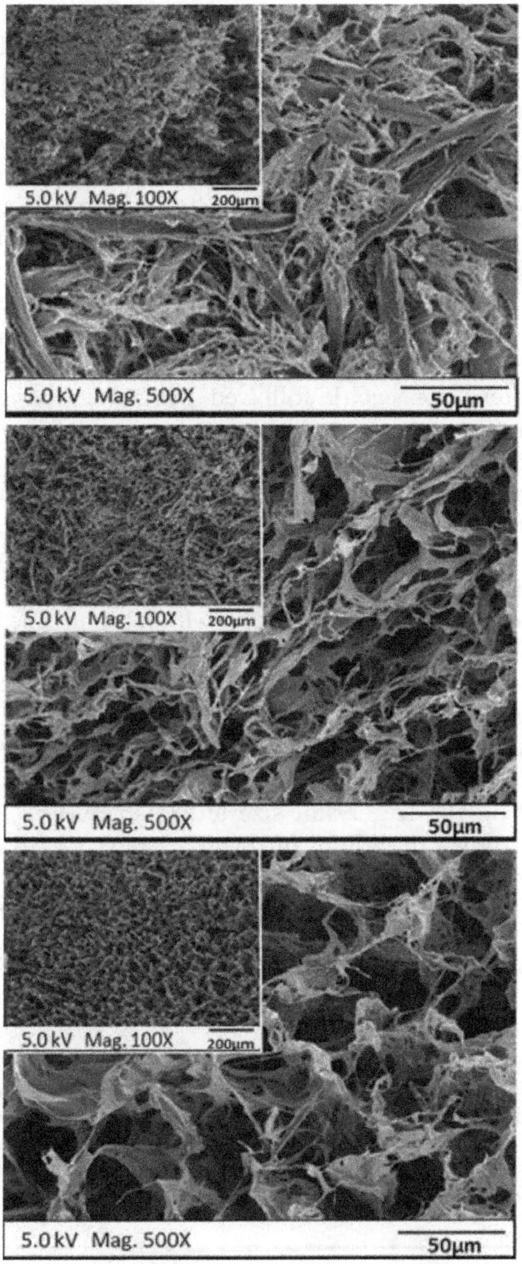

Figure 12.2 SEM images of aerogels prepared from cellulose nanofibrils (CNF) with different degrees of oxidation. From the top to the bottom: DO = 0, 0.1, 0.2. Adapted from *Cellulose*, A fundamental investigation of the microarchitecture and mechanical properties of tempooxidized nanofibrillated cellulose (NFC)based aerogels, **19**, 2012, 1945–1956, T. C. F. Silva, Copyright © Springer Science + Business Media B.V. 2012, with permission of Springer.[30]

applied in a gradient, for example starting from 15%, with incremental steps of 10–20%. Compared with normal freeze-drying from the frozen hydrogel, Ishida *et al.* reported that a tert-butanol solvent exchange followed by freeze-drying resulted in a higher specific surface area of tunicate aerogel (25 *versus* 130 $m^2 g^{-1}$).[34] A similar conclusion has also be reported by Jin *et al.*[35] and Cai *et al.*[31] using different types of cellulose materials. In addition to that, the solvent exchange procedure has also been found to affect the microscopic structure of the aerogel. Sehaqui *et al.* investigated the effects of the nanocellulose surface charge and solvent exchange process on the microscopic properties of the resultant aerogels.[36] Nanocellulose with (TO-CNF) and without (CNF) surface charge were subjected to solvent exchange in one step (water to tert-butanol) or in six steps (water to ethanol then to tert-butanol, three steps for each), followed by freeze-drying. As shown in Table 12.1 and as abovementioned, the introduction of surface charge onto the cellulose surface contributes to a higher specific surface area and thinner fibril diameter, regardless of the solvent exchange due to the strong electrostatic repulsion effect. Multi-stage solvent exchange allows the gradual replacement of water with tert-butanol, which has a lower surface tension and limited capillary effect, consequently minimizing the aggregation of CNF fibrils and thus yielding an aerogel with a higher specific surface area and thinner fibril diameter.

The drying approach is another important factor that affects the properties of aerogels. Currently, freeze-drying and supercritical CO_2 drying are widely used to prepare aerogels. For the freeze-drying, freezing is an important step that directly affects the pore structure and morphology of the aerogel by affecting the ice crystal size and distribution. Generally, a fast freezing, for example, in liquid nitrogen (−196 °C), or liquid propane (−188 °C), produces a smaller average pore size and a more homogenous pore structure.[37] Therefore, control of the freezing rate is one way to tune the aerogel microscopic structure. Jiang and Hsieh prepared nanocellulose aerogels by freezing at either at −20 °C in a freezer for 15 h, or −196 °C in liquid nitrogen for 10 min, followed by freeze-drying.[38] As shown in

Table 12.1 Specific surface area and average pore diameter of CNF aerogels, and estimated average nanofibril diameter based on an assumption of the cylindrical shape of nanofibrils. Adapted from *Composites Science and Technology*, **71**, H. Sehaqui, Q. Zhou and L. A. Berglund, Highporosity aerogels of high specific surface area prepared from nanofibrillated cellulose (NFC), 1593–1599, Copyright 2011, with permission from Elsevier.[36][a]

	Specific surface area ($m^2 g^{-1}$)	Average fibril diameter (nm)	Average pore diameter (nm)
CNF, 1-step	153	17.9	11.2
CNF, 6-steps	249	11.0	16.0
TO-CNF, 1-step	254	10.8	10.9
TO-CNF, 6-steps	284	9.6	11.8

[a]CNF: cellulose nanofibril.

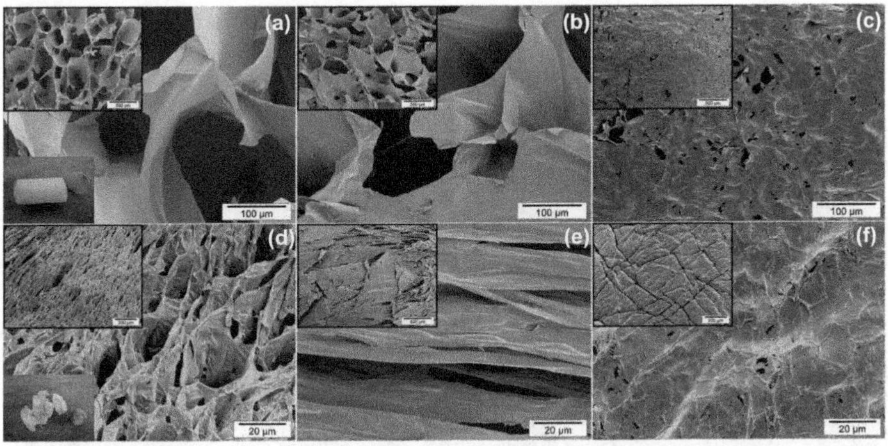

Figure 12.3 CNF aerogels obtained from freezing 0.6% CNF suspensions at −20 °C (a–c) and −196 °C (d–f) followed by freeze drying: (a and d) radial cross-sections, lower left insets are photographs of aerogels; (b and e) longitudinal sections; (c and f) external surfaces. The arrow in (f) points to one of many microfibrils on the surface. Inset scale bar = 500 µm (a, b, c and e) and 200 µm (d and f).
Adapted from ref. 38 with permission from the Royal Society of Chemistry.

Figure 12.3, the aerogels prepared by freezing at −20 °C (a–c) show thicker, larger, and more organized pore walls due to the slower formation of larger ice crystals. However, the aerogels prepared by fast freezing at −196 °C (d–f) have much smaller pores and thinner walls of less organized CNF, due to the rapid and smaller ice nucleation. Similarly, Pääkkö *et al.* compared cryogenic freeze-drying (fast liquid propane freezing at *ca.* −180 °C and drying in a freeze-dryer) and vacuum freeze-drying (freezing and drying in freeze-dryer).[39] As shown in Figure 12.4 and Table 12.2, the cryogenic freeze-drying allows the preservation of the 3D open fibrillar network structure, as well as long and entangled nanofibers with diameters of 5–10 nm with occasional thicker 'fibril bundles' of the aqueous gel, which consequently results in aerogels with only minor shrinkage and higher porosity and specific surface area. By contrast, the vacuum freeze-drying leads to the aggregation of nanofibers into sheet-like structures and higher shrinkage, as well as a much lower specific surface area. Lee and Deng further explored the important role of freezing by using the ice-template method to prepare cellulose aerogels with a unique layer structure and controllable channels.[40] They concluded that an increased unidirectional freezing temperature gradient can lower the ice crystals size and consequently results in a narrow channel structure. They also found that the bridge structures preferred to form between the channels of cellulose microfibrils due to the slower movement of the longer microfibrils (10–30 µm) during freezing, while fewer bridge structures were observed due to the shorter average length of the cellulose nanowhiskers.

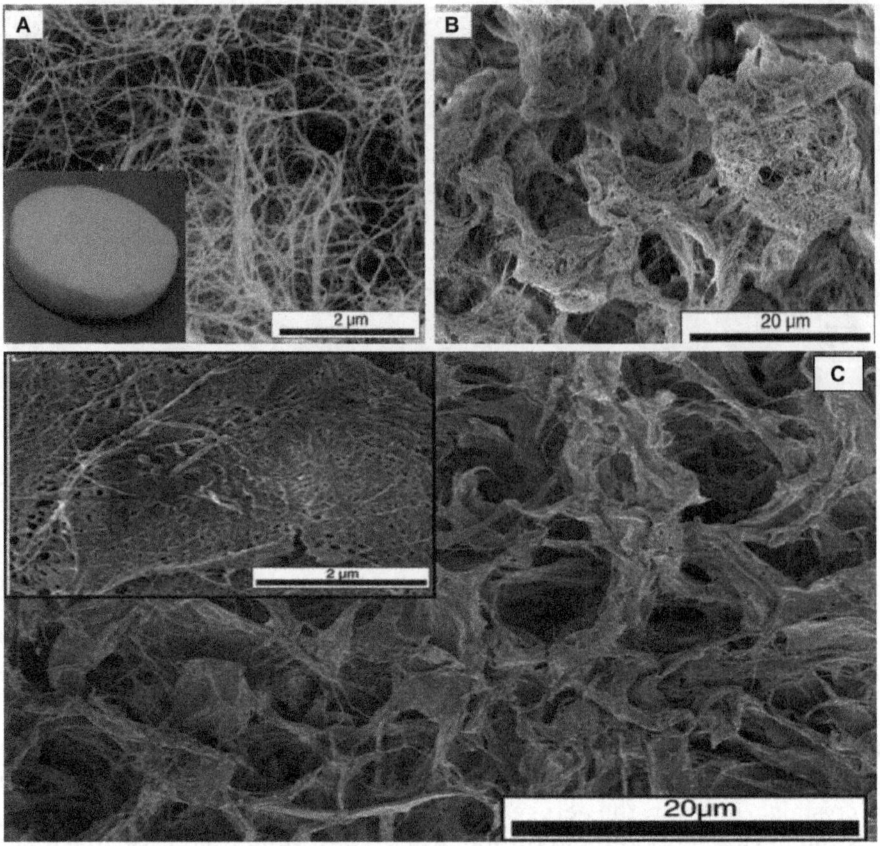

Figure 12.4 SEM images of the aerogels prepared by cryogenic freeze-drying (A, B) and vacuum freeze-drying (C). Adapted from ref. 39 with permission from the Royal Society of Chemistry.

Table 12.2 Compilation of aerogel properties.

Drying method	Drying shrinkage (%)	ρ g^{-1} cm^{-3}	Porosity (%)	BET surface area (m^2 g^{-1})
Cryogenic freeze-drying	7	0.02	~98	66
Vacuum pumping free-drying	20	0.03	~95–98	20

Supercritical drying is another process to transform gels into aerogels without destroying the delicate nanostructured pore network of the material, due to the absence of surface tension and capillary stress, allowing the preservation of the fine cellulose fibrils and their structure and resulting in aerogels with a high specific surface area. Buchtová and Budtova prepared aero-, cryo-, and xerogels by drying the cellulose solution (DMSO/EMImAc) using three different approaches: supercritical CO_2 drying, freeze-drying, and vacuum drying.[41] As shown in Figure 12.5, aerogels prepared from

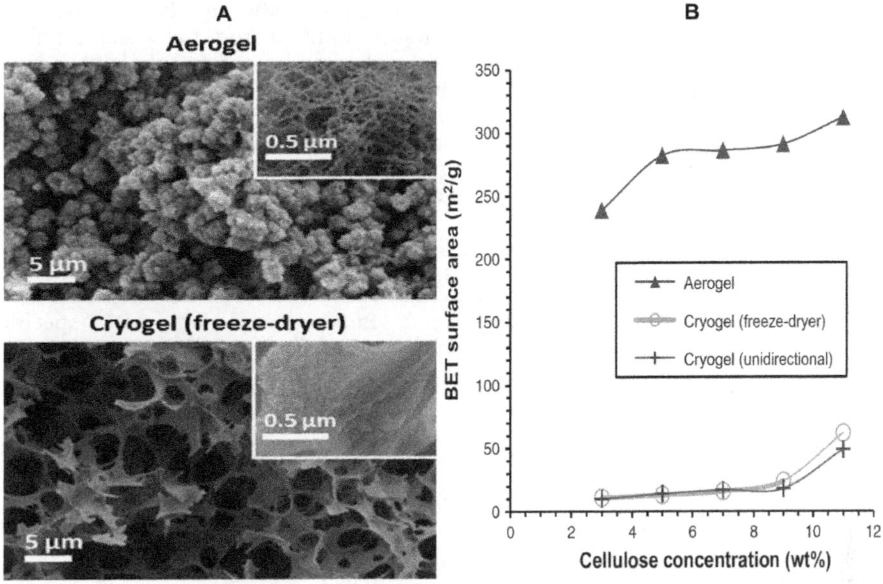

Figure 12.5 (A) SEM images of cellulose aero- and cryo-gel made from 5 wt% cellulose solution. (B) Specific surface area of cellulose aero- and cryogels as a function of cellulose concentration in solution. Adapted from *Cellulose*, Cellulose aero, cryo and xerogels: towards understanding of morphology control, **23**, 2016, 2585–2595, N. Buchtová, © Springer Science + Business Media Dordrecht 2016, with permission of Springer.[41]

supercritical drying have a nanostructured fibrillated texture and *ca.* 20 times higher specific surface area than those obtained from freeze-drying in all concentration ranges from 3–11%.

Control of the solute concentration is another convenient approach and has been widely applied to tune the microscopic structure of aerogels. It has been reported that the microscopic structure of cellulose aerogels can be converted from a 3D web-like network to 2D sheet-like skeletons by carefully controlling the cellulose concentration.[29,42,43] Generally, increasing the concentration of the solute will result in aerogels with a higher bulk density and porosity, while the specific surface area may depend on the cellulose dispersion or hydrogel preparation methods. Table 12.3 summarizes the effect of cellulose concentration on the pore properties of aerogels. Cai *et al.* prepared cellulose aerogels from filter paper by first dissolving cellulose in aqueous alkali hydroxide-urea solution and then regenerating in ethanol, followed by supercritical CO_2 drying.[31] Buchtová and Budtova prepared cellulose aerogels from microcrystalline cellulose using ionic liquid (DMSO/EMImAc) and dried using freeze-drying and supercritical CO_2 drying.[41] Cervin *et al.* prepared cellulose aerogels from carboxymethylated nanocellulose by the normal freeze-drying method.[44] They all reported that with the increase of cellulose concentration, pore porosity decreased and

Table 12.3 Effects of cellulose concentration and preparation methods on the pore properties of aerogels. Adapted and modified from *Cellulose*, Cellulose aero, cryo and xerogels: towards understanding of morphology control, **23**, 2016, 2585–2595, N. Buchtová and T. Budtova, © Springer Science + Business Media Dordrecht 2016, with permission of Springer,[41] *Cellulose*, Ultra porous nanocellulose aerogels as separation medium for mixtures of oil/water liquids, **19**, 2011, 401–410, N. T. Cervin, C. Aulin, P. T. Larsson and L. Wågberg, © Springer Science + Business Media B. V. 2011, with permission of Springer,[44] and from ref. 31 with permission from John Wiley and Sons, Copyright © 2008 WILEY-VCH Verlag GmbH & Co. KGaA, Weinheim.[a]

Sample	Cellulose concentration (%)	Volume shrinkage (%)	Bulk density ($g\,cm^{-3}$)	Porosity (%)	BET specific surface area ($m^2\,g^{-1}$)
Aerogel	3	66	0.126	92	239
	5	49	0.130	91	282
	7	38	0.150	89	286
	9	33	0.175	88	291
	11	21	0.215	86	312
Cryogel (unidirectional freezing)	3	19	0.050	97	10
	5	10	0.073	95	14
	7	7	0.099	93	17
	9	20	0.132	91	18
	11	12	0.163	89	49
Cryogel (freeze-dryer)	3	Sample broken	0.053	96	11
	5	14	0.068	95	13
	7	14	0.099	93	16
	9	23	0.132	91	24
	11	27	0.164	89	62
CNF aerogel	0.5	—	0.004	99.8	42
	1	—	0.008	99.5	16
	1.5	—	0.011	99.3	14
	2	—	0.014	99.1	11
Cellulose aerogel (LiOH/urea)	1	—	0.03	97.0	375
	2	—	0.05	96.5	402
	3	—	0.08	94.6	430
	4	—	0.10	93.4	416
	5	—	0.12	92.1	396
	6	—	0.13	90.4	389
	7	—	0.14	89.7	329

[a]CNF: cellulose nanofibril.

bulk density increased, which has also been reported by Sehaqui *at al.*,[36] Lee *et al.*,[40] Yang and Cranston,[45] Hoepfner *et al.*,[46] and Aulin *et al.*[47] However, the results from Cai *et al.*[31] show that the specific surface area increased until a concentration of 3%, Buchtová and Budtova[41] found that the specific surface area increased with the cellulose concentration, while Cervin *et al.*[44] showed opposite results. Therefore, care should be taken when comparing and interpreting the pore properties of aerogels, especially the specific surface area.

Different from most hydrogel-aerogel pathways, a novel film-hydrogel-aerogel approach was proposed to tune the microscopic structure of nanocellulose aerogels by Liu *et al.*[19,48] The nanocellulose free-standing films were prepared first, followed by swelling of the films to form hydrogels with tunable swelling degree by controlling swelling time and the parameters of the swelling media, for example temperature, ionic strength, and pH. After freeze-drying, nanocellulose aerogels with tunable pore properties and microscopic structures could be prepared. As discussed above and shown in Table 12.4 and Figure 12.6, the surface charge density of the nanocellulose is a key parameter to tune the structure of the aerogels. Changing the pH and ionic strength of the swelling media will cause the association and dissociation of the surface charge (carboxylic groups), which directly controls the swelling degree of the hydrogel, the pore properties and structure of the resultant aerogel. In addition, during the nanocellulose film preparation, hot-pressing will cause irreversible hornification, which will limit the swelling of the hydrogels and thus provide an opportunity to tune the pore properties of the aerogels.

In most case, the removal of solvents from the gel to form an aerogel is a time-consuming process. Strategies to speed up the drying procedure will significantly contribute to the application of aerogels. Judicious control of the gel concentration, freeze-drying or supercritical CO_2 drying conditions, and solvent systems may provide opportunities to shorten the manufacturing time. However, one should keep in mind that these conditions can also directly affect the porous structure of the final product.

Table 12.4 Specific surface area (SSA), pore specific volume (PSV), density, and porosity of the cellulose nanofibrils (CNF) aerogels. Reprinted from *Carbohydrate Polymers*, **148**, J. Liu, F. Cheng, H. Grénman, S. Spoljaric, J. Seppälä, J. E. Eriksson, S. Willför and C. Xu, Development of nanocellulose scaffolds with tunable structures to support 3D cell culture, 259–271, Copyright 2016, with permission from Elsevier.[19]

	SSA ($m^2 g^{-1}$)[a]	PSV ($cm^3 g^{-1}$)	Density ($kg m^{-3}$)	Porosity (%)
HC-WP-25[b]	308.0	0.73	6.4 ± 0.5	99.6 ± 0.03
HC-WP-50	297.6	0.83	4.2 ± 0.2	99.7 ± 0.01
HC-WP-25 (pH3)[c]	279.4	0.96	95.0 ± 2.9	93.5 ± 0.19
HC-HP-25	267.4	0.86	8.8 ± 0.3	99.4 ± 0.02
LC-WP-25	242.1	0.42	17.8 ± 3.7	98.8 ± 0.24
LC-WP-50	203.6	0.46	16.7 ± 3.8	98.9 ± 0.25
LC-WP-25 (pH3)[c]	157.5	0.49	179.6 ± 21.4	88.0 ± 1.43
LC-HP-25	175.9	0.70	18.5 ± 1.3	98.8 ± 0.09

[a]Values were calculated using the skeletal density $\rho_c = 1.50$ which was measured by Helium pycnometry.

[b]The samples were named sequentially according to the charge density of the CNF (low charge and high charge, LC and HC); the CNF film press processing approaches (wet press and hot press, WP and HP); the temperature (25, 37, 50 °C) and pH value (3.0, 7.0, 10.0) of the swelling media.

[c]Samples were swollen in a media with a pH of 3.0. If not specified, other samples were swollen in a neutral environment.

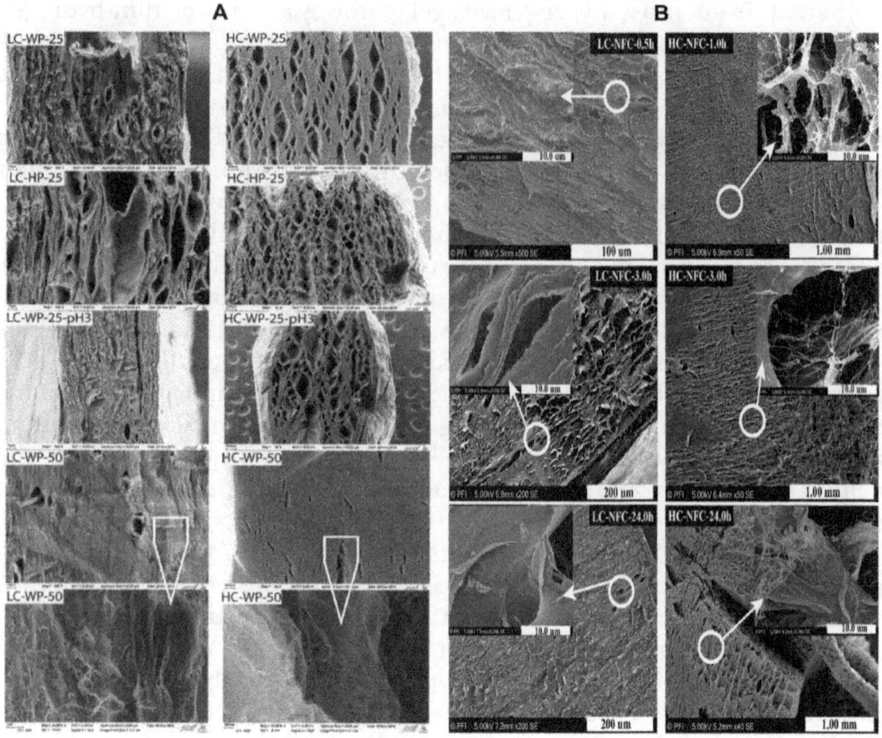

Figure 12.6 Tuning the microscopic structure of nanocellulose by controlling the surface charge density (high charge, HC; low charge, LC), film press technique (wet press, WP; hot-press, HP), film swelling conditions (temperature, pH, and time).
Adapted from *Cellulose*, Hemicellulose-reinforced nanocellulose hydrogels for wound healing application, **23**, 2016, 3129–3143, J. Liu, © Springer Science + Business Media Dordrecht 2016, with permission of Springer,[18] and from *Carbohydrate Polymers*, **148**, J. Liu, F. Cheng, H. Grénman, S. Spoljaric, J. Seppälä, J. E. Eriksson, S. Willför and C. Xu, Development of nanocellulose scaffolds with tunable structures to support 3D cell culture, 259–271, Copyright 2016, with permission from Elsevier.[19]

12.3 Tuning of the Mechanical Properties

The nanocelluloses are known for their high modulus and the aerogels from nanocellulose offers an alternative to conventional aerogels in terms of renewability, being lightweight and flexible, and the possibility of adapting various functionalities. However, insufficient mechanical strength and pore structural stability of nanocellulose aerogels in aqueous conditions limit their application. Enhancing or tuning the mechanical properties of nanocellulose aerogels is of importance for a wide range of applications.

12.3.1 Control of the Density and Surface Charge of the Nanocellulose Aerogels

Generally, the mechanical properties of nanocellulose aerogels directly relate to the density of the aerogels. Sehaqui *et al.* investigated the relationship between the mechanical properties of nanocellulose aerogels and their densities and concluded that the compressive modulus changes by two orders of magnitude with the aerogel density (as shown in Figure 12.7 and Table 12.5).[36,49] Therefore, control of the density provides a simple and efficient approach to tune the mechanical properties of nanocellulose aerogels. The density of nanocellulose aerogels can be controlled by changing the concentration of the nanocellulose dispersion (Table 12.6), by controlling the swelling degree of the hydrogel, by controlling the surface charge density

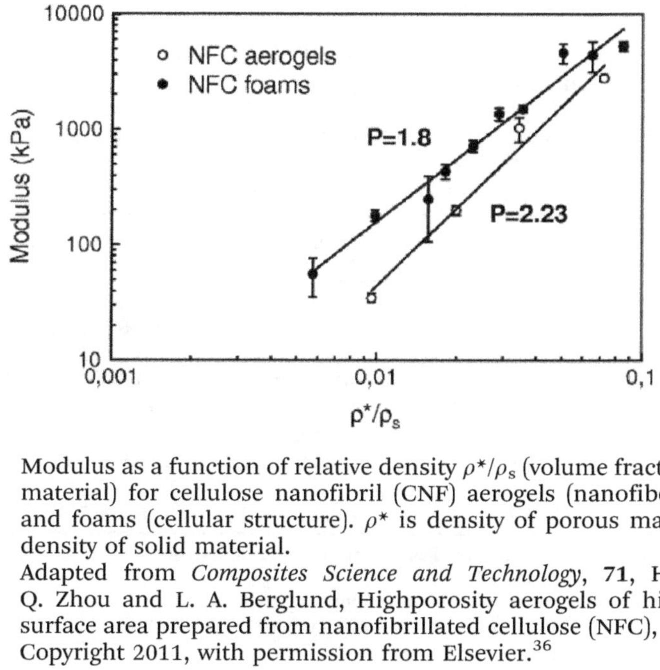

Figure 12.7 Modulus as a function of relative density ρ^*/ρ_s (volume fraction of solid material) for cellulose nanofibril (CNF) aerogels (nanofiber network) and foams (cellular structure). ρ^* is density of porous material, ρ_s is density of solid material.
Adapted from *Composites Science and Technology*, **71**, H. Sehaqui, Q. Zhou and L. A. Berglund, Highporosity aerogels of high specific surface area prepared from nanofibrillated cellulose (NFC), 1593–1599, Copyright 2011, with permission from Elsevier.[36]

Table 12.5 Mechanical properties of cellulose nanofibrils (CNF) aerogels in compression.[a]

Density (kg m^{-3})	14	29	50	105
Porosity (%)	99.0	98.0	96.6	92.8
Modulus (kPa)	34.9 ± 3.0	199 ± 19	1030 ± 240	2800 ± 155
Strength (kPa)	3.20 ± 0.4	24.4 ± 3.9	69.0 ± 8	238 ± 21
Energy absorption (kJ m^{-3})	10.8 ± 0.8	68.0 ± 1	233[a]	720 ± 20

[a]Only one sample reached 70% strain.

Table 12.6 Mechanical properties of microfibrillar cellulose (MFC) foams with different densities. Relative density is $(\rho^*/\rho_{cellulose})$ and the porosity P is $P = 1 - (\rho^*/\rho_{cellulose})$.

MFC concentration (wt%)	0.7	1.2	1.9	2.2	2.8	3.5	4.2	6.0	7.7	10.0
Density (kg m^{-3})	7	12	19	22	28	35	43	61	79	103
Relative density[a]	0.005	0.008	0.013	0.015	0.019	0.023	0.029	0.041	0.053	0.069
Porosity (%)	99.5	99.2	98.7	98.5	98.1	97.7	97.2	95.9	94.7	93.1
Modulus (kPa)	56	180	249	435	718	1360	1510	4650	4450	5310
Yield stress (kPa)	7.8	29.6	26.1	31.9	70.1	92.7	135.1	194	273	515.6
Densification strain (%)	99.4	98.6	98.3	97.9	97.4	96.8	96.1	94.4	92.8	90.5
Energy absorption (kJ m^{-3})	8.4	22.9	36.7	55.5	92	131	210	419.4	538.8	927.7

[a]Calculated using cellulose (1500 kg m^{-3}) as a reference.

Table 12.7 Mechanical properties of the cellulose nanofibrils (CNF) hydrogels and the resultant aerogels. Adapted from Liu *et al.* Reprinted from *Carbohydrate Polymers*, **148**, J. Liu, F. Cheng, H. Grénman, S. Spoljaric, J. Seppälä, J. E. Eriksson, S. Willför and C. Xu, Development of nanocellulose scaffolds with tunable structures to support 3D cell culture, 259–271, Copyright 2016, with permission from Elsevier.[19]

Sample	Density (kg m^{-3})	Compressive Stress (kPa)	Strain (%)	Modulus (kPa)
HC-WP-25	6.4 ± 0.5	28.0 ± 6.6	195.8 ± 10.5	24.7 ± 0.7
HC-WP-50	4.2 ± 0.2	34.2 ± 3.1	251.8 ± 21.1	28.5 ± 3.4
HC-WP-25 (pH3)[a]	95.0 ± 2.9	35.7 ± 5.4	198.1 ± 13.7	47.2 ± 3.6
HC-HP-25	8.8 ± 0.3	31.8 ± 4.9	239.5 ± 5.8	26.5 ± 1.5
LC-WP-25[b]	17.8 ± 3.7	60.3 ± 5.9	91.8 ± 7.8	65.5 ± 3.5
LC-WP-50[b]	16.7 ± 3.8	94.4 ± 14.9	186.7 ± 11.0	100.0 ± 6.2
LC-WP-25 (pH3)[ab]	179.6 ± 21.4	104.41 ± 2.76	179.9 ± 7.6	88.1 ± 5.5
LC-HP-25[b]	18.5 ± 1.3	62.80 ± 5.10	156.4 ± 3.3	66.9 ± 4.0

[a]Samples swelled in media with a pH of 3.0. If not specified, other samples swelled in a neutral environment.
[b]Values were obtained from the measurement of two overlapped specimens to reach the thickness limit (*ca.* > 1.0 mm) of the DMA measurement. If not specified, other samples were determined individually.

of nanocellulose, and by using different drying approaches as abovementioned.[19,49,50]

The effect of the surface charge on the mechanical properties of nanocellulose aerogels has been reported by Silva *et al.*,[30] and Liu *et al.*[19] As shown in Table 12.7 and Figure 12.8, it seems that controversial results were reported. Figure 12.8 shows that the compressive stress increased with the introduction of surface charge density, while Table 12.7 shows that the compressive stress of the aerogels decreased for the higher charged nanocellulose. However, when comparing the effect of nanocellulose surface

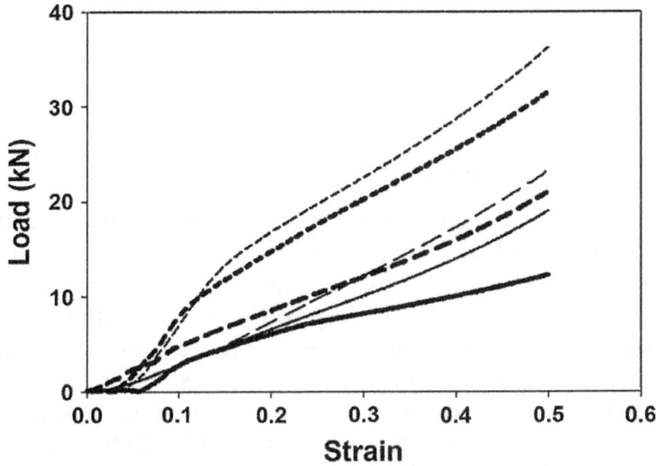

Figure 12.8 Compressive stress–strain of the cellulose nanofibril (CNF)-based aerogel. Bold lines represent the aerogels without HAp addition. Solid line: AG DO = 0; long dashed line: AG DO = 0.1; short dashed line: AG DO = 0.2.
Adapted from *Cellulose*, A fundamental investigation of the microarchitecture and mechanical properties of tempooxidized nanofibrillated cellulose (NFC)based aerogels, **19**, 2012, 1945–1956, T. C. F. Silva, Copyright © Springer Science + Business Media B.V. 2012, with permission of Springer.

Table 12.8 The compressive strength of cross-liked and untreated cellulose nanofibrils (CNF) aerogel in the dry and wet state. Adapted from ref. 50, Copyright 2015 The Authors, under the terms of the CC BY 4.0 licence, https://creativecommons.org/licenses/by/4.0/.[a,b]

	Compressive strength (kPa)	
	Dry state	Wet state
Cross-linked CNF aerogel	411 (44)	180 (16)
Untreated CNF aerogel	317 (35)	120 (13)

[a]S.D. standard deviation.
[b]CNF: cellulose nanofibril.

charge density on the mechanical properties of aerogels, the density of the aerogels should also be taken into consideration. Silva *et al.* concluded that the introduction of surface charge on the nanocellulose leads to more compression-resistance of the aerogels with similar density.[30] Liu *et al.* found that the higher surface charge density on the nanocellulose causes a higher swelling degree of the hydrogel and lower density of the resultant aerogel, consequently leading to lower compression-resistance.[19] Therefore, altering the surface charge density of nanocellulose provides an indirect way to tune the mechanical properties of aerogels, for example by affecting the hydration ability of the hydrogel to achieve the tuning of the aerogel density, and consequently tuning the mechanical properties. However, consideration of the aerogel density should be taken into account when interpreting the mechanical properties.

12.3.2 Chemical Crosslinking

The hydrogen bond between nanocellulose fibrils is the main mechanical resistance contributor of nanocellulose aerogel, which is, however, weak and easy to be destroyed when it comes into contact with water. Crosslinking by forming a covalent or ionic bond under different conditions such as heat, pressure, changes of pH, or radiation treatment has been applied to increase the stiffness and strength of cellulosic materials. Chemical crosslinkers, such as maleic acid anhydride, 1,2,3,4-butanetetracarboxylic acid, polyamide-epichlorohydrin resin, citric acid, and adipic acid dihydrazide have been applied to form different covalent bonds, so as to reinforce nanocellulose aerogels with desired mechanical properties.[45,50–52] As shown in Table 12.8, both the dry and wet compressive strength of the aerogel were improved after crosslinking with maleic acid and hypophosphite, offering the shape and structure integrity of nanocellulose aerogels for the desired application.

12.3.3 Reinforcing with Introducing of Other Composite Components

Reinforcement is another approach to add mechanical properties, for example strength and stiffness, to the nanocellulose aerogels. Cellulose and hemicellulose are closely associated in the plant cell wall and contribute to the rigidity of the cellulose framework in nature.[53] Therefore, hemicelluloses such as xyloglucan (XG) have been applied as a reinforcement component to improve the nanocellulose foams/aerogels by mimicking the plant primary cell wall. As shown in Table 12.9, incorporation of XG significantly increased the modulus and yield stress of the aerogels due to the high affinity of XG with the cellulose. Incorporation of inorganic fillers, such as silica, montmorillonite, and hydroxyapatite, can also enhance the mechanical properties of nanocellulose aerogels (Table 12.10).

Table 12.9 Mechanical properties of microfibrillar cellulose/xyloglucan (MFC/XG) foams (porosity around 98.5%). Adapted from ref. 49 with permission from the Royal Society of Chemistry.

MFC/XG	Density ($kg\,m^{-3}$)	Modulus (kPa)	Yield stress (kPa)
100/0	22.0	440	31.9
90/10	19.8	720	79.6
80/20	20.4	970	97.5
70/30	20.2	1470	112.5

Table 12.10 Properties of the cellulose SiO_2/cellulose aerogels prepared from a water glass solution with different concentrations. Adapted from *Colloids and Surfaces A: Physicochemical and Engineering Aspects*, **439**, S. Liu, T. Yu, N. Hu, R. Liu and X. Liu, High strength cellulose aerogels prepared by spatially confined synthesis of silica in bioscaffolds, 159–166, Copyright 2013, with permission from Elsevier.[54]

Sample	Silica content[a] (wt%)	Density ($g\,cm^{-3}$)	Porosity (%)	Young's modulus[b] (MPa)
CA	—	0.218 ± 0.38	82.21 ± 3.27	49.36 ± 4.17
CWG-5	4.16	0.202 ± 0.18	73.37 ± 4.03	84.67 ± 5.03
CWG-7.5	7.37	0.215 ± 0.22	68.04 ± 2.98	95.67 ± 4.69
CWG-10	9.74	0.212 ± 0.34	66.74 ± 3.68	166.41 ± 7.55
CWG-20	23.96	0.228 ± 0.17	65.11 ± 4.16	204.83 ± 8.31

[a]Data from TG analysis.
[b]Young's modulus of the aerogels.

12.4 Concluding Remarks

Bio-based polymers, such as proteins and polysaccharides, are present in abundant natural resources. Benefits have already been seen when they are used for the design of aerogels to replace petroleum-based chemicals. Due to the inherent properties of bio-based polymers, for example good biocompatibility, these natural polymers shall find high value applications such as in the tailoring of bioactive scaffolds for biomedical treatments. Therefore, considerations need to be taken to meet the requirements of extracellular matrices. For example, microscopic properties and mechanical characteristics are particularly important parameters that need to be controlled. In this chapter, different available approaches to control these properties of aerogels have been elaborated. However, how aerogels that possess these resultant properties would satisfy the application performance is not the focus of the current task, but it is essentially important.

Acknowledgements

The support from the Academy of Finland Project TuneScaffold (0246312-1), EU 6 Multihybrids (IP 026685-2), Nanofire (NMP3-CT 2004-505637) projects, Hungarian Research Found OTKA T049121, Fund of European Union and Hungarian state GVOP/3.1.1.-2004-0531/3.0, Public Benefit Association of Sciences and Sport of the Budapest University of Technology and Economics are acknowledged. This work is part of the activities at the Johan Gadolin Process Chemistry Centre, a Centre of Excellence appointed by Åbo Akademi University.

References

1. A. Liu, L. Medina and L. A. Berglund, *ACS Appl. Mater. Interfaces*, 2017, **9**, 6453–6461.
2. N. Yan, Y. Zhou, Y. Zheng, S. Qiao, Q. Yu, Z. Li and H. Lu, *RSC Adv.*, 2015, **5**, 97467–97476.

3. D. A. Schiraldi, in *Green Polymer Chemistry: Biobased Materials and Biocatalysis*, ed. H. N. Cheng, R. A. Gross and P. B. Smith, American Chemical Society, Washington, DC, 2015, vol. 1192, pp. 471–482.
4. G. Dharunya, N. Duraipandy, R. Lakra, P. S. Korapatti, R. Jayavel and M. S. Kiran, *Biomed. Mater.*, 2016, **11**, 045011.
5. S. V. Murphy and A. Atala, *Nat. Biotechnol.*, 2014, **32**, 773–785.
6. N. C. Hunt and L. M. Grover, *Biotechnol. Lett.*, 2010, **32**, 733–742.
7. D. A. Grande, C. Halberstadt, G. Naughton, R. Schwartz and R. Manji, *J. Biomedical Mater. Res.*, 1997, **34**, 211–220.
8. T. J. Klein, S. C. Rizzi, K. Schrobback, J. C. Reichert, J. E. Jeon, R. W. Crawford and D. W. Hutmacher, *Soft Matter*, 2010, **6**, 5175.
9. S. Nishimoto, M. Takagi, S. Wakitani, T. Nihira and T. Yoshida, *J. Biosci. Bioeng.*, 2005, **100**, 123–126.
10. P. B. Malafaya, G. A. Silva and R. L. Reis, *Adv. Drug Delivery Rev.*, 2007, **59**, 207–233.
11. W. K. Czaja, D. J. Young, M. Kawecki and R. M. Brown, Jr., *Biomacromolecules*, 2007, **8**, 1–12.
12. Y. Jia, J. B. Fei, Y. Cui, Y. Yang, L. A. Gao and J. B. Li, *Chem. Commun.*, 2011, **47**, 1175–1177.
13. V. M. Gaspar, J. G. Marques, F. Sousa, R. O. Louro, J. A. Queiroz and I. J. Correia, *Nanotechnology*, 2013, **24**, 275101.
14. J. D. Fontana, A. M. Desouza, C. K. Fontana, I. L. Torriani, J. C. Moreschi, B. J. Gallotti, S. J. Desouza, G. P. Narcisco, J. A. Bichara and L. F. X. Farah, *Appl. Biochem. Biotechnol.*, 1990, **24–25**, 253–264.
15. D. Wanna, C. Alam, D. M. Toivola and P. Alam, *Adv. Nat. Sci.:Nanosci. Nanotechnol.*, 2013, **4**, 045002.
16. K. Hua, D. O. Carlsson, E. Alander, T. Lindstrom, M. Stromme, A. Mihranyan and N. Ferraz, *RSC Adv.*, 2014, **4**, 2892–2903.
17. Y. R. Lou, L. Kanninen, T. Kuisma, J. Niklander, L. A. Noon, D. Burks, A. Urtti and M. Yliperttula, *Stem Cells Dev.*, 2014, **23**, 380–392.
18. J. Liu, G. Chinga-Carrasco, F. Cheng, W. Xu, S. Willför, K. Syverud and C. Xu, *Cellulose*, 2016, **23**, 3129–3143.
19. J. Liu, F. Cheng, H. Grenman, S. Spoljaric, J. Seppälä, J. E. Eriksson, S. Willför and C. L. Xu, *Carbohydr. Polym.*, 2016, **148**, 259–271.
20. M. W. Tibbitt and K. S. Anseth, *Biotechnol. Bioeng.*, 2009, **103**, 655–663.
21. E. K. F. Yim and K. W. Leong, *J. Biomater. Sci., Polym. Ed.*, 2005, **16**, 1193–1217.
22. P. F. Lu, K. Takai, V. M. Weaver and Z. Werb, *Cold Spring Harbor Perspect. Biol.*, 2011, **3**, a005058.
23. C. J. Chirayil, L. Mathew and S. Thomas, *Rev. Adv. Mater. Sci.*, 2014, **37**, 20–28.
24. T. Saito, S. Kimura, Y. Nishiyama and A. Isogai, *Biomacromolecules*, 2007, **8**, 2485–2491.
25. A. Rees, L. C. Powell, G. Chinga-Carrasco, D. T. Gethin, K. Syverud, K. E. Hill and D. W. Thomas, *BioMed. Res. Int.*, 2015, **2015**, 925757.

26. L. Wagberg, G. Decher, M. Norgren, T. Lindstrom, M. Ankerfors and K. Axnas, *Langmuir*, 2008, **24**, 784–795.
27. N. Lin and A. Dufresne, *Nanoscale*, 2014, **6**, 5384.
28. A. Pei, N. Butchosa, L. A. Berglund and Q. Zhou, *Soft Matter*, 2013, **9**, 2047.
29. W. Chen, Q. Li, Y. Wang, X. Yi, J. Zeng, H. Yu, Y. Liu and J. Li, *ChemSusChem*, 2014, **7**, 154–161.
30. T. C. F. Silva, Y. Habibi, J. L. Colodette, T. Elder and L. A. Lucia, *Cellulose*, 2012, **19**, 1945–1956.
31. J. Cai, S. Kimura, M. Wada, S. Kuga and L. Zhang, *ChemSusChem*, 2008, **1**, 149–154.
32. M. Deng, Q. Zhou, A. Du, J. van Kasteren and Y. Wang, *Mater. Lett.*, 2009, **63**, 1851–1854.
33. O. Aaltonen and O. Jauhiainen, *Carbohydr. Polym.*, 2009, **75**, 125–129.
34. O. Ishida, D. Y. Kim, S. Kuga, Y. Nishiyama and R. M. Brown, *Cellulose*, 2004, **11**, 475–480.
35. H. Jin, Y. Nishiyama, M. Wada and S. Kuga, *Colloids Surf., A*, 2004, **240**, 63–67.
36. H. Sehaqui, Q. Zhou and L. A. Berglund, *Compos. Sci. Technol.*, 2011, **71**, 1593–1599.
37. A. J. Svagan, M. A. S. A. Samir and L. A. Berglund, *Adv. Mater.*, 2008, **20**, 1263–1269.
38. F. Jiang and Y.-L. Hsieh, *J. Mater. Chem. A*, 2014, **2**, 350–359.
39. M. Pääkkö, J. Vapaavuori, R. Silvennoinen, H. Kosonen, M. Ankerfors, T. Lindström, L. A. Berglund and O. Ikkala, *Soft Matter*, 2008, **4**, 2492.
40. J. Lee and Y. Deng, *Soft Matter*, 2011, 7, 6034.
41. N. Buchtová and T. Budtova, *Cellulose*, 2016, **23**, 2585–2595.
42. W. Chen, H. Yu, Q. Li, Y. Liu and J. Li, *Soft Matter*, 2011, 7, 10360.
43. C. Salas, T. Nypelö, C. Rodriguez-Abreu, C. Carrillo and O. J. Rojas, *Curr. Opin. Colloid Interface Sci.*, 2014, **19**, 383–396.
44. N. T. Cervin, C. Aulin, P. T. Larsson and L. Wågberg, *Cellulose*, 2011, **19**, 401–410.
45. X. Yang and E. D. Cranston, *Chem. Mater.*, 2014, **26**, 6016–6025.
46. S. Hoepfner, L. Ratke and B. Milow, *Cellulose*, 2007, **15**, 121–129.
47. C. Aulin, J. Netrval, L. Wågberg and T. Lindström, *Soft Matter*, 2010, **6**, 3298.
48. J. Liu, G. Chinga-Carrasco, F. Cheng, W. Y. Xu, S. Willför, K. Syverud and C. L. Xu, *Cellulose*, 2016, **23**, 3129–3143.
49. H. Sehaqui, M. Salajková, Q. Zhou and L. A. Berglund, *Soft Matter*, 2010, **6**, 1824.
50. C. H. Kim, H. J. Youn and H. L. Lee, *Cellulose*, 2015, **22**, 3715–3724.
51. M. Hamedi, E. Karabulut, A. Marais, A. Herland, G. Nyström and L. Wågberg, *Angew. Chem., Int. Ed.*, 2013, **52**, 12038–12042.
52. W. Zhang, Y. Zhang, C. Lu and Y. Deng, *J. Mater. Chem.*, 2012, **22**, 11642.
53. J. Liu, Ph.D. thesis, Åbo Akademi University, 2016.
54. S. Liu, T. Yu, N. Hu, R. Liu and X. Liu, *ColloidsSurf., A*, 2013, **439**, 159–166.

CHAPTER 13

Applications of Aerogels in Aerospace and Packaging

CINTIL JOSE CHIRAYIL,*[a,b] JITHIN JOY,[a,b] LOVELY MATHEW[c] AND SABU THOMAS[b,d]

[a] Newman College, Thodupuzha East, Idukki, Kerala 685585, India; [b] International and Inter University Centre for Nanoscience and Nanotechnology, Mahatma Gandhi University, Kottayam, Kerala 686560, India; [c] Viswajyothi College of Engineering and Technology, Vazhakulam, Muvattupuzha, Kerala 686670, India; [d] School of Chemical Sciences, Mahatma Gandhi University, Kottayam, Kerala 686560, India
*Email: cintiljose@gmail.com

13.1 Introduction

Aerogel technology provides high added-value lightweight materials with outstanding surface area and open porosity, suitable for loading with active compounds.[1] Efforts have been traditionally focused on aerogel development with a wide range of applications in different fields, for example, aeronautics, biomedicine, construction, environmental remediation or agriculture.[2] Kistler first described the preparation of aerogels from polysaccharides (agar, nitrocellulose and cellulose) in 1931.[3] Since then, many efforts have been focused on aerogel production from polysaccharide-based precursors. However, research on these aerogels addressing biotechnological and pharmaceutical applications has only recently been started. For example, organic aerogels from Federal Drug Administration (FDA) and European Medicines Agency (EMEA) approved bio-based polysaccharides can afford the challenge of acting as a biocompatible plus biodegradable

Green Chemistry Series No. 58
Biobased Aerogels: Polysaccharide and Protein-based Materials
Edited by Sabu Thomas, Laly A. Pothan and Rubie Mavelil-Sam
© The Royal Society of Chemistry 2018
Published by the Royal Society of Chemistry, www.rsc.org

delivery system for the dosing of drugs. Alternatively, the resulting aerogel materials can also meet the performance criteria for other emerging niche markets, for example, the cosmetic, food and biotechnological industry.[4] Fundamentals and processing parameters of the materials manufacturing routes, as well as examples of aerogel systems currently reported in the literature for packaging and aerospace applications will be outlined.

13.2 Aerospace Applications of Aerogels

Aerogels are low density solids with a typically higher surface area and small pore sizes, leading to interesting properties[5] which make them suitable for a variety of aerospace applications such as insulation for astronaut suits and habitats for Mars missions,[6] launch vehicles,[7] inflatable decelerators for entry, descent, and landing operations,[8] and low dielectric substrates for lightweight antennas.[9]

13.2.1 Aerospace: Structural

There is increasing attention on the performance advantages of nanocomposites, and polymer-based nanocomposites in particular, for weight-reducing initiatives. In the interest of sustainability, the specific use of bio-reinforced nanocomposite parts and nanostructured coatings within automotive, aerospace, construction, medical, and packaging applications is accelerating. These "green" nanocomposites can provide high mechanical strength at low density, low weight, and potentially low cost. New efforts are underway to develop and apply sustainable nanocomposites that improve structural properties for applications in the aerospace industry.[10] Typical properties that are improved over existing composites include dimensional stability, structural strength, thermal resistance, chemical resistance, weight reduction, and electrical conductivity. This application falls in the low-volume category because of the high performance requirements for new materials in aerospace, but over the longer term it will likely become a larger application with wide adoption.

13.2.2 Aerospace: Interiors

Much like aero-structures, interiors are being designed with lighter weight materials. Beyond traditional applications in the floorboards, ceiling and panels, monuments and other components, seating has become a major focal point for composites applications.[11] Light weighting of the seats can enable major aircraft weight reductions. Slimmer seat designs also allow operators to make some small increases in the number of seats per aircraft. The use of composites in seats and in a variety of small brackets, clips, trays, plinths, and other structures is increasing. Reducing weight on the aircraft increases fuel efficiency and reduces greenhouse gas emissions.

13.2.3 Astronautical Applications of Aerogels

13.2.3.1 *Hypervelocity Particle Capture*

Researchers anticipated that the kinetic energy absorption characteristics of aerogels would boost the discoveries of extraterrestrial objects in low-Earth orbits.[12] In September 1992, aerogels were sent on the Space Transport System (STS-47) to analyse their ability as a hypervelocity particle capture medium and to test their endurance during launch and reentry. The aerogels successfully survived the launch and reentry and returned without any apparent damage. Generally, the capability of a hypervelocity particle capture is evaluated by how fast it can decelerate the high velocity impacted particles without destroying the latter while being trapped.[13] A wide range of impacts, for example, coorbing flakes, human waste materials, and cosmic dust, were extracted from the aerogel collector to analyse their compositions using scanning electron microscopy with energy dispersive spectroscopy (SEM-EDS) and transmission electron microscopy (TEM), and henceforth their potential origins were suggested.

The most successful aerogel-related mission that has unveiled many scientific discoveries is probably the Stardust Mission. Launched from the Kennedy Space Center in 1999, the function of the mission was to carry a hypervelocity particle collector which would meet up with a known outer solar system body (Comet 81P/Wild 2) to capture coma samples and interstellar dust to be brought back to Earth for laboratory analysis.[14,15]

An aerogel was again selected as a passive detector to capture the micrometeoroids and orbital debris particles. Two transparent tiles of silica aerogels of size $30\times25\times45$ mm^3 were prepared *via* a sol–gel process followed by supercritical drying. The resulting bulk density was 0.087 ± 0.004 g cm^3. An expansive series of impacted particles including metals, glass and mixed oxides ranging from one to several microns were extracted from the aerogels. A study is ongoing to analyse these matter, as mentioned by Woignier *et al.*[16]

Recently, a study was carried out by Jones *et al.*[17] to measure the temperatures experienced by hypervelocity particles during their capture in aerogels. Aggregate projectiles made up of magnetic submicron hematite were employed. The concept used was that when these particles are heated above their Curie temperature (675 °C) during the penetration, they lose their magnetism. Hence, the particles were fired at different velocities to acquire different temperatures upon seizure in the aerogels. Their magnetizations were then observed using atomic and magnetic force microscopy, along with electron paramagnetic resonance. It was found that the heating of these fine particles aggregates highly depended on the location in the capture track where they come to rest. The particles which were fired with velocities up to 6.6 km s^{-1} were still magnetic.

13.2.3.2 *Cryogenic Fluid Containment*

A third function of aerogels is to act as cryogenic fluid containment. This idea was proposed in 2004 when engineers were working on the Satellite

Test of the Equivalence Principle (STEP) mission. The satellite was to be sent into the earth's orbit to probe the underlying foundation of Einstein's theory, the (local) equivalence of gravitational and inertial mass.[18] The test masses and detectors were required to sustain stability from disturbances such as air drag, magnetic field, and solar pressure in order to obtain precise results.[19] This was achieved by placing the measurement instruments in a Dewar containing liquid helium maintained at cryogenic temperatures. The liquid helium coming out of the aerogel would be directed to the spacecraft thrusters throughout the mission. An aerogel control tide was built in the helium storage system.[20]

13.2.4 Aerogels in Aeronautics

The common plastic covers used in aerospace are polyethylene terephthalate (PET), polyvinyl fluoride (PVF), and silicone-coated fiberglass to withstand the high temperature environment. Aerogels, being a superinsulation material and an acoustic shock-absorber, can therefore be considered as a thermal/acoustical insulator for this task. However, a more realistic way to employ aerogels would be by separately exploiting the remarkable properties for thermal insulation, fire protection, and acoustics purposes in different parts of the aircraft.

13.2.4.1 Thermal Barrier

The justification for considering aerogels as a thermal barrier is due to the favourable characteristics in operating temperatures, longevity, chemical (aviation fuels and lubricants) and erosion resistance, and maintenance. A simpler and lighter overall design of the thermal insulation system can be achieved which will consequently reduce the assembly cost as fewer materials are needed. More space will be available for other uses. There will be a rise in the energy efficiency because of the minimization of heat loss and hence fuel will be saved, meaning that the direct operating cost will also decrease. When considering an aero engine, aerogels can be applied in two modes, depending on the temperature and environment requirements. Firstly, it could be sprayed as a thin insulative coating to protect unattainable and uneven substrates from high temperatures. The smooth and uniform layer of insulation will cause little resistance to the airflow. The thermal responses will be improved, which will in turn increase the performance of the engine while the aircraft is cruising at high altitude. Secondly, in compartments where vibration is high, flexible light-weight blankets of aerogels with custom thickness can be used. They can be fastened mechanically to prevent any displacement, and hence interferences problems. Contrary to coatings, blankets are more resistant to contaminations and do not disintegrate easily. Their maintenance cost is also lower than that of coatings.[21]

13.2.4.2 Fire Retardation

Due to being an inorganic and inflammable material with a continuous operating temperature ranging from −273 °C to 650 °C and a high melting point of 1400 °C, silica aerogel makes an excellent firewall compared to the existing combustible organic coatings that cause toxic fumes when burning. The components such as pipes, wires, and electronic accessories within the fire zones of an aero engine can be protected using thin blankets of aerogel, whilst simultaneously enabling weight saving compared to conventional metal sheets.

13.2.4.3 Acoustics

Sound waves are significantly absorbed through aerogels, thus reducing the speed of propagation to 100 m s^{-1}. This is due to their extremely low Young's modulus which is related to the synthesis of the aerogel, more precisely, the interstitial gas type, pressure, and density.[22] Many aerogels are now acknowledged as promising materials for acoustic matching layers of high-sensitivity airborne ultrasonic transducers for boosting of airborne acoustic waves.[23] The aerogel ultrasonic transducer can therefore be incorporated into future aeronautical sensing systems for range findings. Acoustical insulation in aircraft using aerogels can also be considered based on its acoustic absorption. Forest and Gibiat found that the minimum transmission loss in granular aerogel can be 10 dB higher than that of fibre glass with the same thickness.[24]

13.2.4.4 Cost Analysis

The main limitation of preventing aerogels from commercially integrating into the aviation sector is its high cost. One way to achieve this is by reducing the take-off weight, for example, by removing unnecessary items from the aircraft. Despite the existing market competition, it remains difficult to compensate the price of aerogels for other beneficial factors. On the other hand, it is reported that a reduction up to 80% can be attained in the manufacturing cost of Maerogel (standing for Malaysian-made Aerogel), obtained from rice husk.

13.3 Packaging Applications of Aerogels

The highly ordered structure of cellulose nanocrystal improves the mechanical property of film, and the presence of crystalline nanocellulose is believed to increase tortuosity in the film. Furthermore, enormous hydroxyl groups on the cellulose nanocrystal bind the water molecules strongly by hydrogen bonding as a combined result of a slower diffusion process, and hence, reduces the water vapor permeability of the film required for food packaging.[25]

Aerogels prepared from soya protein are brittle, and this limits their potential applications. The mechanical properties of soya protein aerogels could be improved by reinforcement with nanocellulose. Nanocellulosic composite aerogels have a high surface area, low density and good mechanical properties, and all these improved properties make composite aerogels useful candidates for potential applications in packaging, thermal and acoustic insulation, transport media, and so forth.[26]

13.4 Conclusion

Based on the above literature, it can be said that the fundamental synthesis–structure–property relationships of aerogels are now comprehended in the research community after eighty years of tremendous efforts. Different cost-effective manufacturing methods have been developed over time to promote the commercialization of aerogels in various high-tech areas. In the aerospace industry, the capture effectiveness of aerogels as a kinetic energy absorber is already considered to be superior, whilst its potential as a thermal insulator shows great promise for applications ranging from cryogenic temperatures in spacecraft to high temperatures in aero engines. The feasibility tests of insulating space suits with aerogels show that a lower thermal conductivity is required and is yet to be achieved. Science and technology are now continually focusing on green bio-based renewable raw materials and more environmental friendly and sustainable resources. Maerogel, which is an ecological and green-technology material, has been proposed as a promising substitute due its relatively low cost and comparable properties with the conventional silica aerogels being used in aerospace applications.

References

1. I. Smirnova, M. Turk, R. Wischumerski and M. A. Wahl, Comparison of different methods for enhancing the dissolution rate of poorly soluble drugs: Case of griseofulvin, *Eng. Life Sci.*, 2005, **5**(3), 277–280.
2. Y. K. Akimov, Fields of application of aerogels, *Instrum. Exp. Tech.*, 2003, **46**(3), 287–299.
3. S. S. Kistler, Coherent expanded aerogels and jellies, *Nature*, 1931, **127**(3211), 741.
4. D. Renard, F. Van De Velde and R. W. Visschers, The gap between food gel structure, texture and perception, *Food Hydrocolloids*, 2006, **20**(4), 423–431.
5. A. C. Pierre and G. M. Pajonk, Chemistry of Aerogels and Their Applications, *Chem. Rev.*, 2002, **102**, 4243–4265; A. Du, B. Zhou, Z. Zhang and J. Shen, A Special Material or a New State of Matter: A Review and Reconsideration of the Aerogel, *Materials*, 2013, **6**, 941–968.
6. H. L. Paul and K. R. Diller, Comparison of Thermal Insulation Performance of Fibrous Materials for the Advanced Space Suit.,

J. Biomech. Eng., 2003, **125**, 639–647; H. H. Tang; E. S. Orndoff and L. A. Trevino Thermal Performance for Space Suit Elements with Aerogel Insulation for Moon and Mars Exploration, 36th International Conference on Environmental Systems, Norfolk, VA, July 17 – 20, 2006, American Institute of Aeronautics and Astronautics: Reston, VA, 2006, 2006-01-2235.

7. J. E. Fesmire, Aerogel insulation Systems for Space Launch Applications, *Cryogenics*, 2006, **46**, 111–117.
8. J. A. DelCorso; F. M. Cheatwood; W. E. Bruce III; S. J. Hughes; A. M. Calomino Advanced High-Temperature Flexible TPS for Inflatable Aerodynamic Decelerators, 21st AIAA Aerodynamic Decelerator Systems Technology Conference and Seminar, 23–26 May 2011, Dublin, Ireland, American Inst. of Aeronautics and Astronautics, Reston, VA, 2011, 2011-01-2051.
9. M. A. B. Meador, S. Wright, A. Sandberg, B. N. Nguyen, F. W. Van Keuls, C. H. Mueller, R. Rodriquez-Solis and F. A. Miranda, Low Dielectric Polyimide Aerogels as Substrates for Lightweight Patch Antennas, *ACS Appl. Mater. Interfaces*, 2012, **4**, 6346–6353.
10. FMI, The Global Market for Nanocellulose to 2017, Future Markets Inc. Technology Report No. 60, Second Edition, 2012, p. 38.
11. C. Red, Composites in Aircraft Interiors, 2012-2022, Composites World, 2012.
12. F. Hoerz, G. Cress, M. Zolensky, T. H. See, R. P. Bernhard and J. L. Warren, Optical analysis of impact features in aerogel from the orbital debris collection experiment on the MIR station, NASA TM-1999-209372, 1999.
13. P. Tsou, Silica aerogel captures cosmic dust intact, *J. Non-Cryst. Solids*, 1995, **186**, 415–427.
14. H. A. Ishii, J. P. Bradley and Z. R. Dai *et al.*, Comparison of comet 81P/Wild 2 dust with interplanetary dust fromcomets, *Science*, 2008, **319**(5862), 447–450.
15. D. E. Brownlee, P. Tsou and K. L. Atkins *et al.*, Stardust: finessing expensive cometary sample returns, *Acta Astronaut.*, 1996, **39**(1–4), 51–60.
16. T. Woignier, L. Duffours, P. Colombel and C. Durin, Aerogels materials as space debris collectors, *Adv. Mater. Sci. Eng.*, 2013, **6**, 484153.
17. S. M. Jones, M. S. Anderson, G. Dominguez and A. Tsapin, Thermal calibrations of hypervelocity capture in aerogel using magnetic iron oxide particles, *Icarus*, 2013, **226**(1), 1–9.
18. T. J. Sumner, J. Anderson and J.-P. Blaser *et al.*, STEP (satellite test of the equivalence principle), *Adv. Space Res.*, 2007, **39**(2), 254–258.
19. P. Worden, R. Torii, J. C. Mester and C. W. F. Everitt, The step payload and experiment, *Adv. Space Res.*, 2000, **25**(6), 1205–1208.
20. P. W. Worden, STEP payload development, *Adv. Space Res.*, 2007, **39**(2), 259–267.
21. N. Bheekhun, A. R. Abu Talib and M. R. Hassan, Aerogels in aerospace: An overview, *Adv. Mater. Sci. Eng.*, 2013, **2013**, 406065.

22. J. Fricke, Aerogels—highly tenuous solids with fascinating properties, *J. Non-Cryst. Solids*, 1988, **100**(1–3), 169–173.
23. H. Nagahara, T. Suginouchi and M. Hashimoto, Acoustic properties of nanofoam and its applied air-borne ultrasonic transducers, *Proceedings of the IEEE Ultrasonics Symposium*, 2006, pp. 1541–1544.
24. L. Forest, V. Gibiat and A. Hooley, Impedance matching and acoustic absorption in granular layers of silica aerogels, *J. Non-Cryst. Solids*, 2001, **285**(1–3), 230–235.
25. S. S. Costa, J. I. Druzian, B. A. Souza Machado, C. O. de Souza and A. G. Guimaraes, Bi-functional biobased packing of the cassava starch, glycerol, licurinanocellulose and red propolis, *PloS One*, 2014, **11**, e112554, 9.
26. J. C. Arboleda, M. Hughes, L. A. Lucia, J. Laine, K. Ekman and O. J. Rojas, Soyprotein-nanocellulose composite aerogels, *Cellulose*, 2013, **20**, 2417–2426.

CHAPTER 14

Cellulose and Protein Aerogels for Oil Spill Cleaning, Life Science and Food Engineering Applications

HAI M. DUONG,* PENG LIU, THANH X. NGUYEN, SON T. NGUYEN, JINGDUO FENG AND HANLIN CHENG

Department of Mechanical Engineering, National University of Singapore, 9 Engineering Drive 1, EA-07-05, Singapore 117575, Singapore
*Email: mpedhm@nus.edu.sg

14.1 Introduction

Oil spills are considered to be one of the most serious disasters threatening the marine ecosystem. Recently, the explosion of a drilling rig in the Gulf of Mexico caused significant environmental damage.[1] Oil spills are usually related to accidents during oil production, storage, and transportation. As long as fossil fuels are needed, oil spills will remain a significant problem that human beings will have to face.[2–8] Therefore, it is essential to solve, or at least alleviate, this environmental problem.

Several methods of oil spill–cleaning have been developed and can be classified as chemical, biological, and physical methods. Dispersion, *in situ* burning, and solidification are considered to be chemical methods that are complicated and expensive.[9–11] The use of microorganisms *via* biological

Green Chemistry Series No. 58
Biobased Aerogels: Polysaccharide and Protein-based Materials
Edited by Sabu Thomas, Laly A. Pothan and Rubie Mavelil-Sam
© The Royal Society of Chemistry 2018
Published by the Royal Society of Chemistry, www.rsc.org

methods is effective but requires a long time, and the microorganisms are affected by environmental factors, such as pH, temperature, and oxygen content.[12] Moreover, oil spills cleaned using chemical and biological methods are difficult to recover, and recoverability is a crucial factor for oil spill–cleaning applications. With regards to the physical methods, booms and skimmers are often used, but cannot remove oil from the sea effectively.[13] Of these methods, sorption has been considered one of the most effective ways for oil-spill cleaning, as it enables the collection and complete removal of oil from oil-spill sites.[2,11,14–20]

Several materials have been used as sorbents for oil-spill cleaning in both research and practical applications. The oil absorbents can be categorized into inorganic minerals, natural organic materials, and synthetic organic materials.[11,16,17,19] Inorganic materials, such as clay, vermiculite, exfoliated graphite, diatomite, and fly ash, have low oil absorption capacities (4–20 $g\,g^{-1}$).[21–23] Furthermore, some of these inorganic materials, such as clay and vermiculite, are harmful when inhaled by human beings under windy conditions, due to the loose structures of these materials. Natural organic materials from plant and animal residues, such as kapok fibre, sugar-cane bagasse, rice husk, coconut husk, cotton, wool, sawdust, and chitosan, have been examined for oil absorption capabilities.[24–26] However, most of the materials have low oil absorption abilities (3–15 $g\,g^{-1}$), and also absorb water.

On the other hand, synthetic organic materials, such as polypropylene, polystyrene, and polyurethane, possess a high affinity with oil and high oil absorption capacities (4.5–100 $g\,g^{-1}$), but cause waste problems after use due to their slow rates of degradation.[11,16,19] Therefore, there is high demand for new environmentally friendly absorbents with a high oil absorption capacity, good selectiveness, and low cost for oil-spill removal.

A combination of an aerogel structure and recycled cellulose fibres from paper waste can be used to form an advanced material, called a recycled cellulose aerogel, which is cost-effective and a promising material for oil absorption. Although some studies have investigated the use of cellulosic materials for oil absorption, none have fabricated aerogels from paper-waste cellulose fibres and investigated them as absorbents for crude oil-spill cleaning.[2,5,27–31]

The increase in paper consumption has created huge amounts of paper waste, which accounts for 25–40% of global municipal solid waste.[32] In 2004 alone, 360 million tonnes of paper-related waste was generated worldwide. Moreover, paper and paperboard consumption will increase the amount of waste produced by 2.1% each year until 2020, which suggests that more than 500 million tonnes of paper waste can be expected in 2020.[33] In addition, incineration or landfill of the paper waste could damage the environment further with toxic emissions and groundwater contamination.

Recycling paper waste will help to preserve forests and solve the environmental problem. Therefore, it is important to recycle or convert this enormous amount of waste into useful products. Several efforts have been made to solve this problem. For instance, in 2010, 63% of paper waste was recycled in the US.[34] Paper waste has also been investigated as a raw material for the production of bioethanol, polymer precursors, particleboard, and so forth.[32,35,36] Commercially, recycled paper is mainly converted into other paper commodities of lower quality grades than the original products.[37] In addition, the maximum conversion rate from paper waste to other paper products is only approximately 65%.[37] This low conversion rate is due to the length degradation of the cellulose fibres during the recycling processes, which also compromises the quality of the end product.[37] In addition, waste fibres of short lengths generated during recycling are discarded, as they are not suitable for further recycling.[37] It is therefore necessary to develop alternative commodities from paper waste. Although some studies have examined the use of cellulosic materials for oil absorption, none have covered the fabrication of aerogels from paper-waste cellulose fibres, and investigated the aerogels as absorbents for the cleaning of crude-oil spills.[2,5,27–31]

Recycled cellulose fibres from paper waste are a cheap and abundant resource; the price of scrap paper was approximately $100/ton in 2015.[38] A combination of the aerogel structure and recycled cellulose fibre constitute a new material – called a recycled cellulose aerogel – which is cost-effective and has the potential for oil absorption.[39–41] The recycled cellulose aerogel and its silica composite aerogel can also potentially be used as thermal-insulation materials for buildings.[41] Therefore, all the practical applications developed may contribute to the recycling of paper-related waste. This book chapter presents the basic facts about cellulose materials, and comprehensive information about cellulose aerogels and silica–cellulose composite aerogels. Both the fabrication methods and properties of cellulose aerogels and silica–cellulose aerogels are discussed in detail.

In recent years, the production of natural protein-based aerogels has become a highly attractive subject in materials chemistry due to the requirement of biodegradability and biocompatibility for pharmaceutical, medical and food applications.[42] Several types of proteins, including whey protein,[43–45] silk fibroin,[46,47] egg white protein[48] and soy protein,[49–52] have been exploited for the formation of aerogels. The effects of various synthesis conditions, such as drying methods,[43] pH values,[48] ionic strengths[48] and precursor concentrations[47] have been investigated for optimizing the porous structure and multi-properties of the resultant aerogels. These natural aerogels are promising as drug carriers and encapsulation materials.

14.2 Recycled Cellulose Aerogels Using Kymene Binder for Oil Spill–Cleaning Applications

14.2.1 Introduction

This section describes the successful development of an advanced and cost-effective method for the fabrication of recycled cellulose aerogels.

This novel method synthesizes the recycled cellulose aerogels from paper waste by using Kymene as a cross-linker, instead of using sodium hydroxide and urea as in previous reports.[53–55] This method can significantly reduce the toxicity of raw materials, and reduce the entire synthesis duration fromnine days, as in previous methods, to three days.[40,41] After being freeze dried and coated with methyltrimethoxysilane (MTMS) *via* chemical vapour deposition, the recycled cellulose aerogels exhibited ultra-flexibility, high porosity, super-hydrophobicity, and outstanding oil absorption capability.

14.2.2 Synthesis of Cellulose Aerogels Using a Kymene Binder

Recycled cellulose fibres were directly purchased from the market because the established raw–paper waste recycling methods are mature, and using commercial recycled cellulose fibres is cost-effective and time-saving. The recycled cellulose fibres from paper waste (0.075–0.3 g) and Kymene (5–20 μl) were first dispersed in 30 ml of DI water by sonicating the mixtures for 10 min. The suspensions were then placed in a refrigerator at −18 °C for more than 12 h to allow the gelation. The cellulose aerogels were obtained by freeze drying the obtained gels at −98 °C for two days using a Scan Vac CoolSafe 95-15 Pro freeze dryer (Denmark). Thereafter, the cellulose aerogels were further cured at 120 °C for another 3 h to completely cross-link the Kymene molecules.

As the development of the recycled cellulose aerogels was hydrophilic, the highly porous networks of the as-prepared cellulose aerogels were coated with MTMS, to form super-hydrophobic cellulose aerogels for oil absorption and thermal insulation. The proposed coating mechanism[53,54] of the silanation reaction between the cellulose and MTMS is illustrated in Figure 14.1.

The cellulose aerogels and open glass vials containing MTMS (0.5 ml) were placed in big containers. The containers were then capped and heated at 70 °C for 3 h for the silanation reaction. After the aerogel structure was coated completely, excessive MTMS was removed by placing the aerogel in a vacuum oven until the pressure decreased to below 0.03 mbar.

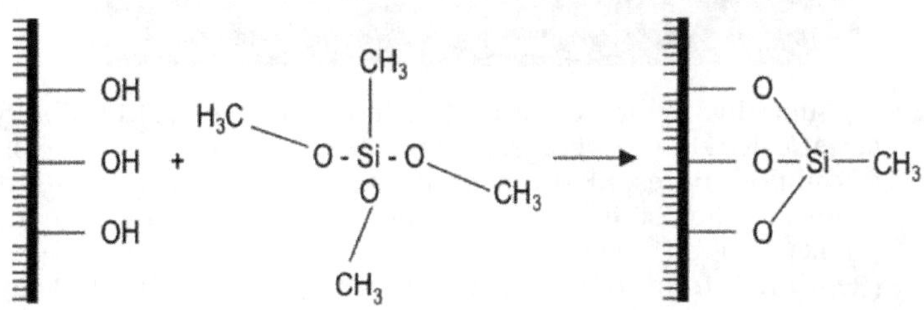

Figure 14.1 The proposed silanation reaction between cellulose and MTMS, which results in super-hydrophobic cellulose aerogels.

14.2.3 Morphology and Hydrophobicity of the Recycled Cellulose Aerogels

In this section, the morphologies and hydrophobic properties of the recycled cellulose aerogels are discussed. The recycled cellulose aerogels exhibited macropore structures. Moreover, the aerogel with the higher cellulose concentration (1.0 wt.%) had a more compacted network and lower porosity. The aerogel with a high stability was also observed to have super-hydrophobicity.

14.2.3.1 Effects of the Cellulose Concentrations

The photographs and SEM images of the developed recycled cellulose aerogels are shown in Figure 14.2. The aerogel sample in Figure 14.2a had dimensions of 45 mm (diameter)×11 mm (thickness), and the same shape as that of its reaction container. As reported previously, the recycled cellulose aerogels were formed *via* hydrogen bonding between the self-assembled cellulose fibres.[40,54]

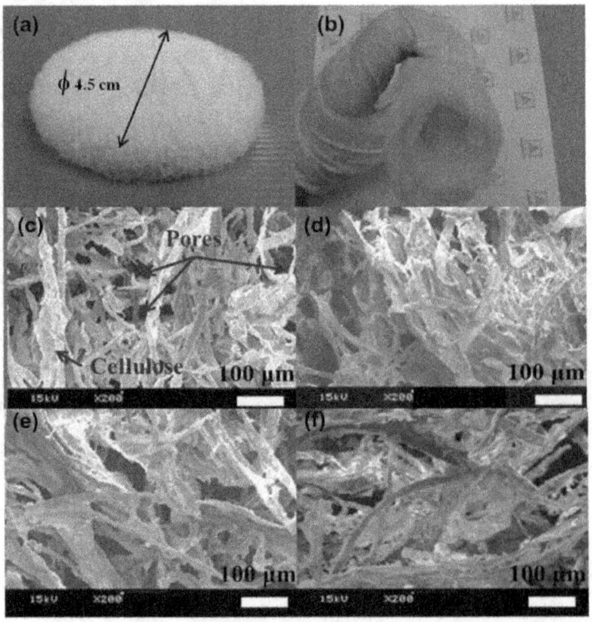

Figure 14.2 (a) Super-hydrophobic recycled cellulose aerogel. (b) Flexibility of the large-scale cellulose aerogel (38 cm×38 cm×1 cm) containing 0.60 wt.% of cellulose fibres, SEM images of the cellulose aerogels with different ratios of cellulose fibres (wt.%) and Kymene (μl): (c) 0.25:5, (d) 1.00:5, (e) 0.60:5, and (f) 0.60:20.
(Reprinted from *Chemical Engineering Journal*, **270**, J. Feng, S. T. Nguyen, Z. Fan, H. M. Duong, Advanced fabrication and oil absorption properties of superhydrophobic recycled cellulose aerogels, 168–175, Copyright 2015,with permission from Elsevier).

In addition, Kymene strengthened the cellulose aerogels by providing a protection mechanism and reinforcement mechanism.[56] For the protection mechanism, some Kymene molecules reacted with other Kymene molecules, and the formed Kymene networks wrapped the cross-linking points between the cellulose fibres to improve the strength of the cellulose network.[56] Moreover, Kymene molecules also bonded with the cellulose fibres to enhance the strength of the cellulose network, providing the reinforcement mechanism.[56] The utilization of Kymene as a cross-linker ensured that the resultant aerogels had a robust structure.[56]

In contrast to the mesopores (2–70 nm) of the aerogels formed by the cellulose nanofibres, the cellulose aerogels with macropores (>50 nm) had a highly porous structure, which can be clearly observed in the SEM images in Figure 14.2c–f.[57–59] Their macropores were possibly caused by the larger size of the recycled cellulose fibres, obtained from the paper waste.[40] Figure 14.2c and d show the morphologies of the cellulose aerogels with cellulose concentrations of 0.25 and 1.00 wt.%, respectively. The aerogel with the higher cellulose concentration (1.0 wt.%) had a more compacted network and lower porosity. However, an increase in the amount of Kymene from 5 to 20 μl in a 30 ml reaction mixture did not significantly impact the aerogel structures, as shown in Figure 14.2e and f, as the amount of Kymene was small compared to that of the cellulose fibres, and the possible minor structure changes might not have been observed.

14.2.3.2 Hydrophobicity of the Cellulose Aerogels

In order to investigate the super-hydrophobicity of the developed cellulose aerogels, the water contact angles were measured on both the external surface and the cross-section of the MTMS-coated cellulose aerogels. As shown in Figure 14.3a and b, large contact angles of 153.5° and 150.8°, respectively, were obtained, thus proving that the hydrophobic coating successfully covered the whole aerogel network. The water contact angle

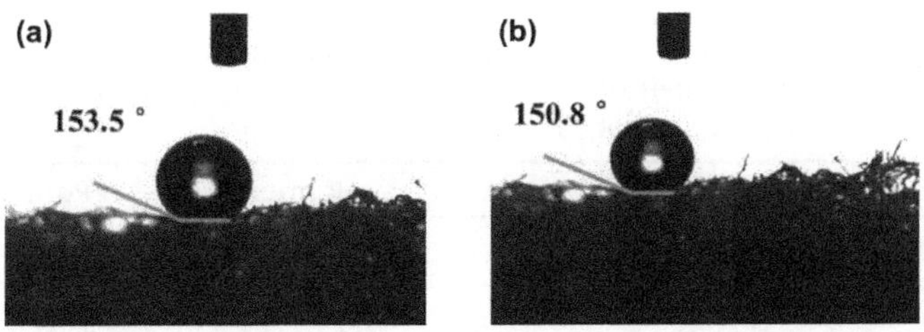

Figure 14.3 Water contact angles on (a) the external surface and (b) the cross-section of the super-hydrophobic recycled cellulose aerogel. (Reprinted from *Chemical Engineering Journal*, **270**, J. Feng, S. T. Nguyen, Z. Fan, H. M. Duong, Advanced fabrication and oil absorption properties of superhydrophobic recycled cellulose aerogels, 168–175, Copyright 2015,with permission from Elsevier).

values of the external surface were slightly higher than those of the cross section, possibly due to the greater accessibility of the external surface.

To examine the hydrophobic stability of the cellulose aerogels, they were then exposed to the normal ambient atmosphere for five months. Their water contact angles of 145–155° over this period were examined. Interestingly, all the cellulose aerogels exhibited similar water contact angles of approximately 150°, regardless of their cellulose concentrations or Kymene amounts. It is well known that the water contact angles strongly depend on the functional groups on the aerogel surfaces. Therefore, in this case, such a small variation in the water contact angles may likely have been a consequence of the identical functional groups (–Si–O–CH_3–) induced by the MTMS coating.[40] Furthermore, the water contact angles of the cellulose aerogels on the external surface and the cross-section did not show any obvious change with time. The hydrophobic properties of the aerogels discussed in this section demonstrate the excellent performance of the MTMS coating on different types of cellulose aerogels.

14.2.3.3 Oil Absorption Properties of the Cellulose Aerogels

The oil absorption properties of the recycled cellulose aerogels are discussed in this section. Several factors, such as the type of oil, the initial cellulose fibre concentration, the temperature, and the seawater effect with different pH values, were investigated with regard to their effects on the oil absorption capacity of the cellulose aerogels. The absorption kinetics and the activation energy values of cellulose aerogels are also investigated in detail in this section.

14.2.3.3.1 Absorption Capacities with Different Oils. A 5w40 motor oil was used to investigate the oil absorption capabilities of the recycled cellulose aerogels listed in Table 14.1. This section focuses on motor oils

Table 14.1 Chemical compositions of the various recycled cellulose aerogels. (Reprinted from *Chemical Engineering Journal*, **270**, J. Feng, S. T. Nguyen, Z. Fan, H. M. Duong, Advanced fabrication and oil absorption properties of superhydrophobic recycled cellulose aerogels, 168–175, Copyright 2015, with permission from Elsevier).

Sample label	Cellulose fibres (wt.%)	Kymene (μl)	Porosity (%)
Sample A	0.25	5	99.4 ± 0.0
Sample B	0.50	5	98.9 ± 0.0
Sample C	0.75	5	98.1 ± 0.0
Sample D	1.00	5	97.2 ± 0.1
Sample E	0.60	5	98.4 ± 0.0
Sample F	0.60	20	98.4 ± 0.0
Sample G	1.00	10	97.4 ± 0.0
Sample H	2.00	20	96.9 ± 0.0
Sample I	4.00	40	96.1 ± 0.3

instead of crude oils, as it aims to show the excellent absorption properties of cellulose aerogels with oil products containing additives.

When the Kymene amount was kept at 5 μl and the cellulose concentration was increased from 0.25 to 0.50 to 0.75 to 1.00 wt.%, the measured absorption capacities of the aerogels (Samples A, B, C, and D in Table 14.1) were 95, 73, 58, and 49 $g\,g^{-1}$, respectively, at 25 °C. The maximum absorption capacity of 95 $g\,g^{-1}$ was achieved with the 0.25 wt.% cellulose aerogel, because it had the lowest density (9×10^{-3} $g\,cm^{-3}$) and the highest porosity (99.4%).

The absorption capacities of all the MTMS-coated cellulose aerogels were one order greater than those of the natural sorbents, two to ten times greater than those of the commercial polypropylene sorbents, and five times greater than those of the recycled cellulose aerogels (approximately 20 $g\,g^{-1}$) reported in previous works that used the sodium hydroxide/urea method.[11,19,25,40,41] The significant enhancement of the absorption capacity may be largely ascribed to the reduced densities and increased porosities of the cellulose aerogels.

The cellulose aerogels fabricated with a Kymene binder can achieve a high porosity (up to 99.4%), while the cellulose aerogels using sodium hydroxide and urea can achieve only 98.0% porosity. When the cellulose amounts were further decreased in the syntheses mixture using sodium hydroxide and urea, rigid aerogels could not be successfully formed.

Temperature is also a major factor affecting the viscosity and the diffusion capability of the oils into the porous aerogel structures. Therefore, the absorption behaviour of the different oils with each aerogel was examined at three different temperatures of 25, 50, and 70 °C. As shown in Table 14.2 and Figure 14.4, the maximum oil absorption capacity increased when the temperature was increased from 25 to 50 °C, but then decreased when the temperature was further increased from 50 to 70 °C. This trend holds for the absorption behaviour of all the oils with the 0.50, 0.75, and 1.00 wt.% cellulose aerogels.

The explanation for this may be that the temperature increase reduced the oil viscosities (shown in Table 14.3), which in turn facilitated oil penetration into the porous aerogel networks. However, the lower viscosities of the oils also had a negative effect on their ability to anchor to the pore walls, reducing the amounts of oil retained in the porous absorbents during drainage. Comparing the maximum oil absorption capacity with the tested temperatures, it can be concluded that 50 °C is the optimum temperature for maximizing the oil absorption performance of the recycled cellulose aerogels.

Besides the temperature effects, the porosity of cellulose aerogels also significantly affects their oil absorbency. Table 14.2 and Figure 14.4 show that the oil absorption capacity of the aerogels was reduced when the initial cellulose concentration increased from 0.50 to 1.00 wt.%. This can be explained by the porosity of the cellulose aerogels. Table 14.1 shows that the aerogel porosity reduced from 98.9 to 97.2% when the cellulose

Table 14.2 Summary of the maximum oil absorption capacities and the absorption rate constants of the cellulose aerogels at different temperatures, with various cellulose fibre concentrations, using the pseudo-first-order and pseudo-second-order models. (Reprinted from *Chemical Engineering Journal*, **270**, J. Feng, S. T. Nguyen, Z. Fan, H. M. Duong, Advanced fabrication and oil absorption properties of superhydrophobic recycled cellulose aerogels, 168–175, Copyright 2015,with permission from Elsevier).

	Cellulose concentration (wt.%)		0.50			0.75			1.00		
	Temperature (°C)		25	50	70	25	50	70	25	50	70
5w50 motor oil	Maximum absorption capacity, Q_m (g g^{-1})		62.6	64.9	59.2	48.1	53.8	46.3	45.9	46.5	46.2
	Pseudo-first-order	R^2	0.986	0.970	0.972	0.977	0.954	0.966	0.986	0.993	0.944
		K_1	0.293	0.324	0.355	0.254	0.319	0.367	0.218	0.287	0.357
	Pseudo-second-order	R^2	0.996	0.996	0.998	0.995	0.998	0.996	0.996	0.995	0.998
		K_2	0.006	0.008	0.009	0.008	0.010	0.015	0.005	0.007	0.013
Singer machine oil	Maximum absorption capacity, Q_m (g g^{-1})		59.3	61.0	58.2	46.1	48.8	47.6	40.4	43.1	42.4
	Pseudo-first-order	R^2	0.983	0.978	0.980	0.981	0.956	0.994	0.978	0.970	0.987
		K_1	0.297	0.311	0.351	0.273	0.306	0.343	0.278	0.377	0.425
	Pseudo-second-order	R^2	0.994	0.999	0.999	0.996	0.995	0.999	0.993	0.994	0.996
		K_2	0.008	0.010	0.012	0.008	0.011	0.015	0.007	0.009	0.016

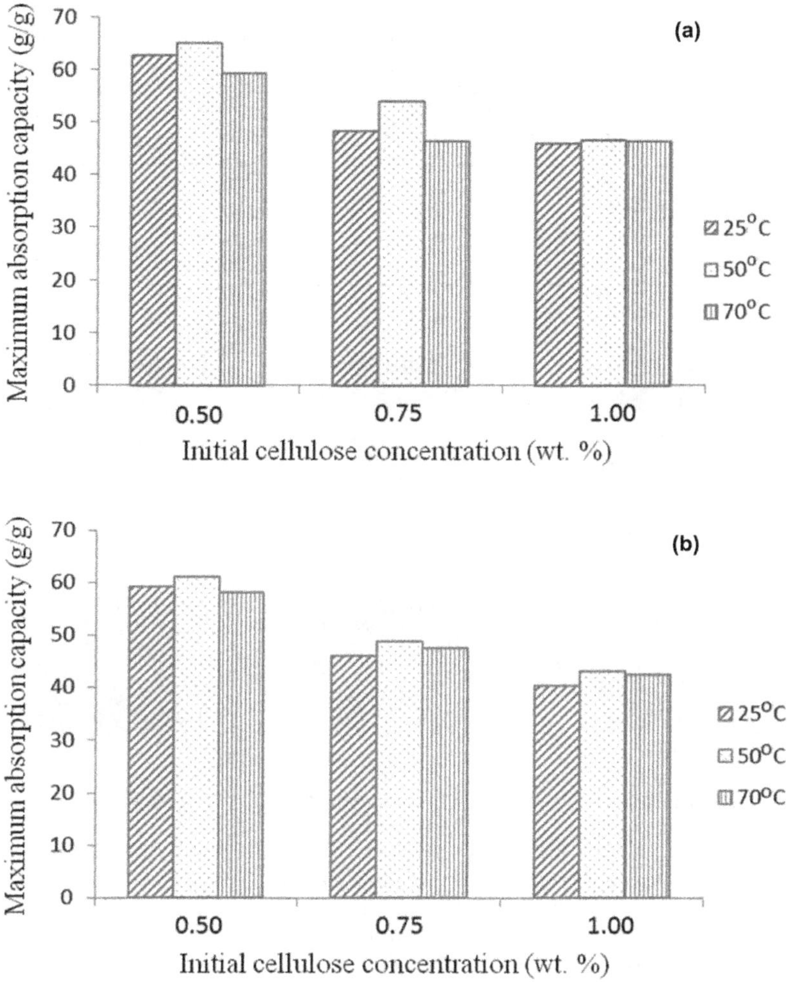

Figure 14.4 Maximum absorption capacities, Q_m, of (a) the 5w50 motor oil and (b) the Singer machine oil with the recycled cellulose aerogels with various cellulose fibre concentrations of 0.50, 0.75, and 1.00 wt.% at 25, 50, and 70 °C.
(Reprinted from *Chemical Engineering Journal*, **270**, J. Feng, S. T. Nguyen, Z. Fan, H. M. Duong, Advanced fabrication and oil absorption properties of superhydrophobic recycled cellulose aerogels, 168–175, Copyright 2015,with permission from Elsevier).

concentration increased from 0.50 to 1.00 wt.%. As the aerogel porosity was lower, there was less space in the aerogel network for oil occupation, and therefore the oil absorbency was lower.

To evaluate the practical oil absorption performance of the recycled cellulose aerogels in the sea, a 3.5% NaCl solution was prepared to imitate seawater. Figure 14.5 illustrates the absorption process over several minutes.

Table 14.3 The relevant viscosities of the tested oils at different temperatures. (Reprinted from *Chemical Engineering Journal*, **270**, J. Feng, S. T. Nguyen, Z. Fan, H. M. Duong, Advanced fabrication and oil absorption properties of superhydrophobic recycled cellulose aerogels, 168–175, Copyright 2015, with permission from Elsevier).

Viscosity (Pa.s)	25 °C	50 °C	70 °C
5w40 motor oil	0.140	n.a.	n.a.
5w50 motor oil	0.160	0.054	0.029
Singer machine oil	0.026	0.009	0.006

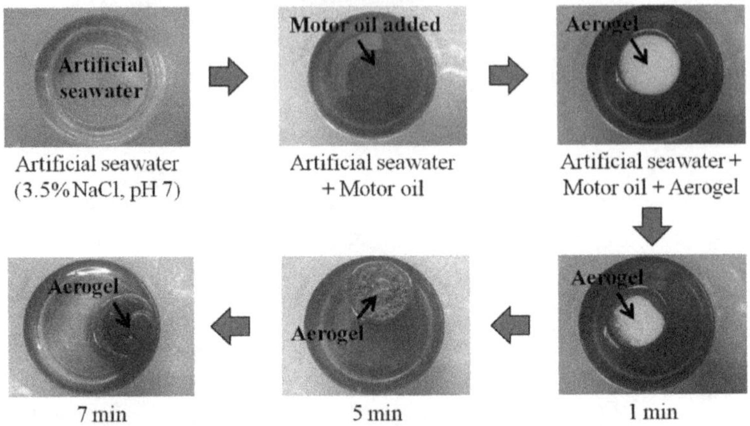

Figure 14.5 Oil absorption process of the recycled cellulose aerogel with 0.5 wt.% of cellulose fibres in artificial seawater (3.5 wt.% NaCl and pH = 7) mixed with 5w40 motor oil and dyed with Sudan Red G before testing. (Reprinted from *Chemical Engineering Journal*, **270**, J. Feng, S. T. Nguyen, Z. Fan, H. M. Duong, Advanced fabrication and oil absorption properties of superhydrophobic recycled cellulose aerogels, 168–175, Copyright 2015,with permission from Elsevier).

The cellulose aerogel was loaded on top of the mixture, and quickly absorbed most of the motor oil within 7 min.

In addition, the effects of pH on the oil absorption capacities of the cellulose aerogels were investigated using different pH values prepared from HCl or NaOH. The oil absorption capacities of the 0.50 wt.% cellulose aerogels under pH = 3, 5, 7, and 9 environments were measured to be 63.00, 62.85, 63.06, and 62.98 $g\,g^{-1}$, respectively. The absorption results indicated pH-insensitive behaviour of the aerogels during the oil absorption tests, possibly because the oil capacities of aerogels are mostly controlled by their porosities and tested oil viscosities, both of which are independent of environmental pH values.

14.2.3.3.2 Absorption Kinetics with Different Oils. The absorption kinetics of the 5w50 motor oil and Singer machine oil on recycled cellulose aerogels were investigated, and are summarized in Table 14.2. Although sample A, with a cellulose fibre concentration of 0.25 wt.% (Table 14.1), had the highest oil absorption capacity, it possessed a less rigid structure and was easy to disintegrate after repeated draining and absorbing of oil. Therefore, three different aerogel samples, marked B, C, and D (Table 14.1) and with cellulose fibre concentrations of 0.50, 0.75, and 1.00 wt.% respectively, were prepared for the oil absorption kinetics tests at 25 °C. As shown in Figure 14.6, the sorption capacity of each oil on the cellulose aerogels was plotted as a function of absorption time. The sorption rate was fast during the first 10 s, and the absorption reached the equilibrium state at 30 s for both of the two oils.

Figure 14.7 shows the absorption kinetics[60,61] of the 5w50 motor oil and Singer machine oil on the 0.50 wt.% cellulose aerogel at 25 °C, respectively. The plots of $\ln[Q_m/(Q_m - Q_t)]$ *versus* time t using the pseudo-first-order model pseudo-second-order models yield the sorption rate constants k_1 and k_2, and the correlation coefficient R^2 are presented in Table 14.2. It can be observed that the correlation coefficient values of the pseudo-second-order model are higher than those of the pseudo-first-order model for both tested oils. Therefore, the pseudo-second-order model could better predict the oil absorption behaviour. Most of the absorption rate constants (k_1 and k_2) for the Singer oil were bigger than those for the 5w50 motor oil. In addition, the absorption rate constants at a higher temperature were larger than those of the same oil and sample at a lower temperature. These suggest the oil absorption processes of the Singer oil, and those at the higher temperature occur faster. Figures 14.7a and b show the experimental absorption kinetics data and the two fitted model curves for the absorption of the two oils on the 0.50 wt.% cellulose aerogels, which show a good agreement.[60,61] The values of k_1 and k_2 are generally much larger than those reported. This phenomenon indicates that the absorption speed of the cellulose aerogels described in this section, with 5w50 motor oil and Singer oil, were much higher than those of the cellulose aerogels described with crude oils. This may be explained by the high porosities of the cellulose aerogels.

The activation energy, E_a, is an important parameter in a thermodynamic study.[62] For example, during a successful absorption process, the activation energy must be overcome by an absorbate to interact with functional groups on the sorbent surface. The activation energy, E_a, can be determined from the change in the absorption rate constant, k, with temperature, T (K), using the Arrhenius equation:[62,63]

$$\ln k = \ln A - \frac{E_a}{RT} \tag{14.1}$$

in which A is the pre-exponential factor and R is the gas constant ($8.314\ \mathrm{J\,mol^{-1}\,K^{-1}}$). By plotting $\ln[k]$ against $1/T$, E_a can be calculated from the slope.

The activation energy values are presented in Table 14.4. It can be observed that the activation energy values of the pseudo-second-order model

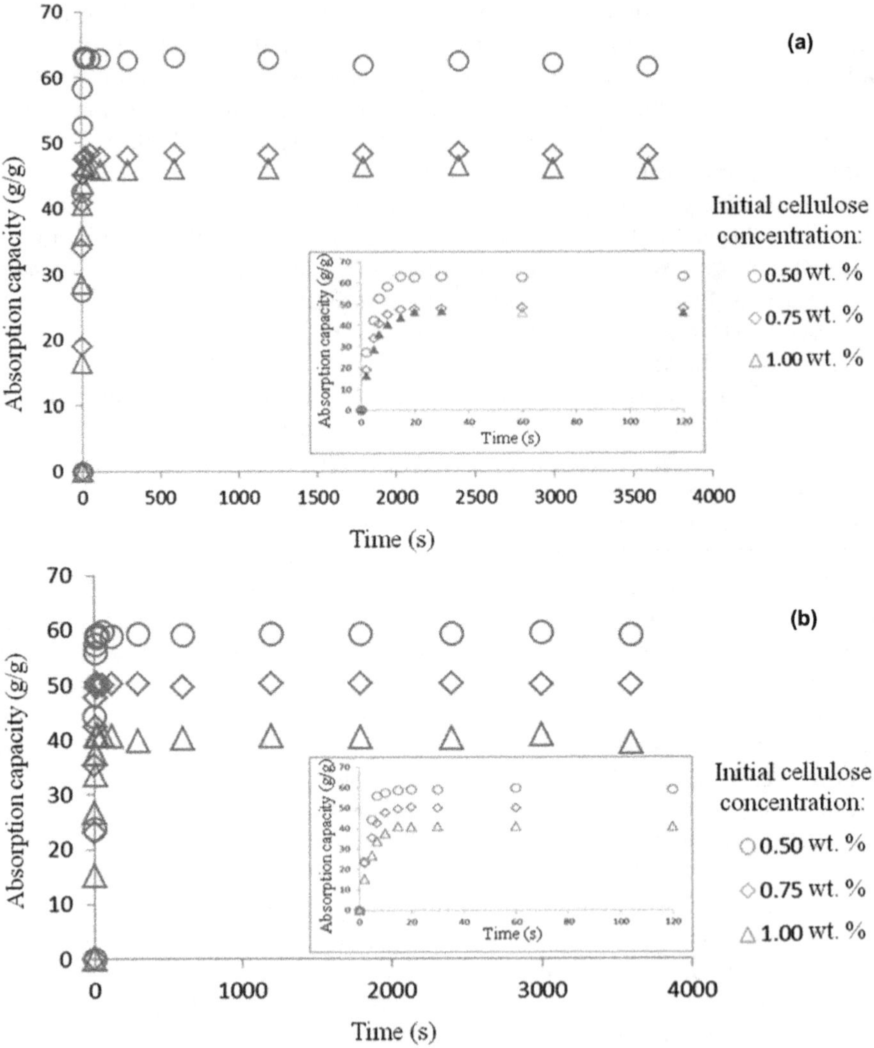

Figure 14.6 Absorption kinetics of (a) the 5w50 motor oil and (b) the Singer machine oil on the recycled cellulose aerogels with various cellulose fibre concentrations of 0.50, 0.75, and 1.00 wt.% at 25 °C. The magnified images show the absorption kinetics of the initial 2 min of the absorption processes.
(Reprinted from *Chemical Engineering Journal*, **270**, J. Feng, S. T. Nguyen, Z. Fan, H. M. Duong, Advanced fabrication and oil absorption properties of superhydrophobic recycled cellulose aerogels, 168–175, Copyright 2015,with permission from Elsevier).

are higher than those of the pseudo-first-order model. This is because the pseudo-second-order model was used for the absorption process controlled by chemi-sorption, which involves higher forces than in physic-sorption.[64]

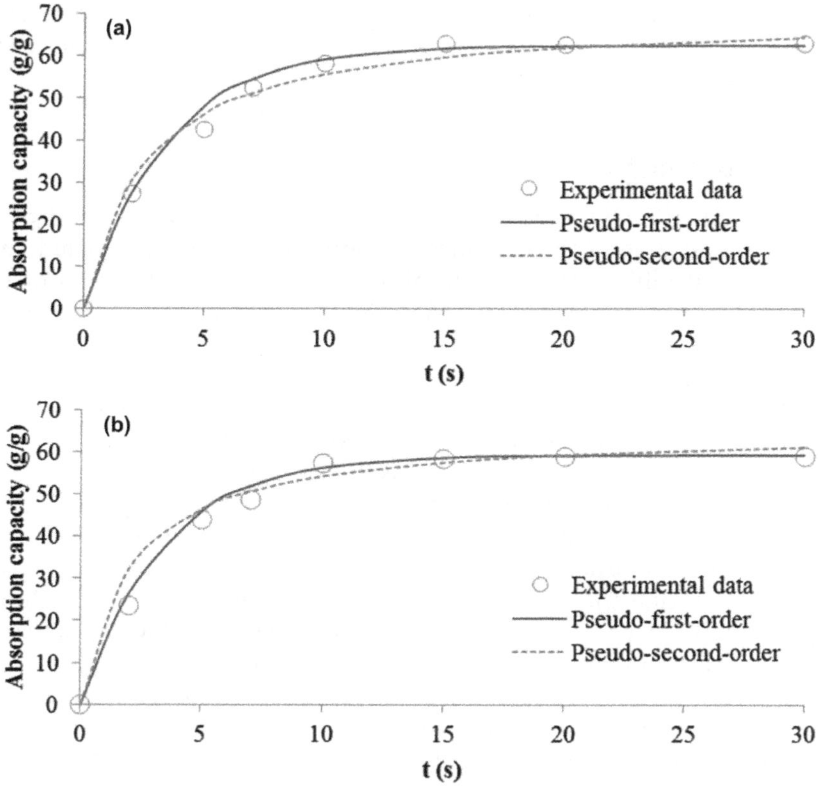

Figure 14.7 Experimental data fitted with the pseudo-first-order and pseudo-second-order models for the absorption kinetics of (a) the 5w50 motor oil and (b) the Singer machine oil on the aerogel with 0.50 wt.% of cellulose fibres at 25 °C.
(Reprinted from *Chemical Engineering Journal*, **270**, J. Feng, S. T. Nguyen, Z. Fan, H. M. Duong, Advanced fabrication and oil absorption properties of superhydrophobic recycled cellulose aerogels, 168–175, Copyright 2015,with permission from Elsevier).

Table 14.4 Activation energies of the absorption of the cellulose aerogels with various cellulose concentrations on the oils, using the pseudo-first-order and pseudo-second-order models. (Reprinted from *Chemical Engineering Journal*, **270**, J. Feng, S. T. Nguyen, Z. Fan, H. M. Duong, Advanced fabrication and oil absorption properties of superhydrophobic recycled cellulose aerogels, 168–175, Copyright 2015, with permission from Elsevier).

Activation energies (J mol^{-1})	Initial cellulose concentration (wt.%)	0.50	0.75	1.00
5w50 motor oil	Pseudo-first-order	3609.6	6984.6	9331.6
	Pseudo-second-order	10682.7	13120.3	18612.6
Singer machine oil	Pseudo-first-order	3090.5	4248.0	8130.2
	Pseudo-second-order	9354.9	10610.3	15240.4

In addition, compared with the 5w50 motor oil, the Singer machine oil had lower activation energy values, which made the oil absorption on the cellulose aerogels more effective.

14.2.4 Summary

In conclusion, the advanced and cost-effective fabrication method of the recycled cellulose aerogels was further improved. The MTMS-coated cellulose aerogels exhibited stable hydrophobicity during a test period of over five months. Their excellent oil absorption capacities were demonstrated with motor oil and Singer oil. It was found that the initial cellulose fibre concentration significantly affected the oil absorption capability of the developed cellulose aerogels. The 0.25 wt.% cellulose aerogel yielded a maximum absorption capacity of 95 $g\,g^{-1}$ with the 5w40 motor oil. The maximum absorption capacity of the cellulose aerogels could be reached at 50 °C, regardless of the pH values of the seawater/oil suspensions, and decreased with an increase in the cellulose fibre concentration.

The pseudo-first-order and pseudo-second-order kinetics models were applied to describe the oil absorption behaviour of the recycled cellulose aerogels for the first time. The pseudo-second-order model was more suited to the oil absorption kinetics study of the aerogels, due to its chemi-sorption nature. Moreover, the recycled cellulose aerogels displayed excellent flexibility: the large-scale cellulose aerogel could be easily bent or rolled without damaging its shape. All of the tested cellulose aerogels could also be compressed to up to 70% strain. The experimental results demonstrate that the super-hydrophobic recycled cellulose aerogels could be very promising sorbents for oil-spill cleaning.

14.3 Cellulose-based Aerogels for Heat-insulation Applications

14.3.1 Introduction

The greenhouse effect is gradually warming up the earth and potentially threatening human life. It was found that CO_2 emissions from buildings contributed approximately 31% of global greenhouse-gas emissions in 2010.[65,66] Improving thermal insulation of buildings is one of the most effective solutions for this issue. Therefore, many efforts have been made to develop new insulation materials.[67] Silica aerogels have been investigated as insulation materials for buildings.[68] However, they are very brittle. A flexible, aerogel-based insulation material has been developed by Aspen Aerogels (USA), but it is much more expensive than conventional insulation materials.[68] As a result, there is considerable need for insulation materials with reasonably low thermal conductivities and costs. These materials should also have high thermal stability for fire safety.

This section focuses on the thermal properties, such as thermal conductivity and thermal stability, of the recycled cellulose aerogels and their silica composites.[69] This is the first time that the benchmark data of the thermal properties of recycled cellulose-based aerogels have been reported. Of the recycled cellulose-based materials, the recycled cellulose aerogels using sodium hydroxide–urea aqueous solutions showed the lowest thermal conductivities (0.032 W mK^{-1}); however, the cellulose aerogels display a continuous weight loss from below 100 °C, during the thermogravimetric (TGA) test. The cellulose–silica composite aerogels displayed the best thermal stability; however, their thermal conductivities (0.039–0.041 W mK^{-1}) are much higher than those of the cellulose aerogels using sodium hydroxide–urea aqueous solutions (0.032 W mK^{-1}). In summary, the recycled cellulose aerogels using a Kymene binder (the aerogels with the lowest cost), exhibit lower thermal conductivities (0.034–0.037 W mK^{-1}) than those of the composite aerogels (0.039–0.041 W mK^{-1}), and a higher thermal stability (with a decomposition temperature of approximately 300 °C) surpassing that of the cellulose aerogels using sodium hydroxide–urea aqueous solutions. The recycled cellulose aerogels are therefore the most promising thermal insulation material.

14.3.2 Synthesis of Silica–Cellulose Aerogels

The catalyst solution was prepared by mixing 10.250 g of ammonium hydroxide and 0.927 g of ammonium fluoride in 50 ml of deionised (DI) water. After that, 6 ml of MTMS solution was mixed with 11 ml of ethanol and stirred for 5 min (forming an MTMS/ethanol solution). Then, 7 ml of DI water, 11 ml of ethanol, and 0.5 ml of the catalyst solution were mixed in another beaker (forming a water/ethanol/catalyst solution). While the MTMS/ethanol mixture was still being stirred, the obtained DI water/ethanol/catalyst solution was poured slowly into the MTMS/ethanol mixture and stirred for another 15 min to form the sol. The sol was poured into a mould that contained the cellulose aerogel, and the gelation and aging processes were conducted at room temperature (25 °C) for three days. After solvent exchange between the gel and DI water, the obtained hydrogel was frozen and dried using a Scan Vac CoolSafe 95-15 Pro freeze dryer (Denmark) for 24 h. Different silica–cellulose aerogels were synthesized from the different cellulose aerogel matrixes with varied cellulose fibre concentrations inside the initial cellulose aqueous suspensions. The fabrication method of pure recycle cellulose aerogel can be found in the section 14.2.2.

14.3.3 Thermal Properties of the Cellulose-based Aerogels

14.3.3.1 Thermal Properties of the Recycled Cellulose Aerogels Using a Kymene Binder

The thermal conductivities of the aerogels were measured under ambient conditions using a C-Therm TCi Thermal Conductivity Analyzer (C-Therm

Table 14.5 Effects of cellulose fibre concentrations on the thermal conductivities of the cellulose aerogels using a Kymene binder. (Reprinted from *Colloids and Surfaces A: Physicochemical and Engineering Aspects*, **506**, J. Feng, D. Le, S. T. Nguyen, V. T. C. Nien, D. Jewell, H. M. Duong, Silica cellulose hybrid aerogels for thermal and acoustic insulation applications, 298–305, Copyright 2016, with permission from Elsevier).

Cellulose fibre concentration (wt.%)	Density (g cm^{-3})	Thermal conductivity (W mK^{-1})
1.0	0.0392 ± 0.0005	0.0340 ± 0.0003
2.0	0.0473 ± 0.0004	0.0353 ± 0.0006
4.0	0.0592 ± 0.0032	0.0366 ± 0.0003

Technologies, Canada), with the MTPS method. Using this approach, the thermal conductivities of the recycled cellulose aerogels fabricated with a Kymene binder were measured. The porosity of the cellulose aerogels using a Kymene binder significantly affected their thermal conductivities. Table 14.5 shows that the thermal conductivity of the aerogels using a Kymene binder increased from 0.034 to 0.037 W mK^{-1} when the initial cellulose fibre concentration increased from 1.0 to 4.0 wt.%. Increases in the initial cellulose fibre concentration led to decreases in the porosity of the resultant cellulose aerogels. When the porosity is lower, there is more solid substance to promote thermal conduction and reduce thermal insulation.

The thermal stability of the cellulose aerogels using a Kymene binder was much better than that of the cellulose aerogels fabricated from the sodium hydroxide–urea aqueous solution. A weight loss of 6.5 wt.%, corresponding to the water release, was observed in the temperature range of 25–150 °C for the cellulose aerogel using a Kymene binder. Slow decomposition of the material was observed between 150 and 300 °C, and 86.3 wt.% remained of the cellulose aerogel using a Kymene binder at 300 °C. In comparison, 40.7 wt.% remained of the cellulose aerogel fabricated from the sodium hydroxide–urea aqueous solution at the same temperature. The obvious differences suggest that the thermal stability of the cellulose aerogel using a Kymene binder was much better than that of the cellulose aerogel synthesized *via* the sodium hydroxide–urea route. The differences between the thermal stabilities of these two cellulose aerogels could possibly be explained by two major factors: (1) there was no urea involved in the fabrication of the cellulose aerogels *via* the Kymene route, and so there was no urea residue;[39] and (2) the sonication method applied with the Kymene route, as a mechanical approach, had an advantage over the chemical treatment methods regarding thermal stability of the cellulose materials.[70,71]

14.3.3.2 Thermal Properties of the Silica–Cellulose Aerogels

The silica-cellulose composite aerogels were fabricated by immersing the cellulose matrixes inside the solutions containing a silica precursor. The cellulose matrixes were the recycled cellulose aerogels (obtained after hydrophobic coating) fabricated with a Kymene binder. The main purpose of the development of the silica–cellulose aerogels was to further improve the thermal stability and mechanical strength of the pure cellulose aerogels. The different silica–cellulose aerogels were synthesized from the varied cellulose matrixes with different initial recycled cellulose fibre concentrations (1.0–4.0 wt.%) in the initial cellulose suspensions.

14.3.3.2.1 Morphology and Hydrophobicity of the Silica–Cellulose Aerogels. The morphologies of the silica–cellulose composite aerogels were investigated with FE-SEM. Figure 14.8a–c display the SEM images of the silica–cellulose aerogels fabricated with the cellulose matrixes with different cellulose fibre concentrations in the initial cellulose aqueous suspensions. The images in Figure 14.8a–c suggest that the cellulose matrix served as the three-dimensional supporting frame.

The intersection points between the cellulose fibres were strengthened by a great number of hydrogen-bond links, which helped the formation of the strong supporting frame.[72] The strong supporting frame restricted the silica particles firmly within it by a confinement effect.[73] Meanwhile, the interconnected silica particles reinforced the cellulose matrix by attaching themselves to the matrix. As a result, a more rigid structure was formed. The structures of the silica–cellulose aerogels using different cellulose aerogel

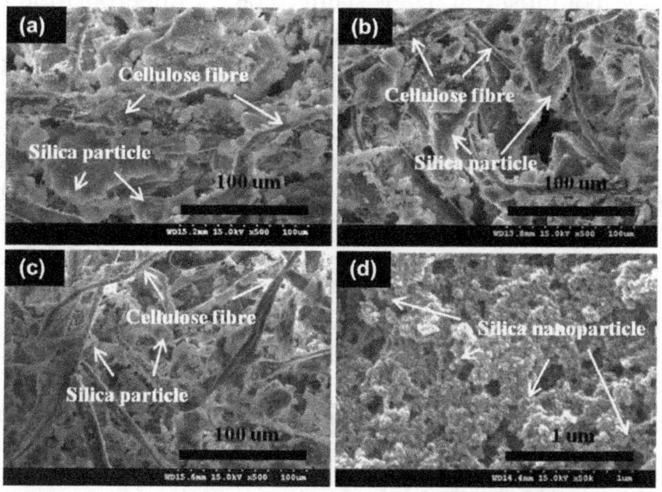

Figure 14.8 SEM images of the silica–cellulose aerogels fabricated with the cellulose matrixes with different cellulose fibre concentrations (a) 1.0 wt.%, (b) 2.0 wt.%, and (c) 4.0 wt.% in the initial cellulose aqueous suspensions. (d) A typical image of the zoomed-in silica region of the composites. (Reprinted from *Colloids and Surfaces A: Physicochemical and Engineering Aspects*, **506**, J. Feng, D. Le, S. T. Nguyen, V. T. C. Nien, D. Jewell, H. M. Duong, Silica cellulose hybrid aerogels for thermal and acoustic insulation Applications, 298–305, Copyright 2016, with permission from Elsevier).

matrixes were compared. When the cellulose content was higher, silica particles with a smaller particle size and more uniform distribution were found. This phenomenon might be explained by the following speculation. Before the freeze drying, the cellulose fibres were embedded in the silica hydrogel. During the freeze drying (including the freezing pre-treatment), the silica particles further away from the cellulose fibres experienced less force from the cellulose matrix than the silica particles next to the cellulose fibres. The denser cellulose matrix led to a more uniform distribution of small silica particles because of the more uniformly distributed force exerted by the cellulose matrix. On the other hand, the loose cellulose matrix with many large pores yielded a number of large silica particles located away from the cellulose fibres because of the small force exerted by the cellulose matrixes. At the same time, inside the loose cellulose matrix, the large silica particles located away from the cellulose fibres coexisted with the small silica particles that formed near the cellulose fibres, yielding a less uniform distribution.

The mesostructure of the silica particles was similar to the findings of other groups that developed silica–cellulose aerogels.[73,74] The BET surface areas of the silica–cellulose composite aerogels in Table 14.6 were between approximately 198 and 296 $m^2 g^{-1}$, comparable to the surface areas found in similar studies conducted by Demilecamps *et al.* (90–170 $m^2 g^{-1}$)[75] and Litschauer *et al.* (220–290 $m^2 g^{-1}$).[76]

As can be observed in Figure 14.9, the X-ray diffraction patterns of the silica–cellulose composite aerogels seem to be the superposition of those of the pure cellulose aerogels and the pure silica aerogels, which is similar to the findings of Cai *et al.*[77] This XRD finding implies that the extent of the chemical reaction between the cellulose fibres and the silica components was quite limited, as no new compound was detected. The XRD results suggest that it was reasonable for the mesoporous structure of the silica–cellulose aerogels to be controlled only by the silica components as the cellulose matrix did not possess a detectable BET surface area, and no new compound was formed. In addition, the XRD results indicate that the attachments between the cellulose fibres and the silica particles observed in the SEM images were physical instead of chemical.

According to Shi *et al.*, the hydrophobicity of the composite could improve the stability of the heat-insulation performance of the material.[74] To date, research on hydrophobic modifications of silica–cellulose composite aerogels

Table 14.6 Morphology studies of the cellulose–silica composites.

Cellulose fibre concentration inside the initial suspension for cellulose matrix fabrication (wt.%)	Density ($g\,cm^{-3}$)	SiO_2 in composite aerogel (wt.%)	BET surface area ($m^2 g^{-1}$)	Porosity (%)
1.0	0.149 ± 0.005	79 ± 4	296 ± 31	93.8 ± 0.0
2.0	0.146 ± 0.005	73 ± 5	248 ± 28	93.8 ± 0.0
4.0	0.138 ± 0.002	60 ± 3	198 ± 24	93.7 ± 0.0

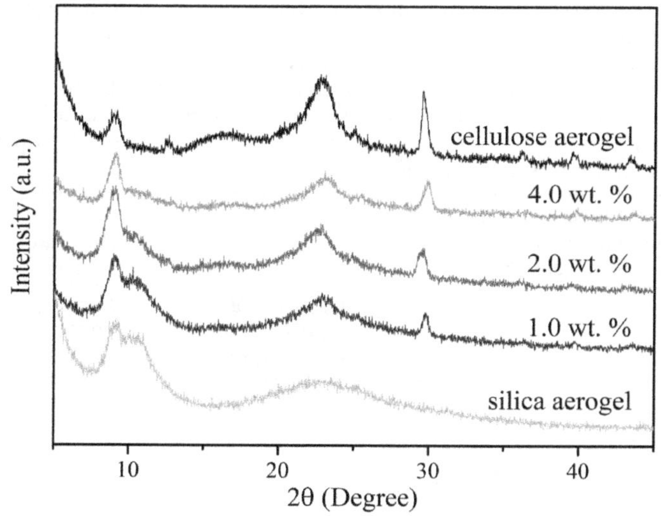

Figure 14.9 XRD patterns of the silica aerogel, the cellulose aerogel, and the silica-cellulose aerogels fabricated from cellulose matrixes with different cellulose fibre concentrations (1.0, 2.0, and 4.0 wt.%) inside the initial suspensions.
(Reprinted from *Colloids and Surfaces A: Physicochemical and Engineering Aspects*, **506**, J. Feng, D. Le, S. T. Nguyen, V. T. C. Nien, D. Jewell, H. M. Duong, Silica cellulose hybrid aerogels for thermal and acoustic insulation Applications, 298–305, Copyright 2016, with permission from Elsevier).

has been limited. In 2013, Shi *et al.* applied a CCl_4 cold-plasma modification approach using gas ionization.[74] In the same year, Sai *et al.* proposed a solvent-immersion method involving a 24 h aging followed by freeze drying.[73] Both methods might be considered uneconomical because of either the expensive equipment or the large amount of chemicals and the long duration involved. Moreover, the water contact angles of the modified composite aerogels from the above two methods were similar (132° and 133°), indicating that the aerogels were not super-hydrophobic.[73,74] However, the silica–cellulose composite aerogels exhibited super-hydrophobic properties. The average water contact angle was approximately 151° for all three composites. The excellent water-repelling property of the composites was inherited from both the hydrophobic cellulose matrix and the silica precursor (MTMS). Meanwhile, the stable methyl group of MTMS was responsible for the excellent hydrophobicity of the silica components.[78] Further modification of the fabricated composite aerogels was not required to achieve the super-hydrophobicity, which could be considered a remarkable advantage.

14.3.3.2.2 Thermal Conductivity of the Silica–Cellulose Aerogels. The thermal conductivities of the silica–cellulose aerogels (0.0385–0.0410 $W\,mK^{-1}$) increase with the increase in density (0.138–0.149 $g\,cm^{-3}$). With references to Table 14.5 and Table 14.7, the thermal conductivities and the densities of

Table 14.7 The thermal conductivities of the silica–cellulose aerogels fabricated from cellulose matrixes with different cellulose fibre concentrations in the initial suspensions. (Reprinted from *Colloids and Surfaces A: Physicochemical and Engineering Aspects*, **506**, J. Feng, D. Le, S. T. Nguyen, V. T. C. Nien, D. Jewell, H. M. Duong, Silica cellulose hybrid aerogels for thermal and acoustic insulation Applications, 298–305, Copyright 2016, with permission from Elsevier).

Cellulose fibre concentration in the initial suspension for cellulose-matrix fabrication (wt.%)	Density ($g\,cm^{-3}$)	Thermal conductivity ($W\,mK^{-1}$)
1.0	0.149 ± 0.005	0.0410 ± 0.0003
2.0	0.146 ± 0.005	0.0390 ± 0.0006
4.0	0.138 ± 0.002	0.0387 ± 0.0003

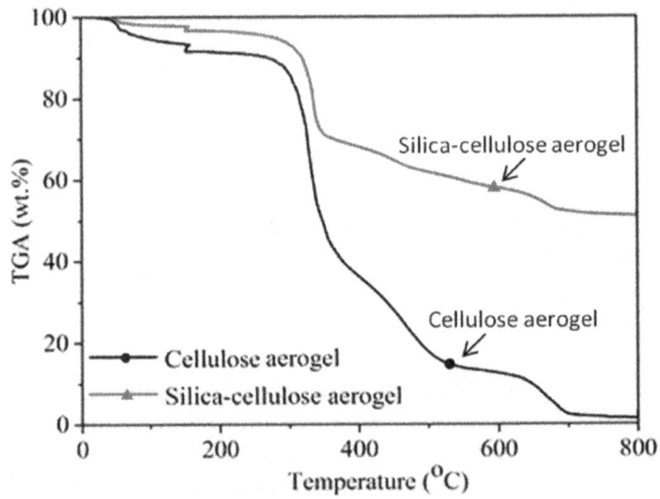

Figure 14.10 The thermogravimetric analysis (TGA) curve of the silica–cellulose aerogel, compared with that of the cellulose aerogel using a Kymene binder. The composite aerogels were fabricated from cellulose matrixes with different cellulose fibre concentration of 4.0 wt.% in the initial suspensions.

the silica–cellulose aerogels (0.0385–0.0410 $W\,mK^{-1}$; 0.138–0.149 $g\,cm^{-3}$) were higher than those of the cellulose aerogels using a Kymene binder (0.034–0.037 $W\,mK^{-1}$; 0.039–0.059 $g\,cm^{-3}$). One possible reason is that the thermal conductivities increase with density.[79,80] The thermal conductivities of the silica–cellulose aerogels (0.0385–0.0410 $W\,mK^{-1}$) were lower than those of the silica–cellulose composites fabricated by other groups (0.15 $W\,mK^{-1}$), and comparable to those of conventional insulation materials, such as polyurethane foams (0.02–0.04 $W\,mK^{-1}$) and insulation boards (0.035–0.16 $W\,mK^{-1}$).[81]

As shown in Figure 14.10, it is clear that the thermal stability of the composite was better than that of the pure cellulose aerogel without the silica embedment. A small weight loss of approximately 2 wt.% of the composites before 150 °C corresponds to the water release. The cellulose aerogel using a Kymene binder showed an approximately 7 wt.% weight loss up to 150 °C, which is greater than that of the composite. This smaller weight loss from the desorbed water was due to the super-hydrophobicity of the composite.

A 25 °C delay in the thermal degradation of the cellulose component of the composite was observed at 325 °C, compared with 300 °C before the silica modification. This increase in the degradation temperature might have been due to an interaction between the silica and the cellulose matrix at higher temperatures, and the interaction showed a positive effect on the thermal resistance of the composite.[81,82] At 300 °C, 93 wt.% silica–cellulose aerogels remained, as opposed to 86 wt.% of cellulose aerogels using the Kymene binder, demonstrating a better thermal stability. The slow mass loss of the residual substances, followed the rapid oxidative decomposition stage, after which the oxidation of the charred residue occurred and continued to the plateau. At the plateau, 51 wt.% of the silica–cellulose aerogel was preserved, mainly containing the silica.

14.3.3.2.3 Mechanical Properties of the Silica–Cellulose Aerogels. The silica modification did not impede the ductile behaviour of the silica–cellulose aerogels. The compressive strain–stress curves of the silica–cellulose aerogels fabricated from different cellulose matrixes are displayed in Figure 14.11.

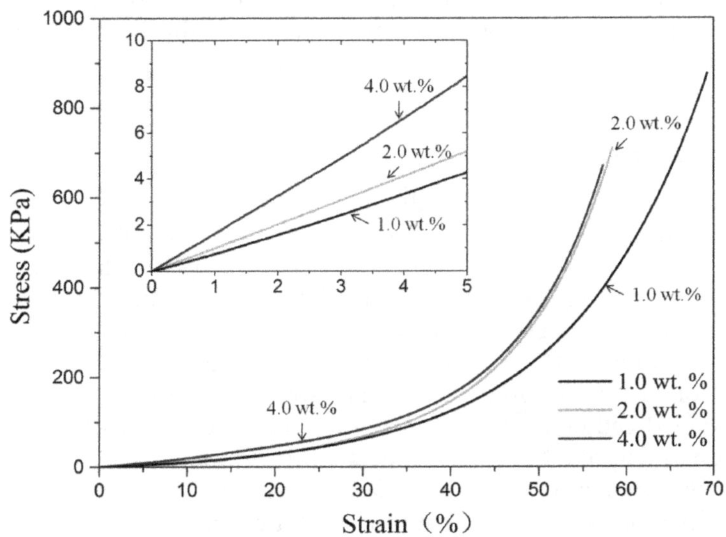

Figure 14.11 The compressive strain–stress curves of the silica–cellulose aerogels fabricated from the cellulose matrixes with different cellulose fibre concentrations (1.0, 2.0, and 4.0 wt.%) in the initial suspensions. The magnified section shows the compressive curves at the low strain (up to 5%) for the composite aerogels.

Table 14.8 The Young's modulus of the silica–cellulose composites and their cellulose aerogel matrixes. (Reprinted from Reference [69] with permission from Elsevier).

Cellulose fibre concentration in the initial suspension for cellulose matrix fabrication (wt.%)	Young's modulus of cellulose matrix (KPa)	Density of composite ($g\,cm^{-3}$)	Young's modulus of composite (KPa)
1.0	13 ± 1	0.149 ± 0.005	86 ± 3
2.0	21 ± 1	0.146 ± 0.005	104 ± 3
4.0	39 ± 2	0.138 ± 0.002	169 ± 5

Silica embedment improved the mechanical strength of the aerogels dramatically. The compressive Young's moduli of the composites were three to five times greater than those of their cellulose aerogel matrixes, as shown in Table 14.8. One contributing factor to the high mechanical strength of the silica–cellulose aerogels was their high densities compared with those of the cellulose matrixes, which is typical for many material systems, as the materials are more rigid and stiff, with a higher density.[83,84] From a microcosmic aspect, the silica particles restrained the bending of the cellulose fibres, improving the mechanical strength of the silica–cellulose aerogels.

For the silica–cellulose aerogels, the Young's modulus increased with slight decreases in density, as suggested in Table 14.8. This might possibly be explained by the increased cellulose content of the silica–cellulose aerogels. The cellulose matrix reinforces and constrains the silica granules, and the higher cellulose content leads to a more rigid cellulose matrix; thus the Young's modulus of the composite increases with the increase in the cellulose content.

14.3.4 Summary

In summary, the three recycled cellulose-based aerogels with three different thermal property combinations have been discussed. The thermal conductivity of the cellulose aerogel using a sodium hydroxide–urea aqueous solution was 0.032 $W\,mK^{-1}$, which was the lowest among the recycled cellulose fibre-based aerogels. However, the cellulose content of this material begins to degrade at 150 °C; therefore the material will require substantial modifications if the thermal stability is a major concern.

The recycled cellulose aerogels using a Kymene binder with cost-effective fabrication displayed better thermal stability than the cellulose aerogels using a sodium hydroxide–urea aqueous solution. However, the thermal conductivities of the aerogels using a Kymene binder were between 0.034 and 0.037 $W\,mK^{-1}$. These thermal conductivity values are slightly higher than those of the cellulose aerogels using a sodium hydroxide–urea aqueous solution.

To improve the thermal stability of the recycled cellulose-based aerogels further, the silica–cellulose composite aerogels were successfully developed.

A 25 °C delay in the cellulose degradation was also observed for the composites. The developed silica–cellulose aerogels also had better mechanical strengths (compressive Young's modulus: 86–169 KPa) than those of the cellulose aerogels (compressive Young's modulus: 13–39 KPa). Moreover, the inherent super-hydrophobic property of the silica–cellulose aerogels is also desirable for reliable and long-term thermal insulation. The thermal conductivities of the silica–cellulose aerogels were approximately 0.04 $W\,mK^{-1}$, comparable to those of commercial insulation products and lower than those of the cellulose aerogels.

14.4 Protein-based Aerogels and Their Applications

Recently, natural protein-based aerogels have attracted increasing research interest due to their biocompatibility and biodegradability for food engineering and life science applications. A few proteins such as whey protein,[43–45] silk fibroin,[46,47] egg white protein,[48] and soy protein[49–52] have been exploited for the formation of aerogels. This section introduces the fabrication methods and the multi-properties of the different types of protein-based aerogels and their potential applications.

14.4.1 Whey Protein Aerogels

Whey protein is the collection of globular proteins isolated from whey, which is the liquid remaining after milk has been curdled and strained for cheese production. This protein is typically a mixture of beta-lactoglobulin, alpha-lactalbumin, bovine serum albumin and immunoglobulins.[85] Whey protein is widely used in the food industry applications including processed meats, bakery products, pasta, ice cream, and infant foods.[44,86] As a by-product of the manufacture of cheese or casein, the broad availability in western countries makes whey protein promising for environmentally friendly-materials.[44]

Betz *et al.* developed a novel whey protein-based aerogel for drug delivery applications.[43] Previous studies indicated that the water-insoluble whey protein hydrogels were applicable for the encapsulation of drugs.[87,88] The use of supercritical drying and freeze drying techniques resulted in the drying of covalently crosslinked hydrogels and generated a highly porous aerogel. The obtained mechanically stable aerogels possessed a high BET-surface area (310–388 $m^2\,g^{-1}$) and drug loading capacity (9.1–9.5%, w/w for ketoprofen). The covalent disulphide bonds avoided the collapse of the aerogel structure when placed in the presence of aqueous media and led to a pH-controlled swelling behaviour upon rehydration. Due to the water-insolubility of the whey protein aerogels, the drug-loaded aerogels showed a sustained drug release at gastric (pH 1.2) and intestinal (pH 6.8) simulated digestive pH conditions. Later, Chen *et al.* prepared whey protein aerogels with enhanced mechanical properties.[44] By blending with alginate, the compressive moduli of the obtained aerogels were significantly enhanced by crosslinking and further increased with increasing aerogel densities. Ahmadi *et al.* prepared whey protein aerogels blended with nanocrystalline and microcrystalline cellulose particles and achieved increased Young's modulus and elastic character of the aerogels.[45] It was found that the nanocrystalline and microcrystalline cellulose particles impacted the texture of whey protein aerogels to different extents. Blending with cellulose nanocrystalline particles reinforced the texture of protein aerogels and decreased the sub-100 nm pore volume. The obtained protein aerogels were further applied for fish oil encapsulation.

14.4.2 Silk Fibroin Aerogels

Silk fibroin is a natural protein, derived from the Bombyx mori silkworm. Due to their biocompatibility and biodegradability, silk fibroin based porous materials have been extensively investigated for biomedical applications.[47] Marin *et al.* developed silk fibroin aerogels as drug delivery devices for the extended release of ibuprofen, a candidate drug.[46] Ibuprofen, a candidate drug compound, was loaded into the silk fibroin aerogels using supercritical CO_2. The obtained aerogels demonstrated high surface areas of 424 ± 75 $m^2 g^{-1}$ and low densities of 0.058 ± 0.001 $g ml^{-1}$. Phosphate buffer solution soaking studies revealed that the silk fibroin aerogels did not swell, nor exhibited any weight loss for up to 6 h. Release of ibuprofen from SF aerogels was found to be governed by Fickian diffusion.

Later, Mallepally *et al.* applied silk fibroin aerogels as potential scaffolds for tissue engineering applications.[47] Silk fibroin aerogel scaffolds were successfully fabricated using sol–gel and supercritical CO_2 processing protocols. The morphology and textural properties of the silk fibroin aerogels could be tuned by the starting concentration of the silk fibroin aqueous solution. When the aqueous fibroin concentration increased from 2 to 6 wt.%, the surface area of the resultant aerogels increased from 260 to 308 $m^2 g^{-1}$ and the compressive modulus increased from 19.5 to 174 kPa. Silk fibroin cryogels were also synthesised, in order to study the effect of hydrogels pretreatments on the morphological and textural properties.[47] The silk fibroin aerogels demonstrated a surface area of 266 $m^2 g^{-1}$ and pore volume of 1.6 $cc\, g^{-1}$, which was significantly higher compared to the freeze-dried silk fibroin cryogels with a surface area of 45 $m^2 g^{-1}$ and a pore volume of 0.05 $cc\, g^{-1}$. These results make these silk fibroin aerogels promising for cell culture and tissue engineering applications. *In vitro* cell culture studies with human foreskin fibroblast cells showed that the silk fibroin aerogels were cytocompatible with human cells and the aerogel scaffold promoted their propagation.

14.4.3 Egg White Protein Aerogels

As a high-quality protein source with high nutrition, egg white protein offers versatile functional properties and it is widely used in the food industry.[89]

Native egg white consists of about 10 wt.% protein and 90 wt.% water.[89] When egg white is heated, the egg white proteins denature thermally and form a gel whose structure is influenced by pH, ionic strength and the type of added salt.[90] Selmer *et al.* developed egg white protein-based aerogels as a carrier material for food applications.[48] The aerogels were prepared by gelation of pasteurized and spray-dried egg white solutions and subsequent supercritical drying. The porosity and surface area of the aerogels could be tailored by controlling the pH and ionic strength of the solution during the gelation process. The largest BET-surface area of 380 $m^2 g^{-1}$ was achieved at pH 2, while the most mechanically stable aerogels were obtained at alkaline pH. In order to produce mechanically stable aerogels with high surface areas, pH values above the isoelectric point should be used. These protein-based aerogels were suitable as microencapsulation materials for sensitive or unpleasant tasting food additives.

14.4.4 Soy Protein Aerogels

Soy flour, as a great source of protein, contains approximately 42% protein, 20% oil, 33% carbohydrates, and 5% ash on a dry basis. Protein from soy is mainly composed of acidic amino acids such as aspartic and glutamic acids, non-polar amino acids (glycine, alanine, valine and leucine), basic amino acids (lysine and arginine), and less than 1% of cysteine.[91] The soy protein-based aerogels have been designed for different applications.[49–52] Amaral-Labat *et al.* reported the first natural organic aerogels at the 91% level, which were prepared from gelation of tannin–soy flour resin in aqueous solution, followed by supercritical drying in CO_2.[49] The resultant aerogels presented a low density of 0.19–0.25 $g\,cm^{-3}$ and high mesopore volumes up to 2.3 $cm^3 g^{-1}$. The pore structures of the aerogels were tuneable by varying the pH values in the synthesis processes. These soy–tannin aerogels have potential for various applications, such as catalyst supports, adsorption of acidic species in liquids or natural super-insulators.

To improve the mechanical properties of the soy protein aerogels, Arboleda *et al.* developed the soy protein–nanocellulose composite aerogels.[50] The nanofibrillar cellulose loading resulted in the morphology transition of the aerogels from network- to fibrillar-like, with a high density of interconnected cells. With soy protein loadings as high as ca. 70%, the aerogels showed a high compression modulus of 4.4 MPa. The developed aerogels can absorb water and other solvents while maintaining their integrity. Due to their low density and good mechanical properties, these soy protein based aerogels are interesting candidates for applications including packaging, thermal and acoustic insulation, and so forth.

14.5 Conclusions

Paper waste can be successfully converted into cellulose-based aerogels with high oil absorption capacities (95 $g\,g^{-1}$), good thermal-insulation performances (0.032 W mK^{-1}), and excellent water-repellent properties (151°). Recycled cellulose-based aerogels utilizing paper waste are promising and economical candidates for several applications, such as oil-spill cleaning and building thermal insulation. In addition, morphology control of the recycled cellulose aerogels could be efficiently achieved by changing the cellulose concentration. To obtain the hydrophobic coating, a simple but effective chemical vapour deposition method *via* MTMS was developed, and the stability of the hydrophobic coating was tested. The recycled cellulose aerogels generated from both fabrication methods demonstrated stable hydrophobic properties over the tested time span of five months.

The initial cellulose fibre concentration played a critical role in the absorption capacities: the decreases in the initial cellulose fibre concentration led to increases in the absorption capacities due to the reduced porosities. The optimum temperature for high oil absorption capacity was 50 °C (of 25, 50, and 70 °C) because of the favourable viscosities of the oils at this temperature, as the viscosities were low enough for the oils to efficiently penetrate the porous aerogel networks, and high enough for the oils to effectively anchor to the pore walls of the cellulose aerogels. In addition, the pH values of the seawater/oil suspension had negligible effects on the oil absorption capacities of the cellulose aerogels, as the key factors affecting absorption capacity (the porosities of the cellulose aerogels and the viscosities of the oils) were independent of the environmental pH values.

The most cost-effective cellulose aerogels using a Kymene binder showed an observable improvement in thermal stability over the previous cellulose aerogels, because the Kymene method mainly applies a mechanical approach and no urea is involved. To improve the thermal stability further, the silica–cellulose aerogels were successfully developed. The major DTA peak also shifted from 352 °C for the cellulose aerogels to 530–550 °C for the silica–cellulose aerogels, as the major DTA peak of the composites is caused by the oxidation of the methyl group of the silica component. In addition, the silica–cellulose aerogels displayed an inherent super-hydrophobic property and better mechanical strength (Young's modulus: 86–169 KPa) than the cellulose aerogels using a Kymene binder (4–39 KPa). However, the thermal conductivities (approximately 0.04 W mK^{-1}) of the silica–cellulose aerogels were higher than those of the cellulose aerogels using a Kymene binder, due to their high density and low porosity.

Protein-based aerogels are novel biodegradable and biocompatible materials for lightweight food engineering and life science applications. The types of proteins, initial protein concentrations, processing techniques and many environmental conditions demonstrate significant influences on the structures and properties of the resultant aerogels. Further studies are required to tailor the morphological and textural properties of these protein-based aerogels with enhanced mechanical stability for practical applications.

References

1. O. A. Centeno-Chalé, M. L. Aguirre-Macedo, G. Gold-Bouchot and V. M. Vidal-Martínez, Effects of Oil Spill Related Chemical Pollution on Helminth Parasites in Mexican Flounder Cyclopsetta Chittendeni from the Campeche Sound, Gulf of Mexico, *Ecotoxicol. Environ. Saf.*, 2015, **119**, 162–169.
2. D. Li, F. Z. Zhu, J. Y. Li, P. Na and N. Wang, Preparation and Characterization of Cellulose Fibers from Corn Straw as Natural Oil Sorbents, *Ind. Eng. Chem. Res.*, 2012, **52**(1), 516–524.
3. H. Li, L. Liu and F. Yang, Hydrophobic Modification of Polyurethane Foam for Oil Spill Cleanup, *Mar. Pollut. Bull.*, 2012, **64**(8), 1648–1653.
4. V. Singh, R. J. Kendall, K. Hake and S. Ramkumar, Crude Oil Sorption by Raw Cotton, *Ind. Eng. Chem. Res.*, 2013, **52**(18), 6277–6281.
5. J. Wang, Y. Zheng, Y. Kang and A. Wang, Investigation of Oil Sorption Capability of Pbma/Sio 2 Coated Kapok Fiber, *Chem. Eng. J.*, 2013, **223**, 632–637.
6. N. Xiao, Y. Zhou, Z. Ling and J. Qiu, Synthesis of a Carbon Nanofiber/Carbon Foam Composite from Coal Liquefaction Residue for the Separation of Oil and Water, *Carbon*, 2013, **59**, 530–536.
7. K. Zhu, Y.-Y. Shang, P.-Z. Sun, Z. Li, X.-M. Li and J.-Q. Wei *et al.*, Oil Spill Cleanup from Sea Water by Carbon Nanotube Sponges, *Front. Mater. Sci.*, 2013, 7(2), 170–176.
8. Q. Zhu, Y. Chu, Z. Wang, N. Chen, L. Lin and F. Liu *et al.*, Robust Superhydrophobic Polyurethane Sponge as a Highly Reusable Oil-Absorption Material, *J. Mater. Chem. A*, 2013, **1**(17), 5386–5393.
9. D. Cormack, B. Lynch and B. Dowsett, Evaluation of Dispersant Effectiveness, *Oil Chem. Pollut.*, 1986, **3**(2), 87–103.
10. M. F. Fingas, R. Stoodley and N. Laroche, Effectiveness Testing of Spill-Treating Agents, *Oil Chem. Pollut.*, 1990, 7(4), 337–348.
11. A. Bayat, S. F. Aghamiri, A. Moheb and G. R. Vakili-Nezhaad, Oil Spill Cleanup from Sea Water by Sorbent Materials, *Chem. Eng. Technol.*, 2005, **28**(12), 1525–1528.
12. R. M. Atlas, Petroleum Biodegradation and Oil Spill Bioremediation, *Mar. Pollut. Bull.*, , **31**(4), 178–182.
13. D. Dave and A. E. Ghaly, Remediation Technologies for Marine Oil Spills: A Critical Review and Comparative Analysis, *Am. J. Environ. Sci.*, 2011, 7(5), 423.
14. A. A. Al-Majed, A. R. Adebayo, M. E. Hossain and A. Sustainable, Approach to Controlling Oil Spills, *J. Environ. Manage.*, 2012, **113**, 213–227.
15. Y. Chu and Q. Pan, Three-Dimensionally Macroporous Fe/C Nanocomposites as Highly Selective Oil-Absorption Materials, *ACS Appl. Mater. Interfaces*, 2012, **4**(5), 2420–2425.
16. C. Cojocaru, M. Macoveanu and I. Cretescu, Peat-Based Sorbents for the Removal of Oil Spills from Water Surface: Application of Artificial Neural Network Modeling, *Colloids Surf., A*, 2011, **384**(1), 675–684.

17. X. Gui, H. Li, K. Wang, J. Wei, Y. Jia and Z. Li *et al.*, Recyclable Carbon Nanotube Sponges for Oil Absorption, *Acta Mater.*, 2011, **59**(12), 4798–4804.
18. X. Gui, Z. Zeng, Z. Lin, Q. Gan, R. Xiang and Y. Zhu *et al.*, Magnetic and Highly Recyclable Macroporous Carbon Nanotubes for Spilled Oil Sorption and Separation, *ACS Appl. Mater. Interfaces*, 2013, **5**(12), 5845–5850.
19. M. O. Adebajo, R. L. Frost, J. T. Kloprogge, O. Carmody and S. Kokot, Porous Materials for Oil Spill Cleanup: A Review of Synthesis and Absorbing Properties, *J. Porous Mater.*, 2003, **10**(3), 159–170.
20. D. Peng, Z. Lan, C. Guo, C. Yang and Z. Dang, Application of Cellulase for the Modification of Corn Stalk: Leading to Oil Sorption, *Bioresour. Technol.*, 2013, **137**, 414–418.
21. D. Bastani, A. Safekordi, A. Alihosseini and V. Taghikhani, Study of Oil Sorption by Expanded Perlite at 298.15 K, *Sep. Purif. Technol.*, 2006, **52**(2), 295–300.
22. O. Karakasi and A. Moutsatsou, Surface Modification of High Calcium Fly Ash for Its Application in Oil Spill Clean Up, *Fuel*, 2010, **89**(12), 3966–3970.
23. V. Rajakovic, G. Aleksic, M. Radetic and L. Rajakovic, Efficiency of Oil Removal from Real Wastewater with Different Sorbent Materials, *J. Hazard. Mater.*, 2007, **143**(1), 494–499.
24. G. Deschamps, H. Caruel, M.-E. Borredon, C. Bonnin and C. Vignoles, Oil Removal from Water by Selective Sorption on Hydrophobic Cotton Fibers. 1. Study of Sorption Properties and Comparison with Other Cotton Fiber-Based Sorbents, *Environ. Sci. Technol.*, 2003, **37**(5), 1013–1015.
25. R. Wahi, L. A. Chuah, T. S. Y. Choong, Z. Ngaini and M. M. Nourouzi, Oil Removal from Aqueous State by Natural Fibrous Sorbent: An Overview, *Sep. Purif. Technol.*, 2013, **113**, 51–63.
26. X. Yuan and T. M. Chung, Novel Solution to Oil Spill Recovery: Using Thermodegradable Polyolefin Oil Superabsorbent Polymer (Oil–Sap), *Energy Fuels*, 2012, **26**(8), 4896–4902.
27. N. T. Cervin, C. Aulin, P. T. Larsson and L. Wågberg, Ultra Porous Nanocellulose Aerogels as Separation Medium for Mixtures of Oil/Water Liquids, *Cellulose*, 2012, **19**(2), 401–410.
28. J. T. Korhonen, M. Kettunen, R. H. Ras and O. Ikkala, Hydrophobic Nanocellulose Aerogels as Floating, Sustainable, Reusable, and Recyclable Oil Absorbents, *ACS Appl. Mater. Interfaces*, 2011, **3**(6), 1813–1816.
29. K. C. Payne, C. D. Jackson, C. E. Aizpurua, O. J. Rojas and M. A. Hubbe, Oil Spills Abatement: Factors Affecting Oil Uptake by Cellulosic Fibers, *Environ. Sci. Technol.*, 2012, **46**(14), 7725–7730.
30. J. Wang, Y. Zheng and A. Wang, Effect of Kapok Fiber Treated with Various Solvents on Oil Absorbency, *Ind. Crops Prod.*, 2012, **40**, 178–184.
31. J. Wang, Y. Zheng and A. Wang, Coated Kapok Fiber for Removal of Spilled Oil, *Mar. Pollut. Bull.*, 2013, **69**(1), 91–96.

32. A. Nourbakhsh and A. Ashori, Particleboard Made from Waste Paper Treated with Maleic Anhydride, *Waste Manage. Res.*, 2010, **28**(1), 51–55.
33. L. Szabó, A. Soria, J. Forsström, J. T. Keränen and E. Hytönen, A World Model of the Pulp and Paper Industry: Demand, Energy Consumption and Emission Scenarios to 2030, *Environ. Sci. Policy*, 2009, **12**(3), 257–269.
34. Paper Recycling, Available from http://Www.Epa.Gov/Osw/Conserve/Materials/Paper/Basic_Info.Htm.
35. D. Mishra and V. K. Sinha, Value-Based Polymer Precursors from Paper Waste and Its Application in Polyurethane Foams, *J. Cell. Plast.*, 2009, **46**(1), 15–30.
36. L. Wang, R. Templer and R. J. Murphy, High-Solids Loading Enzymatic Hydrolysis of Waste Papers for Biofuel Production, *Appl. Energy*, 2012, **99**, 23–31.
37. Y. Ikeda, E. Y. Park and N. Okuda, Bioconversion of Waste Office Paper to Gluconic Acid in a Turbine Blade Reactor by the Filamentous Fungus Aspergillus Niger, *Bioresour. Technol.*, 2006, **97**(8), 1030–1035.
38. The Paper Stock Report: Paper Recycling Online, Available from http://Www.Recycle.Cc/Freepapr.Htm, 2015.
39. J. Feng, S. T. Nguyen, Z. Fan and H. M. Duong, Advanced Fabrication and Oil Absorption Properties of Super-Hydrophobic Recycled Cellulose Aerogels, *Chem. Eng. J.*, 2015, **270**, 168–175.
40. S. T. Nguyen, J. Feng, N. T. Le, A. T. Le, N. Hoang and V. B. Tan *et al.*, Cellulose Aerogel from Paper Waste for Crude Oil Spill Cleaning, *Ind. Eng. Chem. Res.*, 2013, **52**(51), 18386–18391.
41. S. T. Nguyen, J. Feng, S. K. Ng, J. P. Wong, V. B. Tan and H. M. Duong, Advanced Thermal Insulation and Absorption Properties of Recycled Cellulose Aerogels, *Colloids Surf., A*, 2014, **445**, 128–134.
42. L. Heath and W. Thielemans, Cellulose Nanowhisker Aerogels, *Green Chem.*, 2010, **12**(8), 1448–1453.
43. M. Betz, C. García-González, R. Subrahmanyam, I. Smirnova and U. Kulozik, Preparation of Novel Whey Protein-Based Aerogels as Drug Carriers for Life Science Applications, *J. Supercrit. Fluids*, 2012, **72**, 111–119.
44. H.-B. Chen, Y.-Z. Wang and D. A. Schiraldi, Foam-Like Materials Based on Whey Protein Isolate, *Eur. Polym. J.*, 2013, **49**(10), 3387–3391.
45. M. Ahmadi, A. Madadlou and A. A. Saboury, Whey Protein Aerogel as Blended with Cellulose Crystalline Particles or Loaded with Fish Oil, *Food Chem.*, 2016, **196**, 1016–1022.
46. M. A. Marin, R. R. Mallepally and M. A. McHugh, Silk Fibroin Aerogels for Drug Delivery Applications, *J. Supercrit. Fluids*, 2014, **91**, 84–89.
47. R. R. Mallepally, M. A. Marin, V. Surampudi, B. Subia, R. R. Rao and S. C. Kundu *et al.*, Silk Fibroin Aerogels: Potential Scaffolds for Tissue Engineering Applications, *Biomed. Mater.*, 2015, **10**(3), 035002.
48. I. Selmer, C. Kleemann, U. Kulozik, S. Heinrich and I. Smirnova, Development of Egg White Protein Aerogels as New Matrix Material for Microencapsulation in Food, *J. Supercrit. Fluids*, 2015, **106**, 42–49.

49. G. Amaral-Labat, L. Grishechko, A. Szczurek, V. Fierro, A. Pizzi and B. Kuznetsov *et al.*, Highly Mesoporous Organic Aerogels Derived from Soy and Tannin, *Green Chem.*, 2012, **14**(11), 3099–3106.
50. J. C. Arboleda, M. Hughes, L. A. Lucia, J. Laine, K. Ekman and O. J. Rojas, Soy Protein–Nanocellulose Composite Aerogels, *Cellulose*, 2013, **20**(5), 2417–2426.
51. L. Sun, W. Chen, Y. Liu, J. Li and H. Yu, Soy Protein Isolate/Cellulose Nanofiber Complex Gels as Fat Substitutes: Rheological and Textural Properties and Extent of Cream Imitation, *Cellulose*, 2015, **22**(4), 2619–2627.
52. Y. Zhuang, F. Yu, J. Ma and J. Chen, Facile Synthesis of Three-Dimensional Graphene–Soy Protein Aerogel Composites for Tetracycline Adsorption, *Desalin. Water Treat.*, 2016, **57**(20), 9510–9519.
53. S. Hoepfner, L. Ratke and B. Milow, Synthesis and Characterisation of Nanofibrillar Cellulose Aerogels, *Cellulose*, 2008, **15**(1), 121–129.
54. N. Isobe, S. Kimura, M. Wada and S. Kuga, Mechanism of Cellulose Gelation from Aqueous Alkali-Urea Solution, *Carbohydr. Polym.*, 2012, **89**(4), 1298–1300.
55. J. Cai and L. Zhang, Unique Gelation Behavior of Cellulose in Naoh/Urea Aqueous Solution, *Biomacromolecules*, 2006, 7(1), 183–189.
56. V. Stannett. Mechanisms of Wet-Strength Development in Paper. Surfaces and Coatings Related to Paper and Wood: A Symposium [held At] State University College of Forestry at Syracuse University: Syracuse University Press Syracuse, NY, 1967, pp. 269–300.
57. J. Cai, S. Kimura, M. Wada, S. Kuga and L. Zhang, Cellulose Aerogels from Aqueous Alkali Hydroxide–Urea Solution, *ChemSusChem*, 2008, **1**(1–2), 149–154.
58. W. Chen, H. Yu, Q. Li, Y. Liu and J. Li, Ultralight and Highly Flexible Aerogels with Long Cellulose I Nanofibers, *Soft Matter*, 2011, 7(21), 10360–10368.
59. R. T. Olsson, M. A. Samir, G. Salazar-Alvarez, L. Belova, V. Ström and L. A. Berglund *et al.*, Making Flexible Magnetic Aerogels and Stiff Magnetic Nanopaper Using Cellulose Nanofibrils as Templates, *Nat. Nanotechnol.*, 2010, **5**(8), 584–588.
60. Y. Chen and D. Zhang, Adsorption Kinetics, Isotherm and Thermodynamics Studies of Flavones from Vaccinium Bracteatum Thunb Leaves on Nka-2 Resin, *Chem. Eng. J.*, 2014, **254**, 579–585.
61. B. Wu and M. Zhou, Recycling of Waste Tyre Rubber into Oil Absorbent, *Waste Manage.*, 2009, **29**(1), 355–359.
62. P. Saha and S. Chowdhury. Insight into Adsorption Thermodynamics: INTECH Open Access Publisher; 2011.
63. P. Sharma, B. K. Saikia and M. R. Das, Removal of Methyl Green Dye Molecule from Aqueous System Using Reduced Graphene Oxide as an Efficient Adsorbent: Kinetics, Isotherm and Thermodynamic Parameters, *Colloids Surf., A*, 2014, **457**, 125–133.
64. A. M. Vargas, A. L. Cazetta, M. H. Kunita, T. L. Silva and V. C. Almeida, Adsorption of Methylene Blue on Activated Carbon Produced from

Flamboyant Pods (Delonix Regia): Study of Adsorption Isotherms and Kinetic Models, *Chem. Eng. J.*, 2011, **168**(2), 722–730.
65. S. D. L. R. du Can, L. Price and T. Zwickel, Understanding the Full Climate Change Impact of Energy Consumption and Mitigation at the End-Use Level: A Proposed Methodology for Allocating Indirect Carbon Dioxide Emissions, *Appl. Energy*, 2015, **159**, 548–559.
66. I. P. O. C. Change. Climate Change 2014: Mitigation of Climate Change: Cambridge University Press; 2015.
67. B. P. Jelle, Traditional, State-of-the-Art and Future Thermal Building Insulation Materials and Solutions–Properties, Requirements and Possibilities, *Energy Build.*, 2011, **43**(10), 2549–2563.
68. R. Baetens, B. P. Jelle and A. Gustavsen, Aerogel Insulation for Building Applications: A State-of-the-Art Review, *Energy Build.*, 2011, **43**(4), 761–769.
69. J. Feng, D. Le, S. T. Nguyen, V. T. C. Nien, D. Jewell and H. M. Duong, Silica-Cellulose Hybrid Aerogels for Thermal and Acoustic Insulation Applications, *Colloids Surf., A*, 2016, **506**, 298–305.
70. S. Cho, Y. Choi, D. Park, S. Heo, D. Kim and H. Jin, Cellulose Nanocrystals with High Thermal Stability and Their Nanocomposites with Poly (Lactic Acid), *Int. Conf. Compos. Mater.*, 2011, 3–46.
71. G. Tonoli, E. Teixeira, A. Corrêa, J. Marconcini, L. Caixeta and M. Pereira-da-Silva *et al.*, Cellulose Micro/Nanofibres from Eucalyptus Kraft Pulp: Preparation and Properties, *Carbohydr. Polym.*, 2012, **89**(1), 80–88.
72. D. Klemm, B. Heublein, H. P. Fink and A. Bohn, Cellulose: Fascinating Biopolymer and Sustainable Raw Material, *Angew. Chem., Int. Ed.*, 2005, **44**(22), 3358–3393.
73. H. Sai, L. Xing, J. Xiang, L. Cui, J. Jiao and C. Zhao *et al.*, Flexible Aerogels Based on an Interpenetrating Network of Bacterial Cellulose and Silica by a Non-Supercritical Drying Process, J. Mater. Chem. *A*, 2013, **1**(27), 7963–7970.
74. J. Shi, L. Lu, W. Guo, J. Zhang and Y. Cao, Heat Insulation Performance, Mechanics and Hydrophobic Modification of Cellulose–Sio 2 Composite Aerogels, *Carbohydr. Polym.*, 2013, **98**(1), 282–289.
75. A. Demilecamps, G. Reichenauer, A. Rigacci and T. Budtova, Cellulose–Silica Composite Aerogels from "One-Pot" Synthesis, *Cellulose*, 2014, **21**(4), 2625–2636.
76. M. Litschauer, M.-A. Neouze, E. Haimer, U. Henniges, A. Potthast and T. Rosenau *et al.*, Silica Modified Cellulosic Aerogels, *Cellulose*, 2011, **18**(1), 143–149.
77. J. Cai, S. Liu, J. Feng, S. Kimura, M. Wada and S. Kuga *et al.*, Cellulose–Silica Nanocomposite Aerogels by in Situ Formation of Silica in Cellulose Gel, *Angew. Chem.*, 2012, **124**(9), 2118–2121.
78. A. V. Rao, M. M. Kulkarni, D. Amalnerkar and T. Seth, Superhydrophobic Silica Aerogels Based on Methyltrimethoxysilane Precursor, *J. Non-Cryst. Solids*, 2003, **330**(1), 187–195.

79. H. Simmler, S. Brunner, U. Heinemann, H. Schwab, K. Kumaran and P. Mukhopadhyaya *et al.* Vacuum Insulation Panels. Study on Vip-Components and Panels for Service Life Prediction of Vip in Building Applications (Subtask a). *Final report for the IEA/ECBCS Annex*. 2005;39.
80. M. Kiran, A. Nandanwar, M. V. Naidu and K. C. V. Rajulu, Effect of Density on Thermal Conductivity of Bamboo Mat Board, *Int. J. Agric. Forestry*, 2012, **2**(5), 257–261.
81. S. Sequeira and D. V. Evtuguin, I. Portugal. Preparation and Properties of Cellulose/Silica Hybrid Composites, *Polym. Compos.*, 2009, **30**(9), 1275–1282.
82. N. Jia, S.-M. Li, M.-G. Ma, J.-F. Zhu and R.-C. Sun, Synthesis and Characterization of Cellulose-Silica Composite Fiber in Ethanol/Water Mixed Solvents, *BioResources*, 2011, **6**(2), 1186–1195.
83. K. E. Parmenter and F. Milstein, Mechanical Properties of Silica Aerogels, *J. Non-Cryst. Solids*, 1998, **223**(3), 179–189.
84. A. H. Alaoui, T. Woignier, G. W. Scherer and J. Phalippou, Comparison between Flexural and Uniaxial Compression Tests to Measure the Elastic Modulus of Silica Aerogel, *J. Non-Cryst. Solids*, 2008, **354**(40), 4556–4561.
85. A. Haug, A. T. Høstmark and O. M. Harstad, Bovine Milk in Human Nutrition–a Review, *Lipids Health Dis.*, 2007, **6**(1), 25.
86. S. M. Fitzsimons, D. M. Mulvihill and E. R. Morris, Co-Gels of Whey Protein Isolate with Crosslinked Waxy Maize Starch: Analysis of Solvent Partition and Phase Structure by Polymer Blending Laws, *Food Hydrocolloids*, 2008, **22**(3), 468–484.
87. M. Betz and U. Kulozik, Whey Protein Gels for the Entrapment of Bioactive Anthocyanins from Bilberry Extract, *Int. Dairy J.*, 2011, **21**(9), 703–710.
88. S. Gunasekaran, L. Xiao and M. Ould Eleya, Whey Protein Concentrate Hydrogels as Bioactive Carriers, *J. Appl. Polym. Sci.*, 2006, **99**(5), 2470–2476.
89. Y. Mine, Recent Advances in the Understanding of Egg White Protein Functionality, *Trends Food Sci. Technol.*, 1995, **6**(7), 225–232.
90. T. Croguennec, F. Nau and G. Brule, Influence of Ph and Salts on Egg White Gelation, *J. Food Sci.*, 2002, **67**(2), 608–614.
91. J.-M. Raquez, M. Deléglise, M.-F. Lacrampe and P. Krawczak, Thermosetting (Bio) Materials Derived from Renewable Resources: A Critical Review, *Prog. Polym. Sci.*, 2010, **35**(4), 487–509.

CHAPTER 15

Applications of Polysaccharide and Protein Based Aerogels in Thermal Insulation

R. MUTHURAJ, C. JIMENEZ-SAELICES, Y. GROHENS AND B. SEANTIER*

Institut de Recherche Dupuy de Lome (IRDL)-CNRS UMR 6027, University of South Brittany, 56100 Lorient, France
*Email: bastien.seantier@univ-ubs.fr

15.1 Introduction

World energy consumption is steadily increasing due to both economical and population growth.[1,2] The major energy consuming sectors are building, transport and agriculture.[3] Each sector has its own impact on energy consumption. This impact depends on various parameters, such as the sector by itself, the main energy sources, and the global and local climate. The maximum energy consuming field is building sectors due to human comfort, for example heating, ventilation and air conditioning.[3,4] The increase in energy consumption also increases the CO_2 emissions. In many developed countries, almost one third of the greenhouse gases were emitted from the building sector in 2005.[5] When looking closer at the different sectors, their impact drastically varies depending on the country and region.[3,4] For example, the French residential sector is the most energy consuming.[6,7] Intensive efforts have been made to design and develop buildings that

Green Chemistry Series No. 58
Biobased Aerogels: Polysaccharide and Protein-based Materials
Edited by Sabu Thomas, Laly A. Pothan and Rubie Mavelil-Sam
© The Royal Society of Chemistry 2018
Published by the Royal Society of Chemistry, www.rsc.org

consume less energy. Using wall insulation is one way to reduce the greenhouse gas emissions from the building sector. Generally, conventional thermal insulation materials are thicker or multilayered to provide good insulation properties.[3] The thicker insulating materials are not preferable because they make the building heavier, reduce the living space and are more expensive. In Europe, more specifically in Sweden, several teams have conducted research to find a solution to reduce the energy consumption of buildings.[8,9] Sweden even encouraged the energy sector and the building sector to benchmark this work in order to find new ways to reduce the energy consumption of buildings. By using new materials for thermal insulation to optimize the energy and the space utilization in houses, many solutions have been proposed to reach the main goal for the next generation of buildings: obtaining Zero Energy Buildings.[10] Therefore, developing high performance thermal insulation materials is a crucial need for building insulation applications. Numerous efforts have been made to design and develop new thermal insulators meeting the requirements of superinsulation theory. However, building insulation is not the only way to use the newly designed renewable materials. The developed technologies may be used for clothes, transport insulation, cold chambers and food storage. Yet, most of the commercially available thermal insulators are fossil resources based materials and these resources are depleting rapidly.[3] For environmental safety and sustainable development, it is essential to develop renewable resource based thermal insulators with lower thermal conductivity for thermal insulation applications.[11] However, many challenges have to be overcome in order to use renewable resource based insulating materials for such applications.

An aerogel is produced by drying gels. Owing to the large surface area, high porosity, small pore size, and light weight, aerogels possess good thermal and acoustic insulation properties.[3,5,11,12] These unique properties of aerogels are suitable for building applications.[3,13] The annual growth rate of aerogels in thermal and acoustic insulation applications was 20.2% from 2012 to 2017.[11] Many organic and inorganic materials have been explored to produce aerogels for various applications.[3,14] Aerogel production from renewable sources has gained great attention because of concerns about sustainability. For instance, a great number of researchers have investigated the performances of polysaccharide and protein based aerogels and have found that these aerogels have good thermal insulation properties and good mechanical properties.[11,15–19] Therefore, this chapter aims at describing sustainable (*e.g.*, polysaccharide and protein) aerogel production and their thermal insulation properties for many applications. After describing the physical properties needed to obtain a good insulator and some existing materials, polysaccharide and protein based aerogels will be described and discussed for their thermal insulating properties. Finally, the main challenges that need to be overcome to produce good thermal insulators for buildings and other applications will be discussed.

15.2 Thermal Insulation

A thermal insulator is a material that is able to stop heat flow through it. In order to quantify their ability to stop heat flow and be good thermal insulators, specific properties have to be taken into account. Even if other thermal properties, such as diffusivity or specific heat capacity exist, a large number of studies are focused on determining the thermal conductivity of the insulator. Based on the thermal conductivity value, the materials thermal insulation properties can be defined. Thermal conductivity is measured in watts per kelvin-meter ($W(K\,m)^{-1}$). More precisely, this parameter predicts the rate of energy loss (in Watts, W) through a piece of material having a specific thickness (e, m) separating two mediums characterized by temperature difference (in kelvin, K). The total thermal conductivity of an aerogel (λ_{eff}) quantifies the global thermal behavior of a material and is the value measured by most of the existing techniques.

15.2.1 Thermal Conductivity

In order to better understand the evolution of λ_{eff}, it can be described by the parallel flux model.[20] In this model, the effective thermal conductivity is calculated by eqn (15.1), as the arithmetic sum of the contribution of conduction in a solid phase (λ_{cs}), conduction in gas phase (λ_{cg}) and radiation through pores (λ_{rad}).[21]

$$\lambda_{eff} = \lambda_{cs} + \lambda_{cg} + \lambda_{rad} \tag{15.1}$$

The solid conduction highly depends on the material density and its microstructure. To decrease the solid contribution and increase the thermal insulation properties, a material with low density is needed. In the aerogels, the solid contribution is reduced because of the large quantity of pores, which restricts the propagation of phonons in the aerogel skeleton.[22]

Concerning the gas conduction, it is necessary to reduce contacts between gas molecules. To reach this goal, one of the most effective ways is to form small pores that have a size lower than the free mean path of the gas molecules to confine gas molecules inside pores. This phenomenon is known as the Knudsen effect.[23] Typically, if the pore size of a material is lower than 70 nm, the gas conduction starts to be low enough to reach thermal superinsulating properties.[24]

Heat transfer through radiation is caused by the electromagnetic radiation that is emitted by all surfaces. Aerogels are formed by two phases with different refractive indices. Consequently, they are semi-transparent materials which are capable of diffusing the thermal radiation. These three ($\lambda_{cs} + \lambda_{cg} + \lambda_{rad}$) contributions are schematized on Figure 15.1.

Focusing on energy saving *via* thermal insulation properties, the new materials may be used for various insulation applications such as buildings, transport, and personal protection. Whatever the targeted application, the produced materials should have specific thermal properties with the best

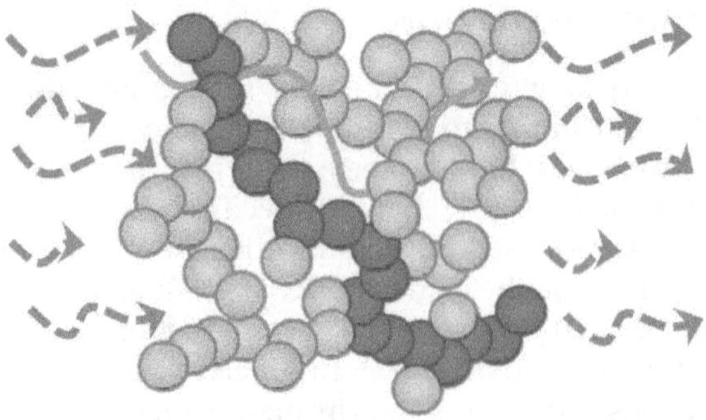

Figure 15.1 Schematic of aerogel structure and heat transfer through the materials. Darker beads represent the solid contribution through the phonons (λ_{cs}). The solid arrow is for the gas contribution (λ_{cg}). Dashed arrows symbolise the heat transfer due to radiation (λ_{rad}).

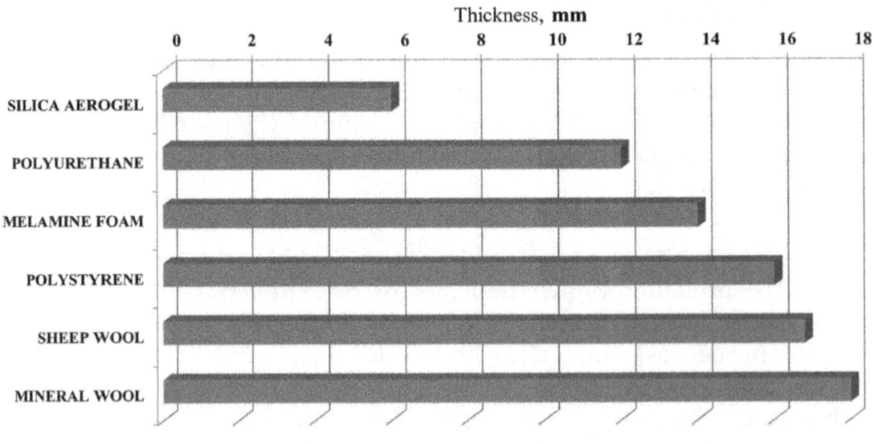

Figure 15.2 Thickness (in cm) of various materials to reach the targeted thermal resistance of 4 $m^2\,kW^{-1}$.

performances. This is the reason why new materials have been developed to obtain better thermal insulation properties. Some materials already exist and the insulation layer thickness to reach a thermal resistance of 4 $m^2\,kW^{-1}$ is represented in Figure 15.2. It can be seen from Figure 15.2, that the silica aerogel needs a lower thickness (~5.9 mm) than other materials because of its lower thermal conductivity.

The evolution of effective thermal conductivity of porous materials with density is shown in Figure 15.3a.[3] When the materials have macropores, the gas conduction contribution is quasi constant over a large range of densities. It is due to this fact that gas is not confined in small pores. Therefore, gas contribution of heat transfer is only due to the gas properties. In the case of air, the gas conduction contribution is the air thermal conductivity

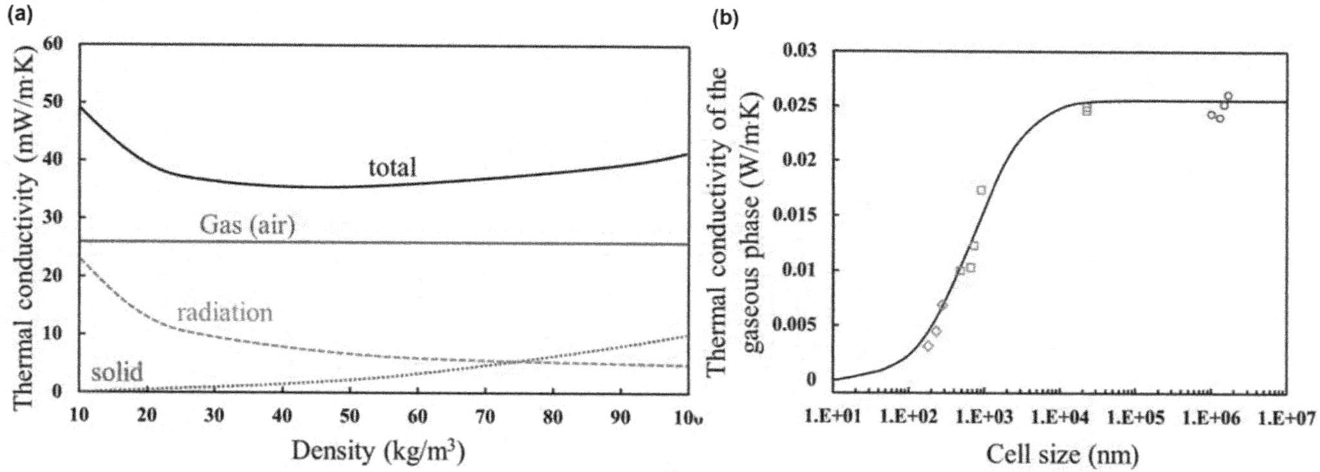

Figure 15.3 (a) Effective thermal conductivity, gas conduction contribution, solid conduction contribution and radiation contribution depending on the density. Reprinted from *Renewable and Sustainable Energy Reviews*, **34**, E. Cuce, P. M. Cuce, C. J. Wood and S. B. Riffat, Toward aerogel based thermal superinsulation in buildings: A comprehensive review, 273–299, Copyright 2014, with permission from Elsevier.[3] (b) Effective thermal conductivity depending on pore size (Knudsen effect). Reprinted from Polymer, 56, B. Notario, J. Pinto, E. Solorzano, J. A. de Saja, M. Dumon and M. A. Rodriguez-Perez, Experimental validation of the Knudsen effect in nanocellular polymeric forms, 57–67, Copyright 2015, with permission from Elsevier.[26]

(0.025 W (K m)$^{-1}$). The solid conduction decreases with a low density aerogel. Indeed, low density means that there is less solid to conduct heat and thus a lower solid contribution. On the other hand, the radiation contribution increases with lower density since radiation will be less attenuated by the solid interaction and will be able to conduct heat through the material. Therefore, the effective heat transfer strongly depends on the radiation and solid conduction for macroporous materials. For instance, expanded polystyrene, extruded polystyrene, mineral wool, loose-fill cellulose and foam glass fall into this category. As gas conductivity is dominating all contributions in the effective thermal conductivity of the aforementioned conventional insulation materials, the effective thermal conductivity is relatively high (between 0.033–0.068 W (K m)$^{-1}$).[25] Figure 15.3b shows the effect of the pore size or cell size on thermal insulation properties (Knudsen effect).[26] It emphasizes the importance to work on the material structure, for example, obtaining small pores (mesopores), in order to reduce the thermal conductivity as low as possible.

Thermal conductivity of materials allows classifying them as thermal conductors ($\lambda_{eff} \geq 0.1$ W (K m)$^{-1}$), insulators (0.1 W (K m)$^{-1}$ $> \lambda_{eff} \geq 0.025$ W (K m)$^{-1}$) and super-insulators ($\lambda_{eff} \leq 0.025$ W (K m)$^{-1}$). Metals are good thermal conductors because of their thermal conductivity values are a few tens to hundreds of W (K m)$^{-1}$. Glass wool, expanded polystyrene, extruded polystyrene, wood, and mineral wool are thermal insulators with the thermal conductivity being between 0.1 to 0.026 W (K m)$^{-1}$.[27,28] Silica aerogels, vacuum insulating panels and vacuum glazing are considered as super-insulating materials because their thermal conductivity is lower than 0.025 W (K m)$^{-1}$.[29]

This short overview of the different thermal contributions highlights the fact that the microstructure and the density of the material have a significant effect on the thermal insulation properties. Currently, there are many thermal insulators (Table 15.1) available and some of them will be described as reference materials to compare with in the next sections.

15.2.2 Existing Thermal Insulation Materials

Among existing thermal insulators, silica aerogel is one of the most extensively studied inorganic thermal insulating materials.[5] As the thermal conductivity of silica aerogel is lower than the thermal conductivity of air (Table 15.1), it is called a thermal super-insulating material.[30] Silica aerogels are commonly used in the fields of space and aeronautics because it is a high added value material that is difficult to afford for common building applications.[31] Clay aerogels are another good inorganic thermal insulator. They have low thermal conductivity (Table 15.1) because of their unique lamellar microstructures, which reduces the heat flow in the aerogels.[32] The heat flow within the 3D aerogel is demonstrated in Figure 15.1. Within the 3D network, the solid conduction is limited due to the high porosity and low density, while the 3D network creates a tortuous path for gas molecules to

Table 15.1 Thermal conductivity of some existing materials.

Existing thermal insulation materials	Total thermal conductivity (λ_{eff}) W (K m)$^{-1}$	References
Silica aerogels	0.012–0.02	3, 27
Clay	0.015–0.05	32
Mineral wool	0.033–0.05	3, 27
Expanded and Extruded polystyrene	0.029–0.055	3, 27
Polyurethane	0.020–0.029	3, 27
Melamine foam	0.035	3
Phenolic foams	0.018–0.025	3, 27
Cork	0.04–0.05	3, 28
Cellulose	0.04–0.05	3
Loose-fill cellulose	0.039–0.042	3
Sheep wool	0.036	3, 41
Foam glass	0.039–0.045	3, 27
Glass wool	0.031–0.043	3, 27
Gypsum foam	0.02	3
Vacuum insulation panels	0.003–0.011	3, 27
Vacuum glazing	0.003–0.008	3, 27
Silica panels	0.023–0.056	29
Silica blankets	0.010–0.023	29

pass through the material, thus limiting gas conduction. Calcium silicate is an abundant material and is produced from limestone. It can also be used for thermal insulation, electric insulating, building boards, and tiles applications.[33]

It has been shown that some inorganic insulating materials are commercially available in the market. However, their insufficient mechanical performances have pushed researchers to develop new materials. One route is to use organic components to produce materials with good mechanical and thermal insulation properties. Expanded polystyrene is widely used as a thermal insulation material because of its lightweight, open-cell structure with low thermal conductivity and hydrophobicity.[34,35] Expanded and extruded polystyrenes have a notable market share around 13% and 8% respectively, in the conventional insulating material family.[3] Another organic powerful insulating material is polyurethane. Polyurethane has the maximum market share (23%) in the conventional insulating materials.[3] The polyurethane foams are used to produce high-resilience seating, insulation panels, and so forth. Melamine foam is a synthetic insulator having good thermal insulation properties. It has been reported that the melamine foam is lightweight, has a good flame resistance, and is a good thermal insulator because of its open cell structure.[36] Melamine foam is used in railway carriages, abrasive cleaners, and sound insulation material for studios, sound stages, and auditoriums.[36,37] Finally, phenolic foams are used for various applications including wall and floor insulation.[3] In recent years, phenolic foams have found increasing applications because of their good fire resistance, with negligible smoke emission, and low levels of toxic gas emission.[38–40]

Still, the production of synthetic organic insulating materials tends to use the most environmentally harmful solvents/chemicals. Therefore, there are some renewable resource based organic materials that have been used to develop environmentally friendly thermal insulating materials, which are described in this section. Due to the low thermal conductivity and lower density, cork is used for thermal insulation applications in house walls, floors, ceilings and facades. Cellulose is also used in good thermal insulating materials because it is the most abundant material with a low thermal conductivity. Like other materials, thermal conductivity of cellulose can vary widely based on the temperature, moisture and density.[3] In terms of the application, cellulose is used to fill various cavities and spaces as a thermal insulation material. Sheep wool is used as a renewable and sustainable thermal insulating material.[41] Sheep wool has many advantages including a fire extinguishing capability, absorption and release of moisture without any change in the thermal properties. Unlike mineral insulating materials, sheep wool is not harmful to human health.[41]

The second route to improve the mechanical properties of inorganic aerogels is to mix the thermal performances of inorganic aerogels with the good mechanical performances of polymers. Many studies have been carried out and over the past few years, many thermal insulators based on silica aerogels reinforced by polymers have been commercialized.[41–44]

In this section of the chapter, the context of the thermal insulation applications has been provided. Now, the new developments of insulating polysaccharide and protein based aerogels will be reviewed.

15.3 Polysaccharide Based and Protein Based Aerogels for Thermal Insulation

Various polysaccharide (*e.g.*, cellulose, pectin, alginate, starch, guar gum, xanthan gum, and chitosan) and protein based aerogels have been studied in the literature with a focus on their thermal insulation properties. The microstructure and their final thermal properties will be correlated to explain why these polysaccharide and protein based aerogel materials have been developed for specific potential applications.

15.3.1 Cellulose

Cellulose is the most abundant natural polymer on Earth. It is present in plants and it can also be produced by bacteria. The cellulose content of vegetable fibers can vary depending on the species. It is the main component of plant cell walls and is responsible for their mechanical strength. As a structural component in all plants, cellulose has a hierarchical structure which is shown in Figure 15.4.[45]

A plant cellulose fiber is formed by bundles of macrofibrils. Cellulose macrofibrils have a length of 1–2 mm with a diameter of 60–360 μm. They are

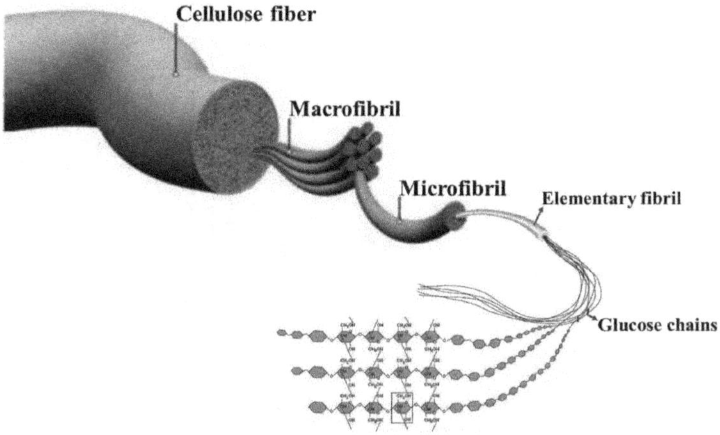

Figure 15.4 Schematic view of the structure of cellulose fibers. Adapted with permission from H. Zhu, Z. Jia, Y. Chen, N. Weadock, J. Wan, O. Vaaland, X. Han, T. Li and L. Hu, *Nano Lett.*, 2013, **13**, 3093–3100, Copyright 2013 American Chemical Society.[45]

the main structural units of the plant cell wall (Figure 15.4).[46] They consist of cellulose microfibril, having a diameter of about 10–30 nm depending on the source of cellulose.[47] Each microfibril is made of elementary cellulose fibrils containing six glucose chains linked together by hydrogen bonds. The diameter of the elementary fibrils typically varies between 1.5 and 3.5 nm,[47,48] and they are considered to be the smallest morphological units in the plant cellulose fiber.[49] It is, therefore, necessary to distinguish different types of aerogels according to the scale of cellulose used. Table 15.2 shows examples of different type of cellulose aerogels and their microstructural characteristics with thermal insulation performances.

15.3.1.1 Cellulose Aerogels

Cellulose aerogels from cellulose II are considered as "third generation" aerogels[50] and they are often called "Aerocelluloses".[51–54] To prepare aerocellulose, cellulose is directly dissolved in a solvent (*e.g.*, *N*-methylmorpholine *N*-oxide (NMMO) monohydrate,[53,55] 8% NaOH–water,[52,56] LiCl/DMAc,[51] calcium thiocyanate[57]) or an ionic liquid,[58–60] regenerated (or coagulated) in a non-solvent (water, alcohol) and then dried in a special way that prevents pore collapse. Aerocelluloses have a wide pore size distribution, from tens of nanometers to several microns, and a high specific surface area of several hundreds of $m^2\ g^{-1}$. Nevertheless, the drying method affects the morphology of the resulting aerocellulose. In the freeze-drying technique, the gel is first frozen and then dried under low pressure by sublimation of the frozen liquid. In this method, the main issue is that the gel network may be destroyed by nucleation and growth of solvent crystals.[61]

Table 15.2 Thermal conductivity and microstructural features of cellulose based aerogels.[a]

Materials	Preparation method	Concentration (wt%)	ρ_{bulk} (g cm^{-3})	Porosity (%)	Pore size (nm)	S_{BET} (m^2 g^{-1})	λ_{eff} (W (K m)$^{-1}$)	References
Amorphous cellulose	SC drying	2–5.6	0.072–0.220	95–86	20–35	222–492	0.024	Ayadi *et al.*[50]
Cellulose	Freeze-drying	3	0.233	84.88	macro	—	0.029	Shi *et al.*[62]
Recycled cellulose	Freeze-drying	—	0.040	94.8	(40–200)·10^3	—	0.032	Nguyen *et al.*[11]
Cellulose - NFC	Freeze-drying	—	0.04	—	—	100	0.024	Seantier *et al.*[18]
NFC TEMPO	SC drying	0.02–2	0.004–0.04	98–99	2–50	500–600	0.018	Kobayashi *et al.*[68]
NFC–chitosan	Freeze-drying	—	0.005	—	—	—	0.015	Chen *et al.*[71]
NFC	SFD	2	0.02	99	40	100	0.018	Jiménez-Saelices *et al.*[16]

[a]SC: supercritical drying, ρ_{bulk}: bulk density; S_{BET}: specific surface area; NFC: nanofibrillated cellulose; SFD: spray freeze-drying; and λ_{eff}: total thermal conductivity.

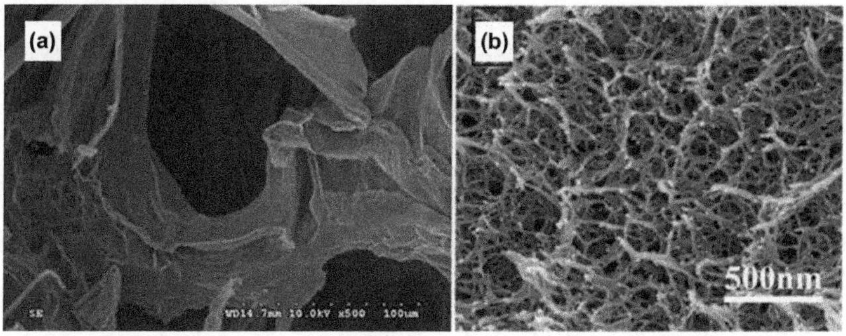

Figure 15.5 (a) SEM image of cellulose aerogel prepared by freeze drying from 5 wt% cellulose. Reproduced from ref. 62 with permission from John Wiley and Sons, Copyright © 2013 Wiley Periodicals, Inc.[62] (b) SEM image of cellulose aerogel prepared by supercritical drying from 2 wt% cellulose.
Reproduced from ref. 50 under the terms of the CC BY 4.0 licence, https://creativecommons.org/licenses/by/4.0/.

This effect tends to create a 2D sheet-like morphology with macropores and a low surface area (Figure 15.5a). Shi and coworkers have prepared aero-celluloses by freeze-drying with a density of around 0.2 g cm^{-3} and a thermal conductivity up to 0.029 W $(K\,m)^{-1}$.[62] Nguyen *et al.* developed a strategy that allowed them to prepare aerogels from recycled cellulose fibers with a density of 0.040 g cm^{-3} and a thermal conductivity of 0.032 W $(K\,m)^{-1}$.[11] Both examples[11,62] are good insulating materials, and their good stable heat insulation performance provides a vast potential in medium and low temperature heat insulation applications. However, these studies could not achieve super-insulating properties. In both aerogels, the pore size was up to 40 μm, which prevents the Knudsen effect occurring.

Supercritical drying can avoid the collapse of aerogel structures during liquid removal by removing the liquid/vapor surface tension and thus achieving structures with nanoporosity.[63,64] Also the specific surface area is much higher, in the order of 100 to 400 $m^2\,g^{-1}$.[52,53,65] However, the lowest thermal conductivities obtained are still higher than the values found for silica aerogels. For example, Ayadi *et al.* have prepared aerocelluloses from different concentrations (2, 4 and 5.6 wt%) of microcrystalline cellulose.[50] These aerocelluloses showed a 3D fibrillar skeleton morphology with a pore size of 20–35 nm and a surface area between 222–492 $m^2\,g^{-1}$ (Figure 15.5b). They have observed the lowest thermal conductivity of 0.024 W $(K\,m)^{-1}$, which is lower than the thermal conductivity of air. This value was achieved for the aerocellulose prepared from 2 wt% microcrystalline cellulose with a density of 0.072 g cm^{-3}. The observed lower thermal conductivity was due to the mesoporous structure of the aerogel, which favors the Knudsen effect and reduces the contribution of gas conduction. The contribution of solid conduction is still high and the pore wall thickness needs to be reduced to improve the thermal insulting properties of the resulting aerogels. Still, the developed aerocelluloses were optically transparent, and their interpenetrating open pore network also permits rapid transport of liquid-phase molecular reactants and nanoscale objects into the aerogel, allowing it to act as a highly porous support for functional inclusions.

15.3.1.2 Multiscale Cellulose Aerogels

Multi-scale cellulose fibers are used to design a composite aerogel. Combining different aspect ratios and varying the formulations, Seantier *et al.* prepared aerogels by freeze-drying.[18] The minimum thermal conductivity was reached when the cellulose nanofillers fraction was between 10 and 20 wt%. The minimal thermal conductivity value obtained for these concentrations of cellulose/nanocellulose fibers (NFC) composite aerogels was 0.024 $W(K\,m)^{-1}$ with a density of 0.040 $g\,cm^{-3}$. The aerogels showed a 2D sheet-like morphology which is a typical morphology of the freeze-drying process. However, the space between the cellulose fibers is reduced due to the formation of a tight 3D network of NFC. These results clearly showed that adding cellulose nanoparticles to the cellulose fiber decreases the apparent thermal conductivity because of the decrease of the pore sizes and the pore wall thickness in the resulting aerogels. This study opens up new opportunities to develop relatively low cost and biodegradable thermal insulators for commercial purposes.

15.3.1.3 Cellulose Nanoparticle Aerogels

Making ultra-light aerogels with cellulose nanofibers, which can be either bacterial cellulose[66] or nanofibrillated cellulose,[67] can be a way to improve the material properties. In both cases the starting material is a network of cellulose I nanofibers filled with water. Water can then be extracted using the same approaches as described for aerocellulose and the morphologies of aerogels are also highly influenced by the drying technique used. With supercritical drying, the resulting material is a network of cellulose I nanofibers. It looks similar to aerocellulose prepared by supercritical drying, but with higher porosities and lower densities. The reason is that the cellulose nanofibers are flexible and they are entangled with each other, creating networks even at very low concentrations. A good example is the NFC aerogel prepared by Kobayashi *et al.*, which reaches a minimal thermal conductivity of 0.018 $W(K\,m)^{-1}$.[68] This value is comparable to the lowest thermal conductivity values reported for silica aerogels (0.015 $W(K\,m)^{-1}$) and is lower than that of the commercially available insulators, such as polyurethane foam, carbonized cork, and mineral wools (0.03–0.05 $W(K\,m)^{-1}$).[69,70] The NFC aerogel has a density of 0.02 $g\,cm^{-3}$ and a 3D fibrillar skeleton morphology. The pore size was 2–50 nm and the specific surface area was around 500 $m^2\,g^{-1}$. The mesoporosity leads to the Knudsen effect and the very low density reduces the contribution of the solid conduction. As a result, the NFC aerogel is a thermal super-insulating material. In addition, these aerogels can be used as a novel high-performance insulator with optical transparency, which could be interesting for window insulation applications.

Generally, the freeze-drying process can produce two kinds of NFC aerogel structures. An open porous structure and a three dimensional fibrillar skeleton morphology are obtained for low concentrations of NFC aerogels. Chen *et al.* have investigated a comparative study of aerogels obtained from differently prepared nanocellulose fibers.[71] They have found better insulating properties for NFC aerogels prepared *via* (TEMPO)-mediated oxidation and ultrasonic treatment, reaching a thermal conductivity of 0.015 $W(K\,m)^{-1}$. Aerogels were prepared from 0.2 wt% NFC suspensions and their density is around 0.005 $g\,cm^{-3}$. The very low density reduces the contribution of solid conduction and the three dimensional fibrillar skeleton morphology results in a small pore size which leads to the Knudsen effect and reduces the contribution of gas conduction. However, to improve the size stability of the NFC aerogels for thermal conductive experiments, chitosan was introduced as a glue to connect the NFC. When the NFC content exceeds 0.5 wt%, the fibers self-aggregated into a 2D sheet-like morphology with macropores.

Recently, Jimenez-Saelices *et al.* have implemented a new drying method to prepare aerogels.[16] They have adapted a spray freeze-drying (SFD) technique, which is an emerging technology utilized in many industries for the production of powders.[72] Their objective was to change the morphology and to reduce the aerogel pore size down to the nanometric scale, even for high NFC concentrations. In their work, they prepared NFC aerogels by conventional freeze-drying and by SFD to compare their properties. While bioaerogels prepared by freeze-drying exhibit a 2D sheet-like morphology with macropores, bioaerogels made by SFD yield a 3D fibril nanostructured skeleton morphology with pore size of few tens of nanometers to a few microns (Figure 15.6). The better insulating properties were found for samples having densities around 0.022 $g\,cm^{-3}$. Aerogels prepared by freeze-drying have a minimal thermal conductivity of 0.024 $W(K\,m)^{-1}$ and the lowest thermal conductivity obtained for SFD bioaerogels is 0.018 $W(K\,m)^{-1}$. This value is similar to the lowest values reported in the literature for NFC aerogels made by supercritical drying, 0.018 $W(K\,m)^{-1}$.[68] Due to the light weight, the contribution of solid conduction was reduced in both the aerogels. However, the macroporous structure of freeze-dried aerogels allows the free circulation of air molecules and the conduction of heat by the gaseous phase. The SFD aerogels have a pore size of around 40 nm, which favors the Knudsen effect and reduces the thermal conductivity.

We highlighted the influence of thermal insulation properties of the cellulose aerogels prepared by changing the preparation methods. In addition to cellulose, there are some other polysaccharides that are used to form aerogels with good thermal insulation properties. Some of them will be described in the coming section.

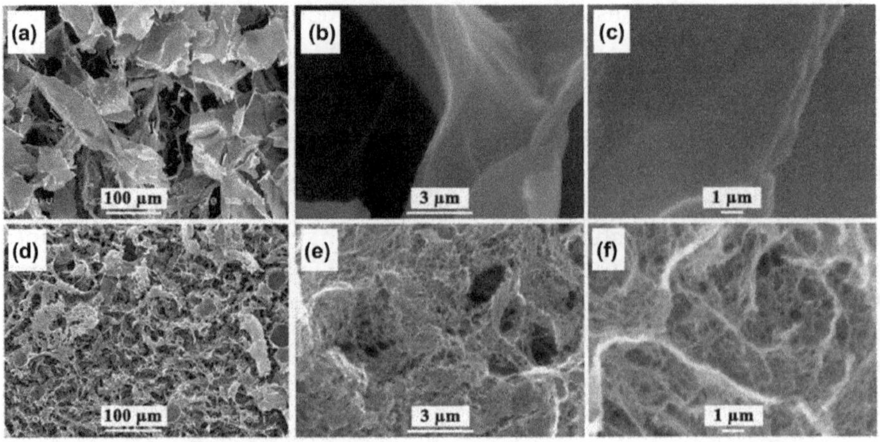

Figure 15.6 SEM images of bioaerogels prepared from 2 wt% NFC solution by (a–c) conventional freeze-drying and (d–f) spray freeze-drying.
Reprinted from *Carbohydrate polymers*, **157**, C. Jimenez-Saelices, B. Seantier, B. Cathala and Y. Grohens, Spray freeze-dried nanofibrillated cellulose aerogels with thermal superinsulating properties, 105–113, Copyright 2017, with permission from Elsevier.[16]

15.3.2 Other Polysaccharide Based Aerogels

Pectin is another abundant and inexpensive commercial heteropolysaccharide. It can be extracted from fruits and is used as a gelling agent, thickening agent, and stabilizer in food.[15] Alginate is an oceanic polysaccharide and is extracted from abundant seaweeds. It is composed of randomly arranged mannuronic acid and guluronic acid units. The alginate has inherent properties such as nontoxicity, biocompatibility, biodegradability[73] and flame-retardant behavior.[72] Starch is also one of the polysaccharides which is composed of a large number of glucose units joined by glycosidic bonds. Chitosan is a kind of polysaccharide and is obtained from the shells of shrimp and other crustaceans with an alkaline substance. The summary of these polysaccharide aerogels thermal insulation properties with microstructural characteristic are presented in Table 15.3. The examples in the Table 15.3 will be discussed in the following section.

15.3.2.1 Pectin Aerogels

Recently, two different sources of pectin (citrus peel and apple pomace) were used to prepare the thermal super-insulating aerogels by the dissolution–gelation–coagulation route and followed by supercritical drying.[15] The detailed thermal insulation properties, porosity, pore size and density of these pectin aerogels are presented in the Table 15.3. Aeropectins showed a fibrillar morphology with long and entangled strands (Figure 15.7).[15] Aeropectin from apple or citrus pectin presents a very similar morphology at the same concentration. For all the aeropectins, specific surface areas vary from

Table 15.3 Thermal conductivity and microstructural features of pectin, starch, alginate, chitosan, xanthan and guar based aerogels.[a]

Materials	Preparation method	Concentration	ρ_{bulk} (g cm^{-3})	Porosity (%)	Pore size (nm)	S_{BET} (m^2 g^{-1})	λ_{eff} (W (K m)$^{-1}$)	References
Citrus peel pectin	SC drying	4 wt%	0.115	~92	~40	300	0.020	Rudaz *et al.*[15]
Apple pomace pectin	SC drying	3 wt%	0.08	~92	~40	250	0.016	
Ammonium alginate	Freeze-drying	5 wt%	0.066	—	—	—	0.025	Shang *et al.*[73]
Sodium alginate	SC drying	4%	0.095	—	~20	367	0.042	Tkalec *et al.*[76]
Starch	Freeze-drying	5 wt%	0.059	—	—	—	0.059	Wang *et al.*[93]
Wheat starch	Freeze-drying	8%	0.12	—	Macropores	—	0.040	Glenn & Irving[80]
	Air dried		0.27				0.044	
	SC drying		0.26				0.037	
Corn starch	Freeze-drying	8%	0.12	—	Macropores	—	0.040	
	Air dried		0.31				0.037	
	SC drying		0.29				0.033	
High-amylose corn starch	SC drying	8%	0.16	—	Macropores	—	0.024	
High methoxy pectin	SC drying	—	0.0861	96.6	19	384	0.023	Tkalec *et al.*[75]
Low methoxy pectin			0.0771	97.0	17	510	0.021	
Xanthan			0.0961	94.9	20	363	0.028	
Alginate			0.1569	95.0	14	147	0.080	
Guar			0.2942	86.9	15	111	0.095	
Chitosan		4 g L^{-1}	0.042	97	10–50	545	0.022	Takeshita & Yoda[17]

[a]SC: supercritical drying, ρ_{bulk}: bulk density; S_{BET}: specific surface area; λ_{eff}: total thermal conductivity.

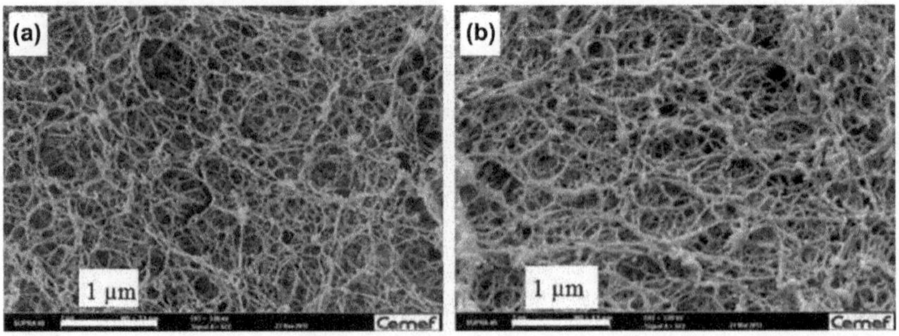

Figure 15.7 SEM images of aeropectins from (a) 3% citrus-based and (b) 3% apple-based.
Reprinted with permission from C. Rudaz, R. Courson, L. Bonnet, S. Calas-Etienne, H. Sallée and T. Budtova, *Biomacromolecules*, 2014, 15, 2188–2195, Copyright 2014 American Chemical Society.[15]

230 to 270 $m^2 g^{-1}$ and the pore size distribution is 11–80 nm. The 3D fibril nanostructured skeleton morphology results in aeropectins with thermal super-insulating properties thanks to the Knudsen effect. The values of thermal conductivity obtained are in the range of 0.016–0.020 $W(K m)^{-1}$. Thermal conductivity increases with increasing pectin concentrations, which is attributed to the increase in the solid phase conduction. Besides the super-insulating properties, this study also demonstrated that the aeropectins are mechanically strong (Young's modulus: 18 MPa with a density of 0.18 $g cm^{-3}$) like aerocellulose.[15] Similarly, Tkalec *et al.* have prepared thermally insulative aerogels from high-methoxyl (HM) pectin and low-methoxyl (LM) pectin using the supercritical drying process.[75] They have found the thermal conductivity of HM-pectin is 0.023 $W(K m)^{-1}$ while the thermal conductivity of LM-pectin is 0.021 $W(K m)^{-1}$.

15.3.2.2 Alginate Aerogels

Several strategies have been used to produce aerogels from alginate.[73–78] Tkalec *et al.* have focused their research on optimizing the alginate gelation process in order to obtain aerogels with better properties.[75] They have used three different types (high, medium and low viscosity) of alginic acid sodium salts to produce aerogels by supercritical drying. The effect of gelation time and gelation solvent on the characteristic properties of the prepared materials was investigated. The best characteristics of alginate aerogels were obtained with high-viscosity alginic acid and with a 24 h gelation time in methanol. High viscous alginate provided dense structure with mesoporosity (Figure 15.8b and c) and a surface area of 367 $m^2 g^{-1}$. On the other hand, the aerogels prepared from low viscosity aliginate showed larger voids between the polysaccharide chains (Figure 15.8a). The alginate aerogel produced with methanol gelation solvent for 24 h gelation showed the lowest thermal conductivity, which was 0.042 $W(K m)^{-1}$.[76] However, this thermal

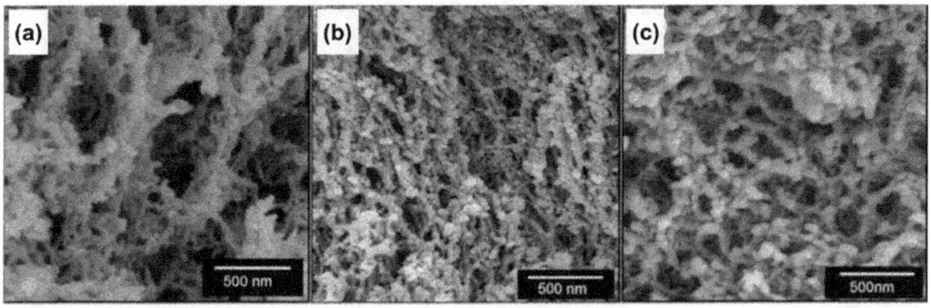

Figure 15.8 SEM image of alginate aerogels prepared from (a) medium viscous ethanol alcogel with 1 h gelation time, (b) high viscous ethanol alcogel with 24 h gelation time and (c) high viscous methanol alcogel with 24 h gelation time.
Reprinted from *Journal of Non-Crystalline Solids*, **443**, G. Tkalec, R. Kranvogl, A. P. Uzunalic, Z. Knez and Z. Novak, Optimisation of critical parameters during alginate aerogels' production, 112–117, Copyright 2016, with permission from Elsevier.[76]

conductivity is still not within the range of some other published results on polysaccharide aerogels. Shang *et al.* prepared ammonium alginate based aerogel by freeze-drying.[73] The prepared alginate aerogel had a density of 0.066 $g\,cm^{-3}$ with a thermal conductivity of 0.025 $W\,(K\,m)^{-1}$. This value is lower than the thermal conductivity of cellulose aerogels, while higher than that of polyurethane.[77,78] Gurikov *et al.* measured the thermal conductivity at ambient conditions of cross-linked and supercritical dried sodium alginate aerogels.[79] The sodium alginate aerogel with different degrees of cross-linking showed a thermal conductivity lower than 0.025 $W\,(K\,m)^{-1}$ which is comparable to the thermal conductivity of air.

15.3.2.3 Starch Aerogels

Starch is also an interesting material to prepare aerogels because it is bio-based and biodegradable. Therefore, Glenn and Irving have prepared unmodified wheat starch (~28% amylose and ~72% amylopectin), corn starch (~28% amylose and ~72% amylopectin), and high-amylose corn starch (~70% amylose and ~30% amylopectin) based aerogels.[80] These aerogels were conditioned at 50% relative humidity for at least 48 h before analysis. At 22.8 °C, the thermal conductivity of these starch aerogels were in the range of 0.024–0.043 $W\,(K\,m)^{-1}$ which is comparable to the commercial beaded polystyrene thermal conductivity. Among these starch based aerogels, the corn starch with a high-amylose content aerogel had a nanostructured morphology, whereas corn starch and wheat starch based aerogels showed 2D sheet-like morphology with macropores. Therefore, the thermal conductivity of corn starch with a high-amylose content aerogel is the lowest (0.024 $W\,(K\,m)^{-1}$) compared to corn starch and wheat starch based aerogels.

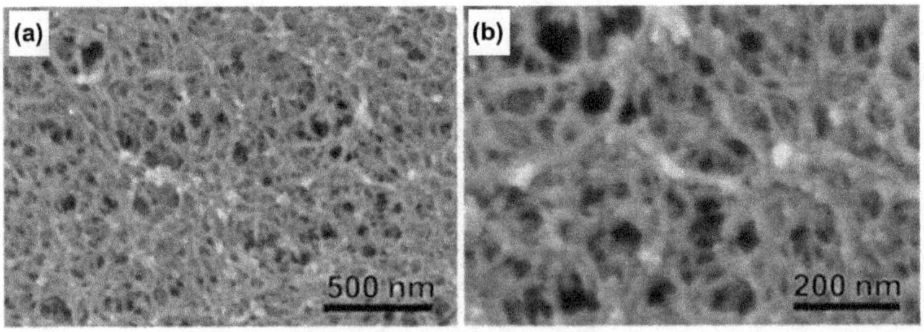

Figure 15.9 SEM images of chitosan aerogel prepared using 4 $g L^{-1}$ chitosan and 7.2 wt% formaldehyde solutions. (a and b) Cross-section at different magnifications.
Reprinted with permission from S. Takeshita and S. Yoda, *Chem. Mater.*, 2015, **27**, 7569–7572, Copyright 2015 American Chemical Society.[17]

15.3.2.4 Chitosan Aerogels

In order to diversify the application of chitosan based materials, transparent and flexible cross-linked chitosan aerogels have been prepared by Takeshita and Yoda.[17] The chitosan aerogel was cross-linked with two different concentrations of formaldehyde, followed by supercritical drying. The cross-linked chitosan aerogels have a 3D fibril skeleton morphology of entangled nanofibers with mesopores of 10–50 nm in size (Figure 15.9) and a specific surface area of 545 $m^2 g^{-1}$. The lowest thermal conductivity was reported for a chitosan (4 $g L^{-1}$) aerogel prepared with 7.2 wt% formaldehyde. This aerogel had a density of 0.042 $g cm^{-3}$ and a thermal conductivity of 0.022 $W (K m)^{-1}$. The thermal conductivity increases with the increasing density because of the higher contribution of solid conduction. Furthermore, a chitosan aerogel with a density of 0.082 $g cm^{-3}$ reaches the thermal conductivity of 0.029 $W (K m)^{-1}$. It can be noticed that the observed thermal conductivity of the cross-linked chitosan aerogels are lower than that of commercial flexible thermal insulators.[17] Based on the performances of this chitosan aerogel, it could be a renewable resource based alternative for thermal insulating applications.

15.3.3 Protein Based Materials

A lot of protein aerogels have been developed recently. Most of the time the protein is coupled with a cross-linker to obtain the best performance aerogels. Yet, only a few protein based aerogels have been prepared for the purpose of obtaining thermal insulating materials. Some trials are reported in the literature with albumin, gelatin and wheat gluten. Here, an overview will be given to describe the various strategies to design protein aerogels with thermal insulating properties. The recent protein based aerogels are summarized in Table 15.4.

Table 15.4 Thermal conductivity and microstructural features of protein based aerogels.[a]

Materials	Preparation method	Concentration (wt%)	ρ_{bulk} ($g\,cm^{-3}$)	Porosity (%)	λ_{eff} ($W\,(K\,m)^{-1}$)	References
Soy protein + polysaccharides + lipids + fibers + tannin + formaldehyde	Sol–Gel process + SC drying	21	0.19–0.25	84–88	0.033–0.034	Amaral-Labat *et al.*[82]
White egg albumin + camphor + formaldehyde	Chemical reaction + Oven curing	40–57	0.26–0.29	77–79	0.040–0.048	Li *et al.*[84]
White egg albumin + camphor + formaldehyde	Chemical reaction + Microwave oven curing	31–65	0.185–0.387	—	0.062–0.069	
Albumin + tannin + additives	Chemical reaction + Oven curing	14–44	0.139–0.417	68.1–90.5	0.0438–0.0956	Lacoste *et al.*[85]
Wheat gluten + TEOS	Wheat gluten gelation + Mixing with TEOS solution + Freeze-drying	50–100	0.127–0.166	87–91	0.039–0.057	Wu *et al.*[86]
Wheat gluten + TEOS + Glutaraldehyde	Wheat gluten gelation + Mixing with TEOS solution + Glutaraldehyde curing	65–74	0.140–0.146	87–91	0.046	
Gelatin + Epoxide	Formation of gelatin fibers from spinning and fiber rowing.	50	0.40–0.60	9–32	0.054–0.060	Stoessel *et al.*[87]

[a]SC: supercritical drying, ρ_{bulk}: bulk density, and λ_{eff}: total thermal conductivity.

15.3.3.1 Soy Protein

Soy flour is one of the protein rich renewable materials. It not only contains protein (~42%), it also contains oil (~20%), carbohydrates (~33%), and ash (~5%) on a dry basis. Soy protein is mainly composed of acidic amino acids (aspartic and glutamic acids), basic amino acids (lysine and arginine), non-polar amino acids (glycine, alanine, valine and leucine), and less than 1% of cysteine.[81] In order to prepare aerogels that are more environmentally friendly, one strategy is to replace part of the chemicals by natural products. In that aspect, one of the first trial was carried out by Amaral-Labat *et al.* with soy flour.[82] The idea was to use the molecules present in the soy flour and the already well-known ability of tannins to bind to proteins to form a 3D network.[83] Indeed, soy flour is composed of proteins, carbohydrates, lipids and fibers that can strongly interact together and with tannins *via* all the functional groups that are at the surface of these molecules. To be able to form an aerogel, the soy flour has been mixed with formaldehyde to produce a gel that was dried by supercritical CO_2. In the obtained aerogel, the protein concentration was 21 wt% of the dry mass and the density was from 0.18 to 0.25 $g\,cm^{-3}$. The most promising application is thermal insulation because of a low thermal conductivity (0.033 to 0.034 $W\,(K\,m)^{-1}$). This is quite good for a material that is up to 91% renewable, inexpensive and a non-irritant. This is a good step towards the development of a natural protein based thermal insulator.

15.3.3.2 Albumin

Albumin also gains a lot of attention for aerogel formation. For example, Li *et al.* used mixtures of albumin, camphor and formaldehyde to develop a strong foam structure.[84] The interactions between the molecules gave rise to an aerogel after curing. Two types of curing were performed for example, oven and microwave oven curing. The obtained structure, having a low density (0.185 to 0.387 $g\,cm^{-3}$), and thermal conductivity of between 0.040 and 0.069 $W\,(K\,m)^{-1}$) are expected to be good insulating materials, but some improvement needs to be done for them to reach thermal super-insulating materials. They still have interesting mechanical properties (good rigidity and elastic properties) that make them promising materials for building insulation.

In another study aimed at replacing camphor by other additives, such as tetra-amines and a larger variety of tannins, different materials were designed varying the preparation protocols (concentration of the molecules, pH, curing process, type of additives and nature of the tannins).[85] Low density porous foams were obtained with some formulations. The results confirm the fact that the thermal conductivity increases linearly with the density. However, the thermal conductivity stays quite high (between 0.042 and 0.098 $W\,(K\,m)^{-1}$). Despite the fact that the materials have low thermal insulation properties, these aerogels are bio-based materials, flexible and elastic, and could be used as thermal insulators for different applications including building and transport.

15.3.3.3 Wheat Gluten

Wu *et al.* chose another strategy to form cellular (87–91% porosity) aerogels from wheat gluten (WG).[86] They prepared flame-retardant and good insulation property aerogels from WG (protein content 77.7 wt%, starch 5.8 wt% and lipid 1.2 wt%) and silica with and without a cross-linker (glutaraldehyde (GA)). After mixing silica precursor (tetraethyl orthosilicate (TEOS)) solution with wheat gluten and a cross-linker, the final mixture was freeze-dried. The obtained material has a pore structure. However, the size of these pores is high (from 4 to 144 μm). As a consequence, the thermal conductivity is high (0.039 to 0.057 $W (K\,m)^{-1}$). Nevertheless, the low thermal conductivity is compensated by good fire resistant properties. The fire resistance performances of the developed materials were compared with polyurethane (PUR, used as a reference) (Figure 15.10). It can be seen that the PUR caught fire after 3 s of fire exposure, whereas WG mixed with 30% of TEOS and 8% GA required more than 3 s to observe an apparent flame.

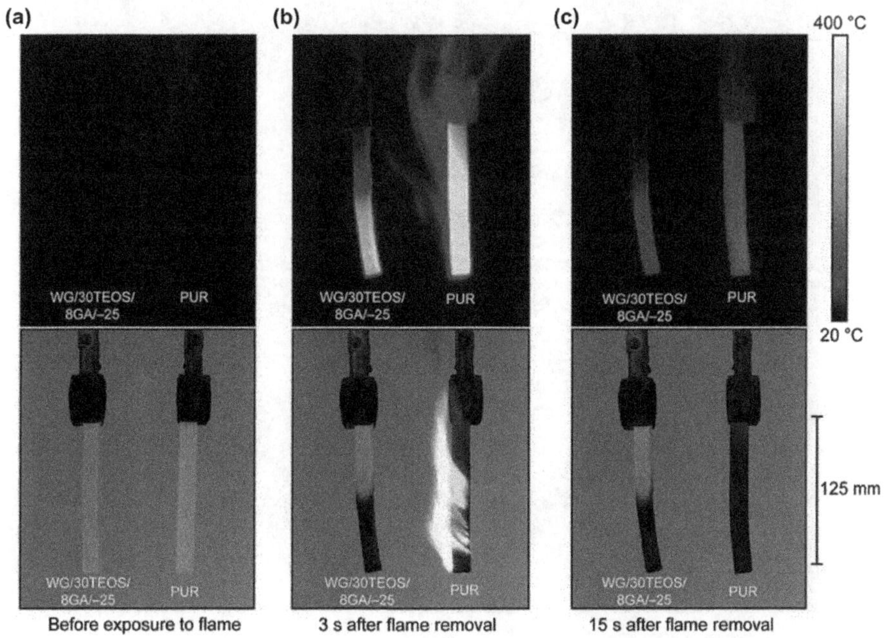

Figure 15.10 Behavior of PUR and WG/TEOS/GA foam against fire. Time-lapse pictures down and corresponding thermal camera pictures taken after (a) 0 s, (b) 3 s and (c) 15 s.
Reproduced from ref. 86 with permission from the Royal Society of Chemistry.

Moreover, a drastic decrease of the time was needed for self-extinguishing to be observed with increasing concentration of TEOS (from more than 68 s to less than a second). Furthermore, this material has good mechanical properties. Besides the fact that the compression moduli decreases when the silica concentration increased, these properties are improved (from 0.5 to 1.5 MPa) with comparable thermal conductivity (0.046 $W(K\,m)^{-1}$) when cross-linked with GA.

15.3.3.4 Gelatin

Not much attention has been paid to producing aerogels from gelatin alone for thermal insulation applications. However, an interesting application was found to fabricate good thermal insulation gloves from gelatin based material.[87] The aim was to produce thermal insulating gloves that are able to protect the user from low temperature. The thermal insulation gloves were obtained from gelatin/epoxide (50/50 wt%) by dry–wet spinning.[87] The structure of these spun fibers is similar to that of an aerogel (Figure 15.11). It looks

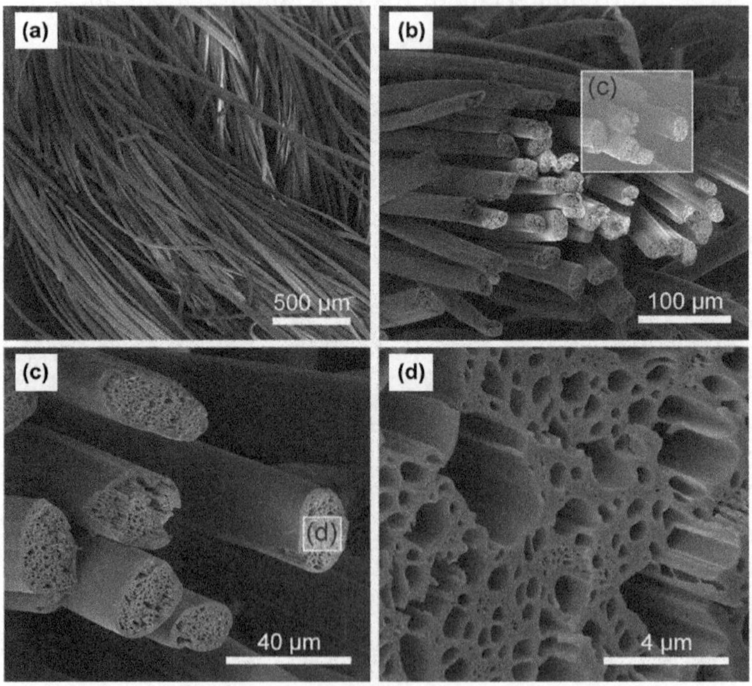

Figure 15.11 SEM pictures of gelatin/epoxy internal structure of electro spun fibers. (a) Top view; (b–d) cross-sections at different magnifications. Reprinted with permission from P. R. Stoessel, U. Krebs, R. Hufenus, M. Halbeisen, M. Zeltner, R. N. Grass and W. J. Stark, *Biomacromolecules*, 2015, **16**, 1997–2005, Copyright 2015 American Chemical Society.[87]

like an aerogel that was stretched in one direction to obtain fibers. In this study, they have developed a process enabling the formation of fabric fibers that can be woven to form gloves. After a cross-linking process, the obtained materials exhibited good mechanical properties and good thermal insulation properties. Although it is not a super thermal insulator, the thermal conductivity of the fibers is as low as 0.054 W $(\mathrm{K\,m})^{-1}$, making it a good candidate for the design of new fabrics in addition to medical applications.[87]

15.4 Challenges

In order to be able to produce and commercialize the developed polysaccharide and protein based aerogels, there are still some improvements that need to be carried out. For example, working on the radiative heat conduction should help with obtaining materials that have even greater thermal insulation properties. It has been shown previously that increasing thermal insulation properties often yields a decreasing performance in mechanical properties. Therefore, new ways to improve mechanical resistance without affecting the thermal properties is crucial for obtaining long-lasting materials. For this goal, many studies have highlighted the fact that moisture can significantly affect the thermal properties. Finding a way to make the developed aerogel more hydrophobic might increase the effectiveness of the materials in time. Fire resistance is another issue that has already been studied but needs to be improved.

15.4.1 Overcoming the Radiative Effect

The thermal conductivity of the polysaccharide and protein based aerogels is still high. Some polysaccharide based aerogels can be called thermal superinsulators, but most of them need to have a lower thermal conductivity. One way to reduce the effective thermal conductivity is to work on the radiative contribution. Electromagnetic radiations that arrive on a surface can be reflected, absorbed or transmitted according to the optical thickness of the material.[88] Heat transfer by radiation through an aerogel therefore depends on this parameter. The optical thickness (τ) may be defined by eqn (15.2) as the ratio between the thickness of the aerogel (d) and the mean free path of photons (l_{ph}):

$$\tau = \frac{d}{l_{\mathrm{ph}}} \tag{15.2}$$

The optical thickness is a statistical measure of the frequency at which a photon interacts with a material at a given distance d. To reduce the radiative transfer in aerogels and improve their thermal insulation properties, so-called infrared opacifiers can be incorporated into the aerogel matrix to enhance the optical thickness. Highly absorbing or scattering particles are integrated into the aerogel structure in a low concentration without significantly increasing the solid thermal conductivity of the backbone.

Carbon black is one example of a suitable opacifier having high absorbing properties.[89,90] Titanium dioxide is also a suitable opacifier to achieve light-scattering properties. These strategies were used to improve thermal performance of silica aerogels by blocking the infrared radiant heat transfer.[91] These methods add strength and improve the high-temperature stability of aerogel panels. However, the use of infrared opacifiers does not seem interesting for polysaccharide based aerogels, since they are not used at elevated temperature. Therefore, other strategies to reduce the radiative contribution of the effective thermal conductivity need to be developed in order to improve the thermal performance of these kinds of aerogels.

15.4.2 Mechanical Properties

Thermal performances should not be the only target for the final application of polysaccharide and protein based aerogels. Once the insulation material is in place, it should stay for a specific period of time without being damaged or losing its properties. Therefore, the developed materials must have good mechanical resistance. Mechanical properties of the pure polysaccharide based aerogels have been studied by many researchers.[15,17,50,68] However, the insufficient mechanical/compressive strength (0.04 MPa) of some of the pure polysaccharide aerogels are limiting to their applications.[92] For example, compressive strength is a key property for aerogels when it is used on the floor. Under the compression load, the porosity of the aerogel is deformed to absorb the energy and then go back to its original form when the load is released if the load is not higher than the elastic modulus of the aerogels. When the aerogel is exposed to high loads, the cell walls will rupture and the amount of porosity will decrease. As already mentioned, changing the structure of the aerogel material will have a drastic effect on the thermal insulation properties. Therefore, it is necessary to develop materials that have high compressive modulus. For the last several years, mechanical properties of the polysaccharides have been improved by different methods. For instance, there have been many reinforcing agents incorporated with polysaccharide aerogels to improve the mechanical properties.[92,93] When 50% clay was incorporated into the starch aerogel, the thermal conductivity (0.053 $W(K\,m)^{-1}$), compressive strength (0.21 MPa), and compressive modulus (7.39 MPa) of the starch/clay aerogel values were increased in comparison with a neat starch aerogel.[93] These increased mechanical and thermal insulation properties are due to the increase density (0.092 $g\,cm^{-3}$) and interaction between the clay and starch, respectively.

Cross-linking with formaldehyde is another strategy to improve the mechanical performances of the polysaccharide aerogels. Therefore, Takeshida and Yoda[17] developed chitosan aerogels by cross-linking with formaldehyde. By this process, they have been able to form flexible, transparent thermal super-insulating materials. They showed that it was possible to get good mechanical properties with a compressive modulus up to 40 MPa

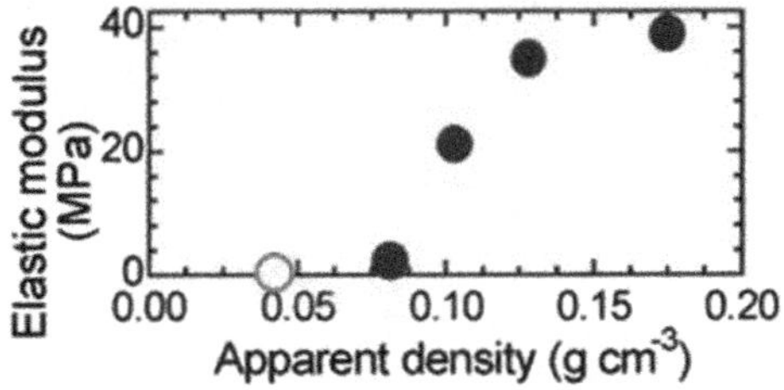

Figure 15.12 Elastic modulus of chitosan aerogels as a fonction of their densities. Open point indicates the elastic modulus of thermal super-insulating aerogel.
Adapted with permission from S. Takeshita and S. Yoda, *Chem. Mater.*, 2015, **27**, 7569–7572, Copyright 2015 American Chemical Society.[17]

(Figure 15.12). However, the best insulators, having a 0.022 $W(K\,m)^{-1}$ thermal conductivity, have low compressive moduli (around 0.35 MPa). Still, these materials are a new class of "environmentally friendly, transparent and flexible" super-insulating materials.

15.4.3 Hydrophobic Character

The hydrophilic behavior of polysaccharide aerogels induces drastic changes on these thermal insulating materials. Indeed, the aerogels made from polysaccharides have high hydrophilic groups with high porosity and specific surface area, which make it easy to adsorb water vapor in the air. Consequently, the thermal insulation performances of the polysaccharide aerogels are deteriorated when exposed to high humidity (water has about 20 times higher thermal conductivity than the air).[62] This is one of the main challenges for the polysaccharide based aerogels for thermal insulation application. Silanes are the most commonly used molecules to improve the hydrophobicity of the hydrophilic surfaces.[12,94,95] However, there are several disadvantages during this modification process, such as the use of expensive reactants, poor efficiency and long modification periods.[96] There are only a few researchers who effectively improved the hydrophobicity of the polysaccharide aerogels by using environmentally friendly strategies including cold plasma modification.[62,92,96] For instance, cellulose and alginate based aerogels were modified using carbon tetrachloride (CCl_4) plasma treatment by Shi *et al.*[62] and Cheng *et al.*,[92] respectively. These studies showed that the hydrophobicity of the resulting aerogels has been improved after plasma treatment. The improved hydrophobicity was confirmed by measuring the contact angle (Figure 15.13). It was observed that the enhanced hydrophobicity of the cellulose aerogel did not affect the thermal conductivity of the resulting cellulose aerogels. The lowest thermal conductivity of both the modified and unmodified 2 wt% cellulose aerogel was around 0.029 $W(K\,m)^{-1}$.[62]

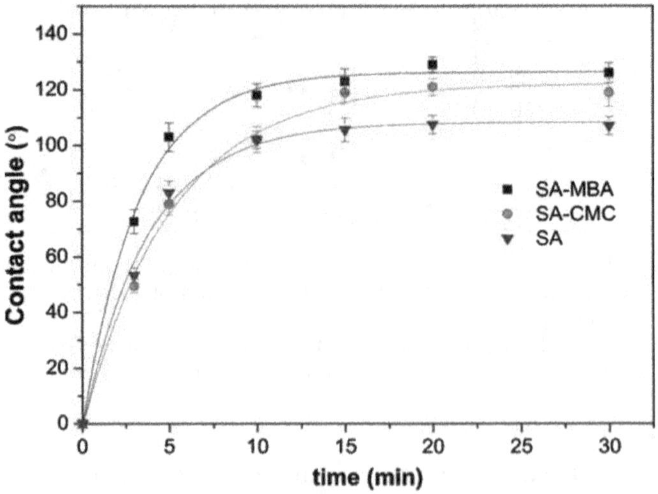

Figure 15.13 Contact angle of aerogels treated by the CCl_4 plasma at 50 W *versus* discharge duration. Samples were labeled as SA aerogel (sodium alginate (SA) without reinforcing agent), SA–CMC aerogel (carboxymethylcellulose (CMC) as the reinforcing agent) and SA–MBA aerogel (*N*,*N*′-methylenebisacrylamide (MBA) as the reinforcing agent), respectively.
Reprinted from *Carbohydrate Polymers*, **88**, Y. Cheng, L. Lu, W. Zhang, J. Shi and Y. Cao, Reinforced low density alginate-based aerogels: Preparation, hydrophobic modification and characterization, 1093–1099, Copyright 2012, with permission from Elsevier.[92]

15.4.4 Fire Resistant Properties

Insulating materials should also be resistant as much as possible to fire. Unlike inorganic materials, the organic aerogels can ignite easily when exposed to a flame. This is a concern, as the industry has to deal with increasing attention due to safety laws. In order to put the new materials on the market and have potential markets for the polysaccharides based aerogels, a large effort has to be done to increase the ignition time and delay the destruction of the thermal insulator. This issue can be addressed by incorporation of a flame retardant additive into the aerogels. The main obstacle is how to finely disperse the flame retardants in the aerogel without collapse of the three-dimensional nanoporous structures in the resulting material. Besides, adding new particles inside the aerogels may increase the solid conduction by increasing the density or closing some pores. As a consequence, the thermal insulation properties of the aerogels might be affected by this process.

There are some commonly used flame retardant additives (clay, metal oxides, graphite, *etc.*) that have been used in polysaccharide aerogels. For example, clay nanoparticles are widely used to improve the fire resistant performances of the polysaccharide aerogels.[93,97–100] Figure 15.14a shows the combustion behavior of Arabic gum aerogels with and without clay

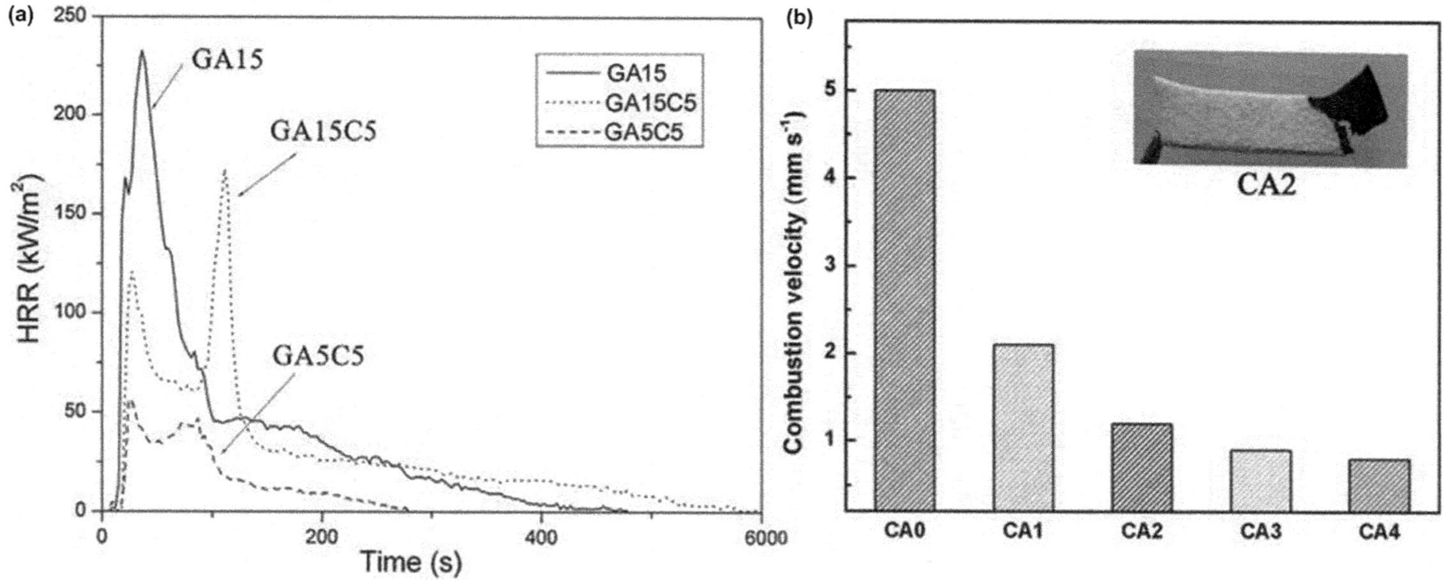

Figure 15.14 (a) Combustion behavior of aerogels: heat release rate (HRR) as a function of time was investigated by cone calorimetry. Sample identification is gum Arabic (GA) and clay (C) used followed by their respective concentration in precursor suspensions, in example, GA15 represents a sample that was prepared from an aqueous suspension containing 15 wt% of gum Arabic, GA15C5 represents a sample that was prepared from an aqueous suspension containing 15 wt% of gum Arabic and 5 wt% of clay, and GA5C5 represents a sample that was prepared from an aqueous suspension containing 5 wt% of gum Arabic and 5 wt% of clay. Reprinted from *Industrial Crops and Products*, **91**, L. Wang, M. Sanchez-Soto and T. Abt, Properties of water-based gum Arabic/clay aerogels, 15–21, Copyright 2016, with permission from Elsevier.[93] (b) Combustion velocity as a function of magnesium hydroxide nanoparticles content. Aerogels with different magnesium hydroxide nanoparticle loadings were coded as CA1 (~40 phr), CA2 (~80 phr), CA3 (~100 phr), and CA4 (~140 phr) in order to distinguish with the aerogel without magnesium hydroxide nanoparticles, which was coded as CA0.
Reprinted with permission from Y. Han, X. Zhang, X. Wu and C. Lu, *ACS Sustain. Chem. Eng.*, 2015, **3**, 1853–1859, Copyright 2015 American Chemical Society.[19]

nanoparticles. With the addition of 5% clay nanoparticles, the fire resistance (heat release rate) performance of the Arabic gum aerogel is enhanced, compared to neat Arabic gum aerogel. In another study,[19] the combustion behavior of a waste cotton fabric based cellulose aerogel was improved by incorporation of magnesium hydroxide nanoparticles from 40 to 140 phr (Figure 15.14b). It was found that the combustion velocity is lower when a higher amount (140 phr) of magnesium hydroxide particles is incorporated. Similarly, the flame resistant properties of cellulose aerogel and alginate aerogels were enhanced with the addition of metal oxide nanoparticles.[19,73,101]

In order to improve the performances of the starch aerogels, a microwave cross-linked starch and clay composite aerogel was prepared by the freeze-drying method.[93] This aerogel has a thermal conductivity of 0.053–0.059 $W\,(K\,m)^{-1}$ with clay content up to 50 wt%. Apparent densities and porosities are not changed after cross-linking the starch. Therefore, there were no significant changes observed in the thermal conductivities of the cross-linked starch aerogel. Moreover, clay addition created more porous structures and hence decreased the thermal conductivity of the cross-linked starch aerogels. The thermal conductivity of a freeze-dried waste cotton fabrics cellulose based aerogel was increased with the addition of magnesium hydroxide nanoparticles from 40 to 140 phr.[19] For example, the thermal conductivity of the cellulose aerogel was 0.056 $W\,(K\,m)^{-1}$ and it was increased to 0.081 $W\,(K\,m)^{-1}$ after incorporation of 140 phr magnesium hydroxide nanoparticles (Figure 15.15).

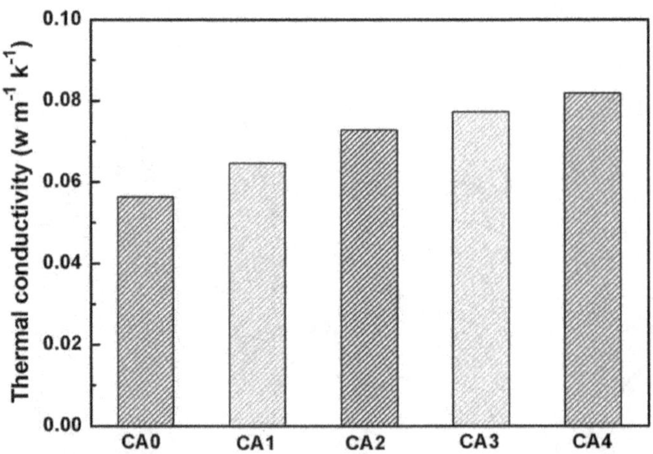

Figure 15.15 Thermal conductivity of cellulose aerogels as a function of magnesium hydroxide nanoparticle content. Aerogels with different magnesium hydroxide nanoparticle loadings were coded as CA1 (~40 phr), CA2 (~80 phr), CA3 (~100 phr), and CA4 (~140 phr) in order to distinguish with the aerogel without magnesium hydroxide nanoparticles (CA0). Reprinted with permission from Y. Han, X. Zhang, X. Wu and C. Lu, *ACS Sustain. Chem. Eng.*, 2015, **3**, 1853–1859, Copyright 2015 American Chemical Society.[19]

The thermal conductivity of the post-cross-linked ammonium alginate aerogel was increased by incorporation of inorganic nanoparticles such as magnesium hydroxide, sodium montmorillonite and Kaolin.[73] For instance, thermal conductivity values of post-cross-linked ammonium alginate aerogel with 50% of magnesium hydroxide, montmorillonite and Kaolin were 0.042, 0.039, and 0.037 $W(K\,m)^{-1}$, respectively. These thermal conductivity values are higher than neat ammonium alginate aerogel (0.025 $W(K\,m)^{-1}$). This is due to the reduction in the porosity of the resulting alginate aerogel with inorganic materials. All these attempts enabled the materials fire-resistant properties to be comparable with the conventional insulating materials such as polyurethane. However, this short review highlights the fact that for the moment, most of the strategies used induced a decrease in the insulating properties. Therefore, a compromise needs to be found between fire resistance and thermal insulation properties.

15.5 Conclusion

In this chapter, the main thermal insulation applications of polysaccharide and protein based aerogels were described. From the thermal insulation context, some expected properties were described to get the most effective thermal insulators. It has been shown that there are many polysaccharide based aerogels exhibiting thermal super-insulating properties. Until now, cellulose, pectin, alginate, starch, and chitosan have been the most studied polysaccharide aerogels for thermal insulation applications because of their abundance, low toxicity, sustainability and renewability. Among these polysaccharides, cellulose, chitosan and pectin showed thermal super-insulating properties with a thermal conductivity lower than air. In addition, few proteins have been used to produce aerogels for thermal insulation applications. As the proteins have a complex structure, it can make the aerogel formation more difficult. Therefore, formation of the final products needs the use of cross-linkers to make interesting insulators from proteins. Polysaccharide and protein based aerogels with good thermal insulation properties can be considered as potential candidates for building or transport thermal insulation applications because they possess better mechanical properties compared to silica aerogel. Some newly designed protein materials are even able to make thermal protective gloves against cold temperatures. However, the developed materials are not suitable to use as structural applications because of their insufficient mechanical properties, hydrophobicity, fire retardancy and so forth. Therefore, investigations are still needed to overcome these technical bottlenecks to diversify their applications. Overcoming all these drawbacks will allow the polysaccharide and protein based aerogels to be used as greener and sustainable thermal insulators for various applications. Overall, designing and developing polysaccharide and protein based aerogels with good thermal insulation, in an eco-friendly manner, mechanically strong, and good fire resistance will open up new opportunities for this new generation of aerogels for thermal insulation applications.

References

1. P. A. Narbel and J. P. Hansen, *Renewable Sustainable Energy Rev.*, 2014, **34**, 91–97.
2. C. Rühl, P. Appleby, J. Fennema, A. Naumov and M. Schaffer, *Energy Policy*, 2012, **50**, 109–116.
3. E. Cuce, P. M. Cuce, C. J. Wood and S. B. Riffat, *Renewable Sustainable Energy Rev.*, 2014, **34**, 273–299.
4. L. Pérez-Lombard, J. Ortiz and C. Pout, *Energy Build.*, 2008, **40**, 394–398.
5. R. Baetens, B. P. Jelle and A. Gustavsen, *Energy Build.*, 2011, **43**, 761–769.
6. A. Bottin, F. Wong, C. Rouquette, D. Reynaud, É. Misak, J.-P. Martin, S. Louati, D. Lepoittevin, J. Lauverjat, C. Hagége, Z. Gong, F.-X. Dussud, K. Delamare, I. Collet, D. Cadin, L. Monnoyer-Smith, G. Mordant, P. Repérant and C. Grosset, *Bilan énergétique de la France pour 2014*, 2015.
7. C. Forman, I. K. Muritala, R. Pardemann and B. Meyer, *Renewable Sustainable Energy Rev.*, 2016, **57**, 1568–1579.
8. M. B. Bulut, M. Odlare, P. Stigson, F. Wallin and I. Vassileva, *Energy Build.*, 2015, **107**, 254–263.
9. A. Kylili and P. A. Fokaides, *Sustainable Cities Soc.*, 2015, **15**, 86–95.
10. S. Deng, R. Z. Wang and Y. J. Dai, *Energy*, 2014, **71**, 1–16.
11. S. T. Nguyen, J. Feng, S. K. Ng, J. P. W. Wong, V. B. C. Tan and H. M. Duong, *Colloids Surf., A*, 2014, **445**, 128–134.
12. N. T. Cervin, C. Aulin, P. T. Larsson and L. Wågberg, *Cellulose*, 2012, **19**, 401–410.
13. P. C. Thapliyal and K. Singh, *J. Mater.*, 2014, **2014**, 1–10.
14. A. Du, B. Zhou, Z. Zhang and J. Shen, *Materials*, 2013, **6**, 941–968.
15. C. Rudaz, R. Courson, L. Bonnet, S. Calas-Etienne, H. Sallée and T. Budtova, *Biomacromolecules*, 2014, **15**, 2188–2195.
16. C. Jiménez-Saelices, B. Seantier, B. Cathala and Y. Grohens, *Carbohydr. Polym.*, 2016, **157**, 105–113.
17. S. Takeshita and S. Yoda, *Chem. Mater.*, 2015, **27**, 7569–7572.
18. B. Seantier, D. Bendahou, A. Bendahou, Y. Grohens and H. Kaddami, *Carbohydr. Polym.*, 2016, **138**, 335–348.
19. Y. Han, X. Zhang, X. Wu and C. Lu, *ACS Sustainable Chem. Eng.*, 2015, **3**, 1853–1859.
20. H.-P. Ebert, in *Aerogels Handbook*, ed. M. A. Aegerter, N. Leventis and M. M. Koebel, Springer New York, New York, NY, 2011, pp. 537–564.
21. *Aerogels*, ed. J. Fricke, Springer Berlin Heidelberg, Berlin, Heidelberg, 1986, vol. 6.
22. J. Fricke and A. Emmerling, in *Chemistry, Spectroscopy and Applications of Sol-Gel Glasses*, ed. R. Reisfeld and C. K. JJørgensen, Springer-Verlag, Berlin/Heidelberg, 1992, vol. 77, pp. 37–87.
23. M. G. Kaganer, *Thermal Insulation in Cryogenic Engineering*, Israel Program for Scientific Translations, Jerusalem, 1969.

24. K. Sakai, Y. Kobayashi, T. Saito and A. Isogai, *Sci. Rep.* 2016, **6**, 20434.
25. D. M. S. Al-Homoud, *Build. Environ.*, 2005, **40**, 353–366.
26. B. Notario, J. Pinto, E. Solorzano, J. A. de Saja, M. Dumon and M. A. Rodríguez-Pérez, *Polymer*, 2015, **56**, 57–67.
27. *Aerogels Handbook*, ed. M. A. Aegerter, N. Leventis and M. M. Koebel, Springer, New York, 2011.
28. B. P. Jelle, *Energy Build.*, 2011, **43**, 2549–2563.
29. BuyAerogel.com | Airloy® X103 Strong Aerogel Large Panels, http://www.buyaerogel.com/product/airloy-x103-large-panels/, (accessed 11 July 2017).
30. C. Buratti and E. Moretti, *Appl. Energy*, 2012, **98**, 396–403.
31. N. Bheekhun, A. R. Abu Talib and M. R. Hassan, *Adv. Mater. Sci. Eng.*, 2013, **2013**, 1–18.
32. E. M. Arndt, M. D. Gawryla and D. A. Schiraldi, *J. Mater. Chem.*, 2007, **17**, 3525.
33. K. Kubo, *Manufacturing aqueous slurry of hydrous calcium silicate and products thereof*, Google Patents, 1970.
34. B. Chen and J. Liu, *Cem. Concr. Res.*, 2004, **34**, 1259–1263.
35. D. Saradhi Babu, K. Ganesh Babu and T. H. Wee, *Cem. Concr. Res.*, 2005, **35**, 1218–1223.
36. F. Björk and T. Enochsson, *Constr. Build. Mater.*, 2009, **23**, 2189–2195.
37. L. Jaouen, A. Renault and M. Deverge, *Appl. Acoust.*, 2008, **69**, 1129–1140.
38. C. Tseng and K. Kuo, *J. Quant. Spectrosc. Radiat. Transfer*, 2002, **72**, 349–359.
39. European Phenolic Foam Association, Properties of Phenolic Foam, http://www.epfa.org.uk/properties.htm.
40. European Phenolic Foam Association, Phenolic Foam Insulation. The ultimate insulation system for the Construction & Building Services Industry, http://www.epfa.org.uk/Pdfs/eng.pdf.
41. J. Zach, A. Korjenic, V. Petránek, J. Hroudová and T. Bednar, *Energy Build.*, 2012, **49**, 246–253.
42. G. Churu, B. Zupančič, D. Mohite, C. Wisner, H. Luo, I. Emri, C. Sotiriou-Leventis, N. Leventis and H. Lu, *J. Sol-Gel Sci. Technol.*, 2015, **75**, 98–123.
43. N. Leventis, *Acc. Chem. Res.*, 2007, **40**, 874–884.
44. N. Leventis and H. Lu, in *Aerogels Handbook*, ed. M. A. Aegerter, N. Leventis and M. M. Koebel, Springer New York, New York, NY, 2011, pp. 251–285.
45. H. Zhu, Z. Jia, Y. Chen, N. Weadock, J. Wan, O. Vaaland, X. Han, T. Li and L. Hu, *Nano Lett.*, 2013, **13**, 3093–3100.
46. O. Lerouxel, D. M. Cavalier, A. H. Liepman and K. Keegstra, *Curr. Opin. Plant Biol.*, 2006, **9**, 621–630.
47. D. Klemm, B. Heublein, H.-P. Fink and A. Bohn, *Angew. Chem., Int. Ed.*, 2005, **44**, 3358–3393.

48. A. N. Fernandes, L. H. Thomas, C. M. Altaner, P. Callow, V. T. Forsyth, D. C. Apperley, C. J. Kennedy and M. C. Jarvis, *Proc. Natl. Acad. Sci.*, 2011, **108**, E1195–E1203.
49. V. A. Frey-Wyssling and K. Mühlethaler, *Makromol. Chem.*, 1963, **62**, 25–30.
50. F. Ayadi, B. Martín-García, M. Colombo, A. Polovitsyn, A. Scarpellini, L. Ceseracciu, I. Moreels and A. Athanassiou, *Carbohydr. Polym.*, 2016, **149**, 217–223.
51. B. J. C. Duchemin, M. P. Staiger, N. Tucker and R. H. Newman, *J. Appl. Polym. Sci.*, 2010, **115**, 216–221.
52. R. Gavillon and T. Budtova, *Biomacromolecules*, 2008, **9**, 269–277.
53. J. Innerlohinger, H. K. Weber and G. Kraft, *Macromol. Symp.*, 2006, **244**, 126–135.
54. R. Sescousse and T. Budtova, *Cellulose*, 2009, **16**, 417–426.
55. F. Liebner, A. Potthast, T. Rosenau, E. Haimer and M. Wendland, *Holzforschung*, 2008, **62**, 129–135.
56. R. Sescousse, R. Gavillon and T. Budtova, *Carbohydr. Polym.*, 2011, **83**, 1766–1774.
57. H. Jin, Y. Nishiyama, M. Wada and S. Kuga, *Colloids Surf., A*, 2004, **240**, 63–67.
58. O. Aaltonen and O. Jauhiainen, *Carbohydr. Polym.*, 2009, **75**, 125–129.
59. M. Deng, Q. Zhou, A. Du, J. van Kasteren and Y. Wang, *Mater. Lett.*, 2009, **63**, 1851–1854.
60. C. Tsioptsias, A. Stefopoulos, I. Kokkinomalis, L. Papadopoulou and C. Panayiotou, *Green Chem.*, 2008, **10**, 965.
61. H. Tamon, H. Ishizaka, T. Yamamoto and T. Suzuki, *Carbon*, 2000, **38**, 1099–1105.
62. J. Shi, L. Lu, W. Guo, Y. Sun and Y. Cao, *J. Appl. Polym. Sci.*, 2013, **130**, 3652–3658.
63. S. S. Kistler, *J. Phys. Chem.*, 1931, **36**, 52–64.
64. S. S. Kistler, *Nature*, 1931, **127**, 741.
65. J. Cai, S. Kimura, M. Wada, S. Kuga and L. Zhang, *ChemSusChem*, 2008, **1**, 149–154.
66. F. Liebner, E. Haimer, M. Wendland, M.-A. Neouze, K. Schlufter, P. Miethe, T. Heinze, A. Potthast and T. Rosenau, *Macromol. Biosci.*, 2010, **10**, 349–352.
67. M. Pääkkö, J. Vapaavuori, R. Silvennoinen, H. Kosonen, M. Ankerfors, T. Lindström, L. A. Berglund and O. Ikkala, *Soft Matter*, 2008, **4**, 2492.
68. Y. Kobayashi, T. Saito and A. Isogai, *Angew. Chem.*, 2014, **126**, 10562–10565.
69. N. Hüsing and U. Schubert, *Angew. Chem., Int. Ed.*, 1998, **37**, 23–45.
70. A. C. Pierre and G. M. Pajonk, *Chem. Rev.*, 2002, **102**, 4243–4266.
71. W. Chen, Q. Li, Y. Wang, X. Yi, J. Zeng, H. Yu, Y. Liu and J. Li, *ChemSusChem*, 2014, 7, 154–161.

72. *Handbook of Industrial Drying*, ed. A. S. Mujumdar, CRC Press, Boca Raton, Fla., 4th edn, 2015.
73. K. Shang, W. Liao, J. Wang, Y.-T. Wang, Y.-Z. Wang and D. A. Schiraldi, *ACS Appl. Mater. Interfaces*, 2016, **8**, 643–650.
74. J. Zhang, Q. Ji, X. Shen, Y. Xia, L. Tan and Q. Kong, *Polym. Degrad. Stab.*, 2011, **96**, 936–942.
75. G. Tkalec, Ž. Knez and Z. Novak, *RSC Adv.*, 2015, **5**, 77362–77371.
76. G. Tkalec, R. Kranvogl, A. Perva Uzunalić, Ž. Knez and Z. Novak, *J. Non-Cryst. Solids*, 2016, **443**, 112–117.
77. F. Fischer, A. Rigacci, R. Pirard, S. Berthon-Fabry and P. Achard, *Polymer*, 2006, **47**, 7636–7645.
78. G. Biesmans, D. Randall, E. Francais and M. Perrut, *J. Non-Cryst. Solids*, 1998, **225**, 36–40.
79. P. Gurikov, S. P. Raman, D. Weinrich, M. Fricke and I. Smirnova, *RSC Adv.*, 2015, **5**, 7812–7818.
80. G. M. Glenn and D. W. Irving, *Cereal Chem.*, 1995, **72**, 155–161.
81. J.-M. Raquez, M. Deléglise, M.-F. Lacrampe and P. Krawczak, *Prog. Polym. Sci.*, 2010, **35**, 487–509.
82. G. Amaral-Labat, L. Grishechko, A. Szczurek, V. Fierro, A. Pizzi, B. Kuznetsov and A. Celzard, *Green Chem.*, 2012, **14**, 3099.
83. I. Mueller-Harvey, *J. Sci. Food Agric.*, 2006, **86**, 2010–2037.
84. X. Li, A. Pizzi, M. Cangemi, P. Navarrete, C. Segovia, V. Fierro and A. Celzard, *Ind. Crops Prod.*, 2012, **37**, 149–154.
85. C. Lacoste, M. C. Basso, A. Pizzi, A. Celzard and M.-P. Laborie, *Ind. Crops Prod.*, 2015, **73**, 41–48.
86. Q. Wu, R. L. Andersson, T. Holgate, E. Johansson, U. W. Gedde, R. T. Olsson and M. S. Hedenqvist, *J. Mater. Chem. A*, 2014, **2**, 20996–21009.
87. P. R. Stoessel, U. Krebs, R. Hufenus, M. Halbeisen, M. Zeltner, R. N. Grass and W. J. Stark, *Biomacromolecules*, 2015, **16**, 1997–2005.
88. *Fundamentals of Heat and Mass Transfer*, ed. T. L. Bergman and F. P. Incropera, Wiley, Hoboken, NJ, 7th edn, 2011.
89. J. Fricke, X. Lu, P. Wang, D. Büttner and U. Heinemann, *Int. J. Heat Mass Transfer*, 1992, **35**, 2305–2309.
90. D. Lee, P. C. Stevens, S. Q. Zeng and A. J. Hunt, *J. Non-Cryst. Solids*, 1995, **186**, 285–290.
91. J. Kuhn, T. Gleissner, M. C. Arduini-Schuster, S. Korder and J. Fricke, *J. Non-Cryst. Solids*, 1995, **186**, 291–295.
92. Y. Cheng, L. Lu, W. Zhang, J. Shi and Y. Cao, *Carbohydr. Polym.*, 2012, **88**, 1093–1099.
93. L. Wang, M. Sánchez-Soto, T. Abt, M. L. Maspoch and O. O. Santana, *Polym. Int.*, 2016, **65**, 899–904.
94. A. Tejado, W. C. Chen, M. N. Alam and T. G. M. van de Ven, *Cellulose*, 2014, **21**, 1735–1743.
95. Z. Zhang, G. Sèbe, D. Rentsch, T. Zimmermann and P. Tingaut, *Chem. Mater.*, 2014, **26**, 2659–2668.

96. R. Lin, A. Li, T. Zheng, L. Lu and Y. Cao, *RSC Adv.*, 2015, **5**, 82027–82033.
97. H.-B. Chen, Y.-Z. Wang, M. Sánchez-Soto and D. A. Schiraldi, *Polymer*, 2012, **53**, 5825–5831.
98. H.-B. Chen, B.-S. Chiou, Y.-Z. Wang and D. A. Schiraldi, *ACS Appl. Mater Interfaces*, 2013, **5**, 1715–1721.
99. L. Wang, M. Sánchez-Soto and T. Abt, *Ind. Crops Prod.*, 2016, **91**, 15–21.
100. L. Wang, D. A. Schiraldi and M. Sánchez-Soto, *Ind. Eng. Chem. Res.*, 2014, **53**, 7680–7687.
101. B. Yuan, J. Zhang, J. Yu, R. Song, Q. Mi, J. He and J. Zhang, *Sci. China: Chem.*, 2016, **59**, 1335–1341.

CHAPTER 16

Biomedical Applications of Polysaccharide and Protein Based Aerogels

CARLOS A. GARCÍA-GONZÁLEZ,* CLARA LÓPEZ-IGLESIAS, ANGEL CONCHEIRO AND CARMEN ALVAREZ-LORENZO*

Departamento de Farmacología, Farmacia y Tecnología Farmacéutica, Facultad de Farmacia, R+D Pharma group (GI-1645), and Health Research Institute of Santiago de Compostela (IDIS), Universidade de Santiago de Compostela, E-15782-Santiago de Compostela, Spain
*Email: carlos.garcia@usc.es; carmen.alvarez.lorenzo@usc.es

16.1 Bio-based Products

Biopolymers have a notable impact on the performance of relevant products for biomedical applications, such as surgical sutures, dressings or implants.[1] The broad portfolio of natural, semisynthetic and synthetic biodegradable polymers with diverse physicochemical (*e.g.*, molecular weight, hydrophilicity, chemical functionalities and degradation profiles), mechanical (elasticity, ductility and compressibility) and biological (biocompatibility, induction of cell responses and microbial growth inhibition) properties opens up possibilities for the processing of innovative drug delivery systems and medical devices with customized properties. The use of implanted medical devices from biodegradable polymers able to degrade *in vivo* at a rate compatible with the functional demand is especially interesting as it avoids interventions or practices for removal or checking of the implant.

Green Chemistry Series No. 58
Biobased Aerogels: Polysaccharide and Protein-based Materials
Edited by Sabu Thomas, Laly A. Pothan and Rubie Mavelil-Sam
© The Royal Society of Chemistry 2018
Published by the Royal Society of Chemistry, www.rsc.org

Polysaccharides and proteins are families of biodegradable natural polymers of special interest due to their high abundance, good biological performance, similar structure to the extracellular matrix and degradability by enzymes present in the body (Table 16.1). Proteins of human (collagen, elastin, fibrin and albumin), animal (silk fibroin and sericin, whey protein) and vegetal (gluten, zein and soy protein) origin, and chemical derivatives (gelatin) are among the most common polypeptides contained in medical devices as the major component (*i.e.* matrix) or as components of polymer blends, building blocks or carriers.[1–4] Similarly, polysaccharides can be obtained from a variety of sources including human and animal (hyaluronic acid, chondroitin sulfate, heparin, and chitin) or vegetal (alginate, pectin, starch, agar, and carrageenan) origin. Synthetic (chemical and chemoenzymatic) routes to obtain modified polysaccharides (chitosan, polysaccharide conjugates and blends) along with cost-effective production (extraction → purification) techniques to obtain biomedical grade-polysaccharides have spread the use of this family of polymers for life sciences applications.[1,5–7] Moreover, protein–polysaccharide combinations have also been largely explored for biomedical applications.[3] Current and future trends of protein and polysaccharide-containing biomedical products lies in the design of the desired chemical composition to obtain the desired bioactivity, and at the proper physical structure and format (microparticles, nanoparticles, fibers, dressings, macroporous, mesoporous or dual-porous materials) to obtain an advanced physiological and biological response from the product at high performance. This chapter seeks to provide an analysis of the state-of-the-art of bio-based aerogels and their biomedical applications.

16.2 Bio-based Aerogels for Drug Delivery

Bio-based aerogels offer attractive features and the high processing versatility needed for the development of emerging nanostructured drug carriers.[26–28] Many polysaccharides and proteins are already components of Federal Drug Administration (FDA)- and European Medicines Agency (EMA)-approved medicines. Indeed, the myriad of bio-based aerogel sources allows the design of carriers with advanced features due to specific drug-carrier chemical interactions, tuneable degradation rates or pH-sensitive responses. Moreover, the aerogel processing conditions can be adapted to render tailor-made internal (*e.g.* specific surface area, overall porosity, density and specific drug loading capacity) and external (particle size and coatings) morphological properties of interest for drug dosage form manufacturing (accurate flow properties for dosing and handling) and functioning (suitable drug release profiles and targeted release).[26–31]

Bio-based aerogels drug carriers may meet the challenges of the oral administration route, notably regarding the biopharmaceutical limitations of poorly-soluble BCS-class II and IV drugs, and the limited stability against GI tract conditions (enzymes and pH) of proteins and polypeptide drugs.[32–35] The proper formulation design for oral administration would avoid the use

Table 16.1 Uses, advantages and limitations of the most common biodegradable natural polymers in biomedical applications.[8–25]

Biopolymer	Biomedical uses	Advantages	Limitations	Other characteristics
Agarose	• *in vitro* cell culture • Scaffolds for neural and cartilage tissue engineering • Phantom material for ultrasound imaging and hyperthermia applications	• Swelling nature and high water content • Stimulate *in vitro* nerve formation	• Poor mechanical properties • Changes in mechanical properties may interfere with biological functions	• Forms thermally reversible gels • Insoluble in water at room temperature
Albumin	• Targeted intracellular delivery of drugs • Additive in tissue culture media • Long-term scaffolds in tissue engineering	• Multiple ligand binding sites • Long circulatory half-life • Interaction with several cellular receptors • Promoter of the viability and proliferation of mesenchymal stem cells • Potential autogenic source of biomaterials for tissue engineering • High water-binding capacity	• Needs enzymatic polymerization to generate solid biomaterials	• The most abundant plasma protein in mammalian blood
Alginate	• Drug and protein delivery in oral and pulmonary administration • Wound dressings • Tissue engineering: blood vessels, bone, cartilage, muscle, nerve, pancreas and liver • Cell culture and transplantation	• High processing versatility • Can be easily modified to obtain specific properties • Similarity to extracellular matrix • Mucoadhesive • Anti-viral and anticoagulant agent	• Lacks cell adhesivity • Limited long-term stability of gels in physiological conditions • Variety of origins and molecular weight, low reproducibility • May contain toxic impurities: requires high purification to be biocompatible • Limited mechanical stiffness • Inflammatory activity	• Anionic • Mild gelation conditions • Water soluble

Table 16.1 (*Continued*)

Biopolymer	Biomedical uses	Advantages	Limitations	Other characteristics
Carrageenan	• Gelling, stabilizing and thickening agent • Induction of experimental inflammation • Controlled release formulations • Wound dressings • Solubility enhancer	• Anti-viral, antioxidant, immune-inflammatory, anticoagulant, anti-hyperlipidemic • Absorbs exudates	• Several types with very different properties • Presents cytotoxicity, specially in parenteral administration • Inflammatory activity: formulation requires special care	• Anionic • Highly available and renewable source • Water soluble
Cellulose (bacterial)	• Bone tissue engineering • Drug delivery systems • Wound dressings • Artificial cornea • Dental grafting • Diagnostic sensors • Synthetic blood vessels	• Water holding ability • Obtained under mild conditions • Higher surface area than plant-obtained cellulose • Moldability (flexible)	• Usually needs chemical functionalization to adapt it for biomedical applications • High production costs	• High purity degree, very low toxicity • Insoluble in water • Water vapor permeability • Inert towards pH variation
Cellulose (plant-derived)	• Membranes for hemodialysis • Enzyme carriers for biosensors • Drug delivery • Scaffolds for bone, cartilage, liver, skin and blood vessels • Wound dressings	• Possibility to modulate biodegradability • Excellent mechanical properties • Can be easily modified to obtain specific properties • Ability to integrate into surrounding tissue	• Variety of sources, low reproducibility and varied costs • Complicated and expensive purification process	• The most abundant polysaccharide in nature • Insoluble in water, although a large number of soluble derivatives are available
Chitosan	• Drug, gene and growth factor delivery • Oral and pulmonary delivery • Tissue engineering as scaffolds for skin, bone, cartilage, liver, nerve and blood vessels • Wound healing	• Biodegradable • Can be easily modified to obtain specific properties • Hemostatic and anti-inflammatory activity • Favors hyaluronan synthesis and scar formation • Antibiotic activity • Mucoadhesive	• Variety of origins, molecular weights, low reproducibility • Difficulty to control processing • May contain toxic impurities • Elevated price	• Cationic • Abundant in nature • High charge density • Soluble at acidic pH but insoluble in water • Variety of molecular weight and deacetylation degree

Collagen	• Vascular grafts • Low load-bearing grafts • Hemostatic pads • Wound dressings • Corneal shields • DNA, protein and cell delivery • Drug targeting in lung and liver fibrosis and psoriasis	• Support the attachment, growth and migration of mesenchymal stem cells and their differentiation to the osteogenic lineage • Collagen-based 3D-structures promotes vascularization in tissues • Biomimetic approach • Mid-term degradation times • High water-binding capacity	• Poor mechanical properties • Production of recombinant human collagen still not possible	• Main component of organic extracellular matrix • The most abundant protein in the human body • 28 human collagen genetic types
Dextran	• Hydrogels for protein, drug and growth factor delivery • Colon-specific or parenteral administration of drugs • Wound dressings • Plasma expander and anti-trombolythic agent • Imaging agent	• Biodegradable by natural enzymes • Possibility to modify hydroxyl groups to obtain specific properties • Anti-coagulant activity • Anti-inflammatory activity	• Needs chain modifications or cross-linking agents to gelify	• Hydrophilic • Water-soluble
Gelatin	• Capsules for oral and rectal administration • Cardiovascular, hepatic, corneal and bone tissue engineering • Skin grafts • Wound dressings • Injectable fillers for drug delivery • Carriers for cancer treatment, ocular drug delivery and gene delivery	• High processing versatility • Enhanced epithelialization in wounds • Promoter of the viability and proliferation of mesenchymal stem cells • Transparency of gelatin gels • Tunable degradation rate of the gels by cross-linking extent	• Material properties largely influenced by collagen source and type • Requires chemical cross-linkers for long-term *in vivo* applications that can pose problems of biocompatibility • Source control is needed to avoid risk of disease transmission	• Reversible thermal gelation below 37 °C • Wide range of molecular weights controlled by the denaturation process used • Mainly obtained from pig skin for biomedical applications

Table 16.1 (*Continued*)

Biopolymer	Biomedical uses	Advantages	Limitations	Other characteristics
Hyaluronic acid	• Dermal fillers • Intra-articular viscosupplements • Corneal and dermal wound repair • Post-surgical adhesion prevention • Cell and molecule delivery • Cell culture	• Versatility for chemical modification • Naturally present in human body; does not induce immune response • High processing versatility • Enhances tissue formation and repair • Lubrication and water retention • Support the attachment, migration and proliferation of cells	• Needs modifications to avoid rapid degradation	• Anionic
Pectin	• Nasal, oral, ocular, colon-targeted and cancer-targeted drug delivery • Gene delivery • Tissue engineering • Wound healing	• Anti-methastatic properties • Ability to immobilize cells, genes, proteins, drugs or growth factors • Multiple gelling mechanisms by changing pH, temperature or using cations • Mucoadhesive	• Different sources and extraction processes, low reproducibility	• Anionic • Variety of molecular weight and degrees of esterification and amidation
Silk fibroin	• Bone tissue engineering • Soft tissue engineering • Wound dressings and other non-implantable biomedical textiles • Carriers for drug delivery • Orthopaedic implants	• Stimulation of the fibroblasts proliferation • Anti-inflammatory activity • Stimulation of glucose transport in adipocytes • Supports the attachment, growth and spreading of mesenchymal stem cells and their differentiation to the osteogenic lineage • Remarkable mechanical properties	• Composition and molecular structure change among different sources • Some allergic reactions linked to the presence of sericin residues • Bone promoting ability varies with the source	• High processing versatility • Hydrophilic • Compatible with common sterilization methods • Tunable biodegradation rate

Soy protein	• Lowering of blood cholesterol levels • Wound dressings • Topical applications • Tissue engineering	• High processing versatility by thermal and chemical treatments • Moisturizing agent • Protection against neurological damage • Antioxidant properties	• Too fast degradation rates for tissue engineering purposes • High denaturation temperature • Purification process can be complex	• Low cost • Variety of protein-to-oil contents • Long-term storage stability
Starch	• Bone cements • Hydrogels for drug delivery • Bone grafts • Electrospun wound dressings • Excipient in oral formulations	• Completely biodegradable by natural enzymes • Mechanical properties can be tailored by chemical modifications • Promoter of cell adhesion and proliferation • Promoter of the expression of osteoblastic markers	• Difficulty to process in pure form • Different sources and compositions, different structures, low reproducibility • Need to be manipulated to obtain certain mechanical properties and moisture sensitivity	• Low cost • Abundant in nature • Hygroscopic • Gelation upon heating
Whey protein	• Carrier for oral and intestine-targeted drug delivery • Excipient in buccoadhesive tablets	• Versatility for chemical modification • Good tableting properties • Mucoadhesive • Protective properties of bioactive compounds • Antioxidant properties • Antibacterial agent • Muscle strength promoter	• Lactose content in whey protein may cause sensitive reactions to intolerants • Purification process can be complex	• Anionic • Low cost • Used in the native, denaturized and reticulated forms • Gels with pH-sensitive swelling behaviour

Table 16.1 (*Continued*)

Biopolymer	Biomedical uses	Advantages	Limitations	Other characteristics
Zein	• Water and moisture barrier in tablets • Drug carriers for intestinal drug delivery • Oral delivery of proteins and peptides • DNA transfection • Vaccine delivery • Scaffolds for tissue engineering • Bioimaging	• Resistance to microbial attack • Mucoadhesive • High drug binding capacity • Water-proof biomaterial • Drug protection against gastric degradation • Reduces the toxicity of quantum dots	• Non-aqueous based processing options • Low mechanical strength • Solubility influenced by the presence of xanthophyll pigments	• High processing versatility • Low solubility in water • Heat and abrasion resistance • Low cost

of more invasive routes like subcutaneous implantation or parenteral (intravenous, subcutaneous and intramuscular injections) and an enhanced drug dosing precision and longer *in vivo* biological half-life of the active agents.

There is an increasing proportion of new drug candidates that are poorly water soluble due to the advent of more complex molecular entities arising from emerging synthetic technologies and the target-based drug discovery approach.[36] According to the modified Noyes–Whitney equation, the dissolution rate of a drug can be improved by (i) increasing the contact surface area of the drug with the dissolution medium, as well as by (ii) increasing the saturation solubility of the drug in the medium. Both parameters can be properly tuned by encapsulation of BCS-class II and IV drugs in bio-based aerogels.

The outstanding specific surface area of aerogels facilitates the exposition of the loaded drugs to the GI-tract medium, thus improving the drug dissolution rate. The high inner surface area of aerogels would promote long-term stability of the small drug particles (for instance. drug nanocrystals) loaded into the carrier, as well as preventing their agglomeration. The external surface area of bio-based aerogels can also be increased by reducing the aerogel particles size by means of milling or emulsion templating. Drugs can be loaded into aerogels during any of the processing steps of the aerogel preparation, for example the gelation step, the solvent exchange step or the supercritical drying, as well as by post-processing (*e.g.*, supercritical and liquid CO_2 impregnation of aerogels) (Figure 16.1).[37–42] The gel precursors of bio-based aerogels are usually obtained in the form of hydrogels, for example in an aqueous medium, so the loading of BCS-class II and IV drugs during the gelation step is usually discarded owing to the intrinsic low water solubility of these drugs, which in turn leads to low drug loading content. Alternatively, the direct gelation of polysaccharides in ethanol and containing BCS-class II and IV drugs has recently been reported.[43,44] The loading of the drug in the gels during the solvent exchange is regarded as an auspicious solution in the case of sufficient drug solubility in the solvent coupled to the high drug partition coefficient towards the bio-based gels. Drug encapsulation during supercritical drying is scarcely used due to the low loading yields obtained, as the drying process usually takes place in the continuous mode with a supercritical CO_2 flow that may solubilize and extract the drug. Finally, the supercritical impregnation of drugs in aerogels is an interesting post-processing approach leading to a solvent-free product with high yields and no need of downstream processes. The main drawback of this approach is the need for a certain solubility of the drug in the supercritical medium to reach relevant drug loadings in a reasonable timeframe. For any of the abovementioned loading strategies, the chemical groups present in the gel backbone play a key role in the affinity of the drug for the aerogel matrix and can be tuned by the choice of polysaccharide/protein source or by derivatization.

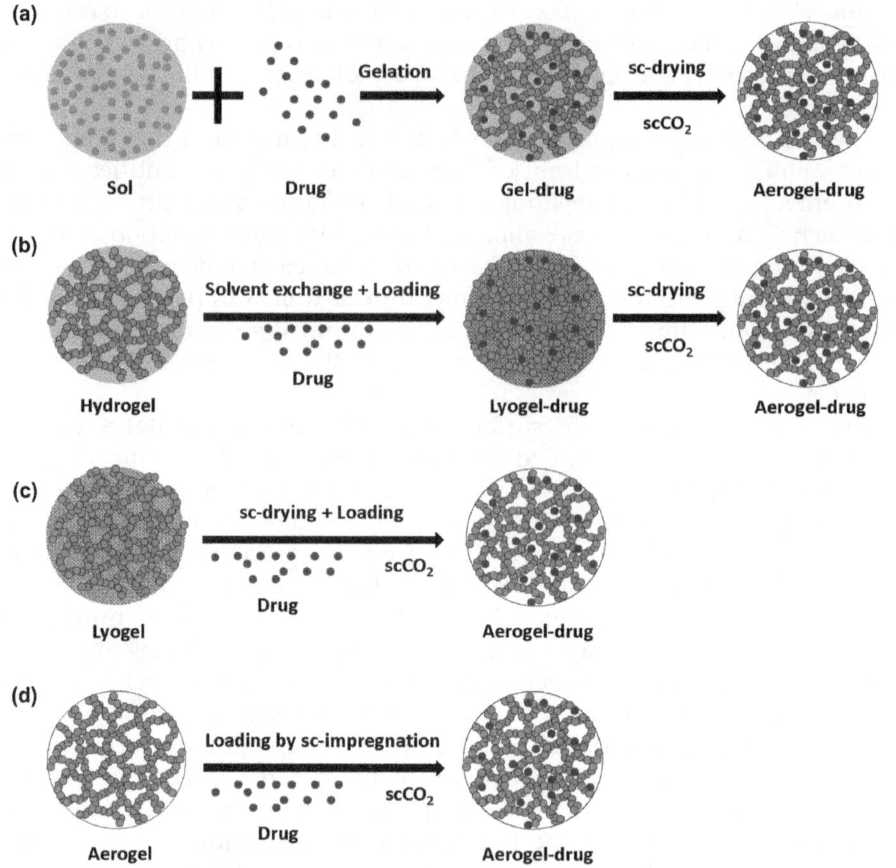

Figure 16.1 Strategies to obtain polysaccharide and protein-based aerogels loaded with drugs: (a) during the sol–gel process (co-gelation); (b) in the lyogel matrix during the solvent exchange; (c) in the lyogel matrix during the supercritical drying process; (d) in the aerogel matrix through supercritical impregnation (post-treatment method).
Adapted from *Carbohydrate Polymers*, **86**, C. A. García-González, M. Alnaief and I. Smirnova, Polysaccharide-based aerogels—Promising biodegradable carriers for drug delivery systems, 1425–1438, Copyright 2011, with permission from Elsevier.[26]

Loading during the solvent exchange step or by supercritical impregnation commonly results in the drug particles being either in the amorphous state or as nanosized crystals. Although there is still controversy in the literature on the actual crystalline form of the drugs loaded in aerogels, both approaches give a significant increase in the apparent drug solubility in aqueous media and subsequently drug dissolution becomes faster.

Bio-based aerogels also appear to be efficient tools for the formulation of therapeutic peptides and proteins in oral products. Several critical aspects regarding the delivery of proteins can be likely overcome using this type of nanostructured carrier. The short half-life of peptides in the GI-tract is related to the physicochemical conditions of the physiological medium, as well as the presence of proteolytic enzymes resulting in conformational changes (denaturation) or degradation.[34] The development of a formulation targeting the lower GI-tract reduces the enzymatic degradation of proteins and peptides. Two strategies for the targeting of bio-based aerogels to the gut have so far been reported in the literature.[33,45,46] The first approach relies on certain bio-based aerogels (pectin, alginate and whey protein) that exhibit pH-dependent drug release (Figure 16.2).[32,47] Namely, pectin, a polysaccharide resistant to the degradation from proteolytic enzymes (such as protease and amylase), may be useful to prepare aerogels that release most of the drug in the colon where pectin would be digested by the local microflora.[33,46] The swelling and degradation rates of pectin aerogels are dramatically affected by the pH (1.2 and 6.5) (Figure 16.3a)[33]. Pectin source (apple or citrus pectin) also influences both the swelling degree and the degradation rate at simulated intestinal pH conditions; apple pectin releasing the drugs (nicotinic acid and theophylline) faster with *ca.* 90% release after 1 h (Figure 16.3b).

Pectin aerogels in the form of microspheres have been further decorated with magnetic (maghemite) nanoparticles for targeted colonic delivery, with slight modification of the textural properties of the aerogels.[46] The incorporation of magnetic nanoparticles opens up the possibility of *in vivo* directing of the drug carrier to the target tissue with real time tracking by imaging.[48] In a different approach, bio-based aerogels were endowed with protection against the gastric medium by means of a multistep gelation process to obtain multi-membrane spheres or of enteric coatings through an aerogel post-processing method.[33,37,45] The former method would entrap the active pharmaceutical ingredient (API) in the inner layers of the aerogel avoiding burst effects and allowing a more controlled release in the lower GI-tract.[33,37] The latter method was tested for starch-alginate aerogel cylinders coated with polymethacrylates and caused a change of the mechanical behaviour of the material from plastic (uncoated) to viscoelastic (coated) and an improvement in stability against humidity.

Overall, bio-based aerogels are quite appealing for oral administration, although there are still some limitations that have to be overcome and challenges that need to be faced. For the targeted release of APIs a new generation of aerogels endowed with smart behavior is still to emerge. Smart aerogels may represent a step-forward from pH-dependent API release to pH-triggered or pH-responsive release. Moreover, the low apparent density of the bio-based aerogels, typically in the range of 0.06 to 0.15 $g\,cm^{-3}$, means a relatively low weight of carrier per volume unit. In the case of hard capsules, the standard capsule sizes for human medicines give a serious limitation to reaching body fill weights above 50–100 mg of aerogel carriers (in the specific favorable case where the apparent density is the same as the tapped density). The specific loading capacity of the bio-based aerogels depends on the natural polymer source, the textural properties and the drug-aerogel interactions, and usually ranges from 0.01–0.25 g drug per gram of aerogel.

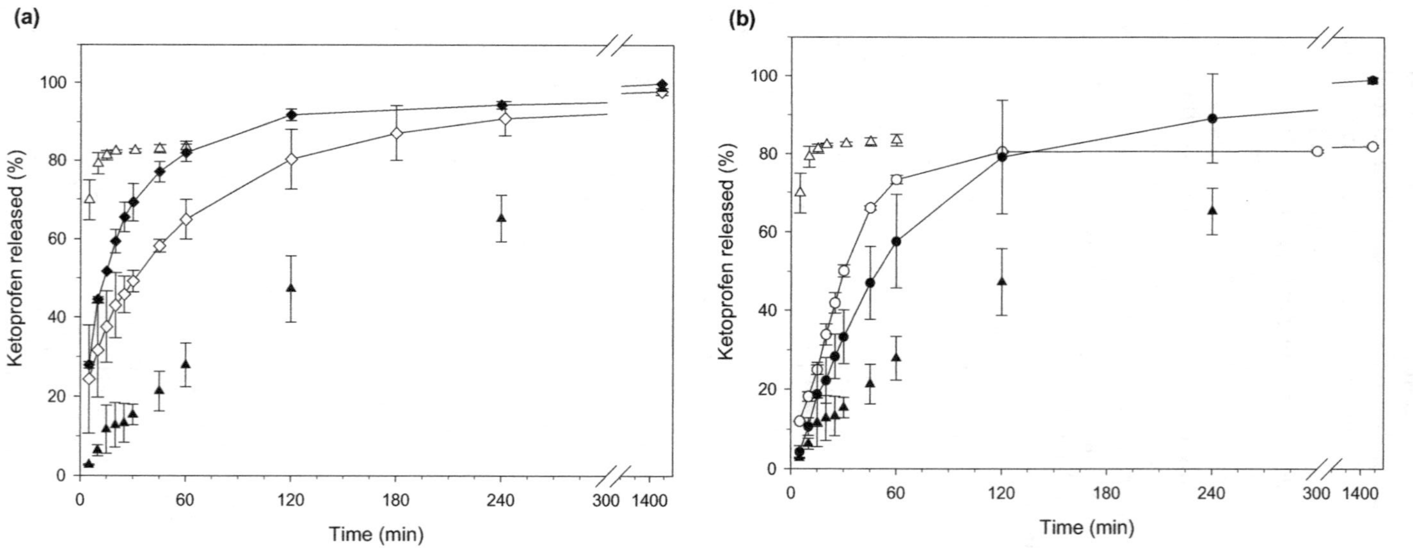

Figure 16.2 Effects of pH and polysaccharide matrix (pectin or alginate) on the release of ketoprofen from polysaccharide aerogels.[32] *In vitro* release profiles of ketoprofen from (a) alginate (diamonds), and (b) pectin (circles) aerogel microspheres at two different pH media (solid lines). Release studies of the aerogels were carried out at 37 °C and pH 1.2 (0.1 N HCl, dark symbols) and 6.8 (0.1 M phosphate buffer saline solution, blank symbols). Release profiles of the raw ketoprofen (dotted lines) are also plotted for the sake of comparison (dark triangles for pH 1.2 and blank triangles for pH 6.8).
Reprinted from *Carbohydrate Polymers*, **117**, C. A. García-González, M. Jin, J. Gerth, C. Alvarez-Lorenzo and I. Smirnova, Polysaccharide-based aerogel microspheres for oral drug delivery, 797–806, Copyright 2015, with permission from Elsevier.

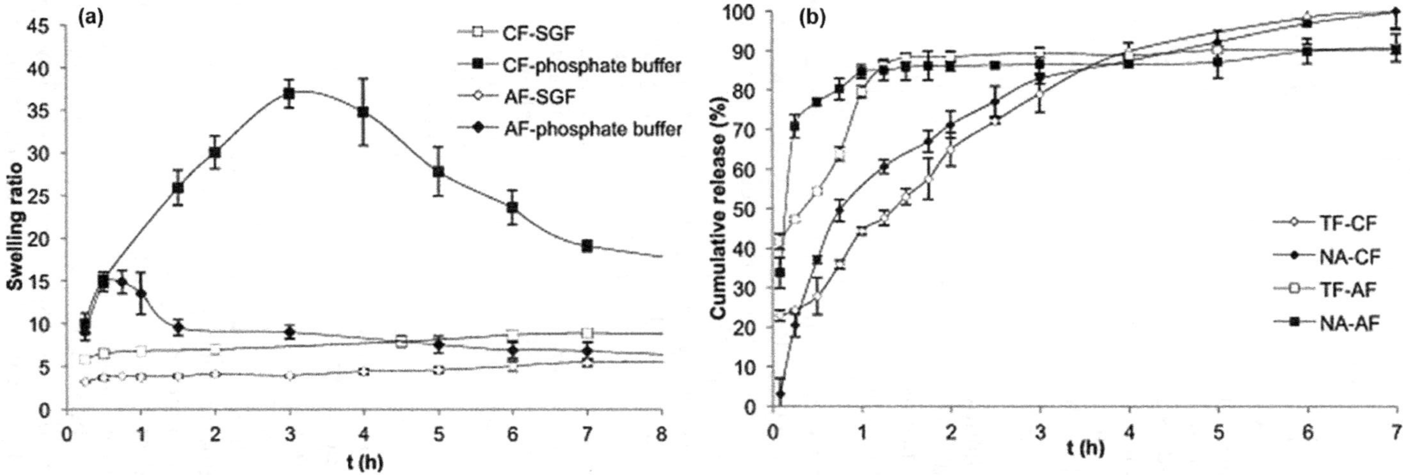

Figure 16.3 Effect of the pectin source (apple –AF– or citrus –CF– pectin) on the use of multi-membrane pectin aerogels for colonic drug delivery:[33] (a) Fluid absorption capacity at two pH conditions (pH 1.2 –SGF–, and 6.5 –phosphate buffer–); (b) Nicotinic acid (NA) and theophylline (TF) release profiles from the aerogels at simulated intestinal conditions (37 °C, pH 6.5). Reprinted from *Carbohydrate Polymers*, **113**, A. Veronovski, G. Tkalec, Ž. Knez and Z. Novak, Characterisation of biodegradable pectin aerogels and their potential use as drug carriers, 272–278, Copyright 2014, with permission from Elsevier.

Therefore, the maximum dosing of capsules using aerogels as drug carriers are in the range of 0.5–25 mg of drug, which might not be sufficient for a number of treatments. In the case of tableting of aerogels, the main fact to be tested is the effect of the compression condition on the preservation of the textural properties of the bio-based aerogels. Previous studies with ordered mesoporous silica have shown that mesopores collapse.[35,49] Blending with excipients (*e.g.*, microcrystalline cellulose) before the compression step can mitigate the loss of integrity of the mesoporous material.[49] Moreover, the flow properties of aerogel powders have been scarcely studied, even though they are of utmost importance for the tableting and capsule filling to facilitate the pneumatic transport of the aerogel, to ensure a precise control of the dosing per tablet/capsule, as well to obtain a proper mixing of the ingredients of the formulation. The effects of the aerogel-containing tablet formulation and of the tableting process itself on the release profile of the drugs loaded in the aerogels have been only faintly studied in inorganic (silica) aerogels[50] and need to be examined. Finally, the storage stability of bio-based aerogels under different ICH climate conditions is still an uncertainty and a critical issue to be assessed for the development of aerogel drug formulations.

Drug-loaded bio-based aerogels may also find applications in orally disintegrating tablets due to the possibility of obtaining prompt drug releases from aerogels, and in floating drug dosage forms because of their low density.

The inherent low density and high porosity mean aerogels exhibit high air flowability, which is especially promising for their use as carriers for the delivery of APIs in the respiratory tract, for example the pulmonary administration route by inhalation, and nasal drug delivery. Pulmonary drug delivery can be applied for the treatment of respiratory disorders, but also for systemic absorption.[36] In the former case, targeted drug delivery to the site of action is pursued in order to optimize the therapeutic effect and minimize the untoward effects. Systemic administration using the pulmonary route provides a painless and non-invasive alternative to reach blood circulation due to the high surface area and permeability of the alveolar epithelium to the flow blood supply through the pulmonary arteries, in contrast to intravenous or transdermal injections and implants, and to the low enzymatic activity if compared with the oral administration route.

For inhalation, the aerodynamic particle size plays a critical role in obtaining an optimum aerodynamic particle flow pattern, avoiding inertial impaction in the oropharyngeal region of large particles and the Brownian motion with air entrainment and exhalation of small particles. An optimum median aerodynamic diameter, for example equivalent geometric diameter of a solid sphere with the same aerodynamic diameter as the studied particles, has been proposed to be in the range of 1–5 μm for pulmonary delivery.[51,52] The median aerodynamic diameter is directly dependent on the geometric diameter and to the square root of the particles bulk density. The high porosity of aerogel particles translates into a much lower particle bulk density than that of a solid particle of the same diameter and from the same material, resulting in lower aerodynamic diameters. According to this, a geometric diameter of 2–10 μm has been proposed for aerogel particles to be administered through dry powder inhalers.[53] This larger particle size has a positive effect on the aerogel performance as less particle aggregation is likely to take place. Moreover, the wetting of aerogels when in contact with aqueous fluids in the mucous membrane or pulmonary surfactant present in the lung tissue is dramatically accelerated. Thus, the aerogels readily dissolve due to the effect of capillary forces and the high specific surface area with respect to solid particles of the same mass, resulting in a faster drug access to the blood stream.[27,53]

Aerogel powders for inhalation in the 0.5–10 μm range have been prepared by jet-milling and dripping.[53,54] Aerogels were loaded with APIs (salbutamol, insulin, morphine, sildenafil citrate) during the gel solvent exchange to ethanol prior to supercritical drying, this approach being compatible with the use of heat sensitive drugs. Insulin-loaded aerogel powder from derivatized trehalose (a natural disaccharide found in several plants, bacteria, fungi and invertebrate animals) in the 0.5–10 μm particle size range was prepared by sol-gel and supercritical drying, followed by jet milling.[53] Low density (0.001–0.1 $g\,cm^{-3}$), *ca.* 95% overall porosity and specific surface areas reaching up to 1200 $m^2\,g^{-1}$ have been reported for the trehalose aerogels. The preservation of the biological integrity and the activity of the insulin loaded into these aerogels was confirmed using the SDS-PAGE technique and with insulin receptor transfected NIH 3T3 fibroblasts cell cultures, respectively. Insulin release from the aerogels on simulated mucous membrane was very fast and uniform. Salbutamol-loaded chitosan aerogels were obtained in the form of particles of *ca.* 10 μm size and tapped densities of *ca.* 0.12 $g\,cm^{-3}$ by a dripping method of a chitosan solution into a sodium tripolyphosphate (TPP) aqueous bath followed by supercritical drying.[54] The physicochemical stability of chitosan aerogels upon storage was confirmed by a three-month test at 25 °C. Prolonged salbutamol release profiles (PBS pH 7.4) longer than 2 hours and dependent on the chitosan source (molecular weight) and TPP concentration were obtained, it was found that the aerogels are suitable for pulmonary drug delivery intended for the treatment of respiratory disorders.

Dry powder inhalation (DPI) devices seem to be the most straightforward way of administration of aerogels for inhalation.[52] Several DPI device options can be considered (unit-dose, multi-unit-dose or multidose DPI) with the administered dose being dependent on the premetered aerogel weight and the specific drug loading per dose. Alternatively, the delivery of the drug-loaded aerogel powder in a room where the patient is located with a total dose dependent on the aerogel concentration in the environment and the exposure duration has been proposed.[53]

The nasal administration route is suitable not only for local treatment, but also for systemic delivery, drug delivery to the central nervous system through the nose-to-brain route, and vaccinations.[55–57] Nasal powder sprayers are regarded as suitable devices for the intranasal delivery of bio-based aerogels. For these devices, the particle size of solid powders should be in the range of 10–50 μm, these being the lower and upper limits defined by the risk of pulmonary deposition of the powders and by the deposition in the anterior nasal cavity with poor absorption properties, respectively.[58] This aerodynamic diameter range translates into a geometric diameter range of 20–100 μm for highly porous materials such as aerogels. The processing of several bio-based aerogels with sizes in this range by means of emulsion-gelation techniques has been reported in the literature.[29,32,59,60] This approach provides not only a fine control on the particle size distribution of the aerogel microspheres, but also a regularity in the shape and dimensions, which is important for obtaining reproducible aerodynamic flow profiles. Interestingly, the inherent gel-forming capacity of the polysaccharides and proteins used as aerogel sources is commonly aligned with mucoadhesive properties.[61,62] Mucoadhesion may prolong the residence time of the drug-loaded aerogel powder in the nasal cavity through attachment onto the nasal mucous layer and through the holding back of mucociliary drainage. The mucoadhesion of certain bio-based aerogels (alginate containing aerogels) to porcin mucin was confirmed through turbidimetric analysis.[59] The cytotoxicity of alginate-based aerogels loaded with ketoprofen, and transmucosal transport of the drug payload through the nasal epithelium have been evaluated.[63] RPMI 2650 cells were used as the best *in vitro* model alternative to excised human nasal mucosa tests (*i.e. ex vivo* model), which is the current common practice, and showed similar results regarding permeability for sodium fluorescein, as well as transepithelial electrical resistance and mucus production. Direct contact studies of ketoprofen-loaded aerogels onto the cells showed not only an absence of cytotoxicity after a one-hour incubation, but also an enhancement in the drug permeation with respect to the pure drug. Polysaccharide aerogel microspheres were loaded with a kyotorphin derivative, a neuropeptide showing analgesic behaviour of interest for nasal delivery.[59,64] An encapsulation efficiency of the kyotorphin derivative of *ca.* 80% was obtained by incorporation of the peptide into the aerogel matrix during the gelation process, although the stability and retained activity of the peptide in the aerogel were not reported.[59]

16.3 Bio-based Aerogels for Tissue Engineering

Tissue engineering has been denoted as a multidisciplinary field aiming to orchestrate the trinomial combination of stem cells, resorbable scaffolds and bioactive molecules.[65] In particular, scaffolds can be used to engineer tissue regeneration using two different strategies: (i) as an *in vitro* cell support for ulterior transplantation, and (ii) as a support to promote *in vivo* cell colonization and tissue growth. Nanostructuration of materials opens novel process windows in the scaffold architecture. On the other hand, materials from bio-based sources may favour cytocompatibility and biomimetic designs of the grafts. Thus, the use of nanostructured bio-based materials in the form of aerogels is an auspicious solution to merge the advantages from both nanostructuration and natural materials domains (Figure 16.4).[27,66–70]

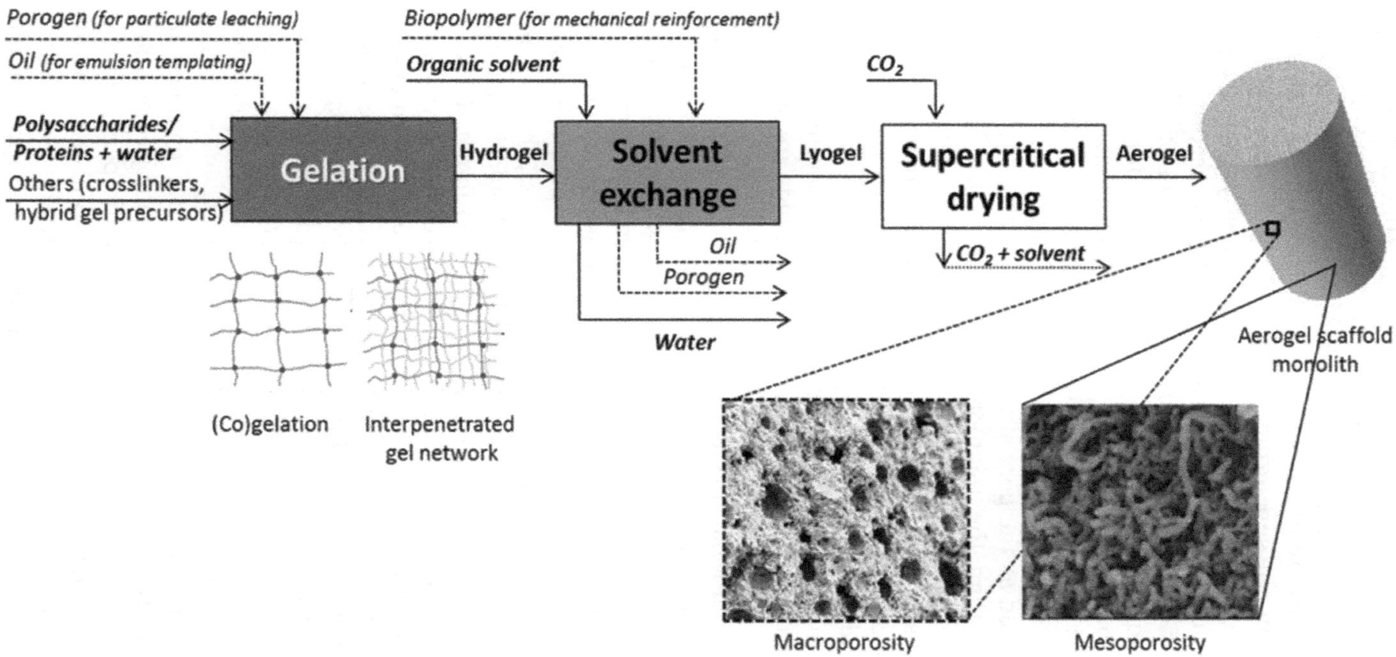

Figure 16.4 Bio-based aerogel scaffold processing steps: gelation, solvent exchange and scCO_2-assisted drying of gels. Gelation step can occur from a single or multiple gel precursors. For multiple gel precursors, the gelation mechanism can be cogelation or interpenetrated gel network formation. The resulting aerogel scaffolds in the form of monoliths (right) have a high porosity formed by interconnected mesopores. Alternatively, advanced properties (macroporosity, mechanical reinforcement) can be conferred to the aerogel scaffolds by combining with other techniques (dashed arrows) during the processing. Emulsion templating or particulate leaching can be used to obtain aerogel scaffolds with interconnected macroporosity (dashed square) by incorporating a hydrophobic dispersed phase and a solid porogen during the gelation step, respectively. The mechanical reinforcement of the aerogel scaffolds can be achieved by impregnation of the aerogels with a biopolymer during the solvent exchange step.

High porosity, full pore interconnectivity and presence of mesopores are important features of aerogel scaffolds to promote cell attachment, angiogenesis and transport of nutrients and waste for tissue repair. Moreover, bioactivity and cell proliferation capacity have been also reported for certain bio-based aerogels.[71–74] Despite using potentially cytotoxic organic solvents during aerogel production, the supercritical drying process has been proven to be effective enough to reduce residual solvents to non-cytotoxic levels. Finally, aerogels processed to be used in tissue engineering need to reach SAL-6 sterility conditions, meaning a possibility in a million of a microorganism survival, as required by regulatory agencies for terminal sterilization. The sterilization of aerogels can be quite straightforward if sterile conditions are preserved during the supercritical drying of the alcogel precursor. If sterilization is still needed, only the external outer surface of the aerogel has to be sterilized as the pore size of aerogels (*i.e.* 2–50 nm mesoporous range) is lower than the microorganisms size (>0.21 μm) thus avoiding their penetration into the aerogel network.

Processing of aerogel scaffolds is a technological approach that is able to give a high amount of control over the purity and the micro- and macrostructure of the material, as well as allowing operation under milder conditions compared to other conventional methods such as solvent casting, melt moulding, or electrospinning. Bio-based aerogel scaffolds are obtained through methods where temperature conditions can be below 45 °C during the whole process, no additional downstream processes are necessary for purification, and the use of organic solvents can be minimized.[26,67] Moreover, CO_2 and organic solvents can be recycled and reused for the sake of process integration and economics. Several attempts have been carried out to optimize the supercritical drying step regarding pressure, temperature and time. From an economical point of view, these three processing parameters should be kept at a minimum value.[29,72] Nevertheless, the feasible operating region to obtain optimum textural properties should be accordingly established for each bio-based aerogel system. As an example, the rule-of-thumb of overestimating the processing time for supercritical drying has been reported to be detrimental for starch aerogels because of the removal of structural water from the aerogel backbone.[29] On the contrary, the underestimation of the supercritical drying time can lead to poor textural properties and aerogels with solvent contents that can be toxic for living cells. Supercritical processing time can also have an impact on the removal of the excess of cytotoxic crosslinkers in the case of bio-based aerogels obtained from chemically cross-linked hydrogels.[75,76]

Bio-based aerogels can undergo intermediate processing methods to promote osteointegration. Bio-based hybrid aerogels obtained by co-gelation or interpenetration of components improve the biological properties of the resulting nanoporous material.[75,77,78] Chemical modifications of bio-based aerogels for bone and cartilage implants, such as derivatization with phosphate groups or oxidation, can promote the bioactivity and resorption of the material.[79,80] Nevertheless, the main biological limitations on the use of aerogels in tissue engineering are related to the intrinsic lack of macroporosity needed to promote cell colonization, proliferation and angiogenesis. Several approaches are reported in the literature to confer macroporosity to the scaffolds with certain limitations in the simultaneous control of pore particle size, overall macroporosity and pore interconnectivity.[67,77,81]

Emulsion templating has been explored to obtain starch aerogel cylindrical monoliths ($L = 30$ mm; $D = 6$ mm) containing round and interconnected macropores of *ca.* 15 µm.[67] In this approach, oil-in-water emulsions are used in which the continuous aqueous phase undergoes gelation of the polysaccharide and the dispersed phase forms oil droplets. After solvent exchange and supercritical drying, the continuous and dispersed phases become the aerogel backbone and the macroporosity of the scaffold, respectively. This method was extended to cellulose aerogels, which showed an anisotropic distribution with interconnected macroporosity in the 100–350 µm range and representing 62% in volume (Figure 16.5).[82] The resulting cellulose aerogel with macropores had a Young's modulus of 2.47 MPa and a lower yield stress at 1% strain than the cellulose aerogel without macropores. The results obtained are rather promising, although further research on the effect of the processing parameters is needed. The optimization of this processing approach should lead to a better control of the macropore size distribution, as well as on the control of the overall macroporosity.

Incorporation of porogens during gelation followed by particulate leaching has been introduced as an alternative method to confer macroporosity to bio-based aerogels.[71,83] As an example, a bed of porogen particles (paraffin or PMMA) of selected particle sizes was formed as a negative template by thermal fusion, and then the voids of the bed were filled with a solution containing the cellulose aerogel precursors. After gelation of the cellulose, the porogen was removed by leaching with an organic solvent (THF for paraffin and acetone for PMMA) and aerogels were obtained by supercritical drying. Cellulose aerogels containing interconnected macropores in the size range of the paraffin particles, with 98% porosity and Brunauer–Emmett–Teller (BET)-specific surface areas in the 136–243 $m^2\,g^{-1}$ range were thus obtained (Figure 16.6a). Results showed low fibroblast viability for cellulose aerogels obtained from paraffin porogen and close to 100% for those obtained from PMMA porogen. A certain cracking of the gel network upon the porogen leaching was noticed in the case of the PMMA porogen. Macroporous aerogels were also attempted by means of the agglomeration of (chitin) gel microspheres followed by supercritical drying.[84] Fusion of the adjacent gel microspheres was promoted using gellan gum and the interparticular voids gave rise to the macroporosity of the resulting aerogel scaffold in the 50–200 µm range (Figure 16.6b). Finally, the macroporosity was also obtained in aerogels through the foaming of CO_2-induced bio-based hydrogels by a controlled CO_2 depressurization rate followed by supercritical drying.[77] After sufficient CO_2 immersion time for gelation had elapsed, the CO_2 was depressurized at different rates. Slow depressurization rates gave

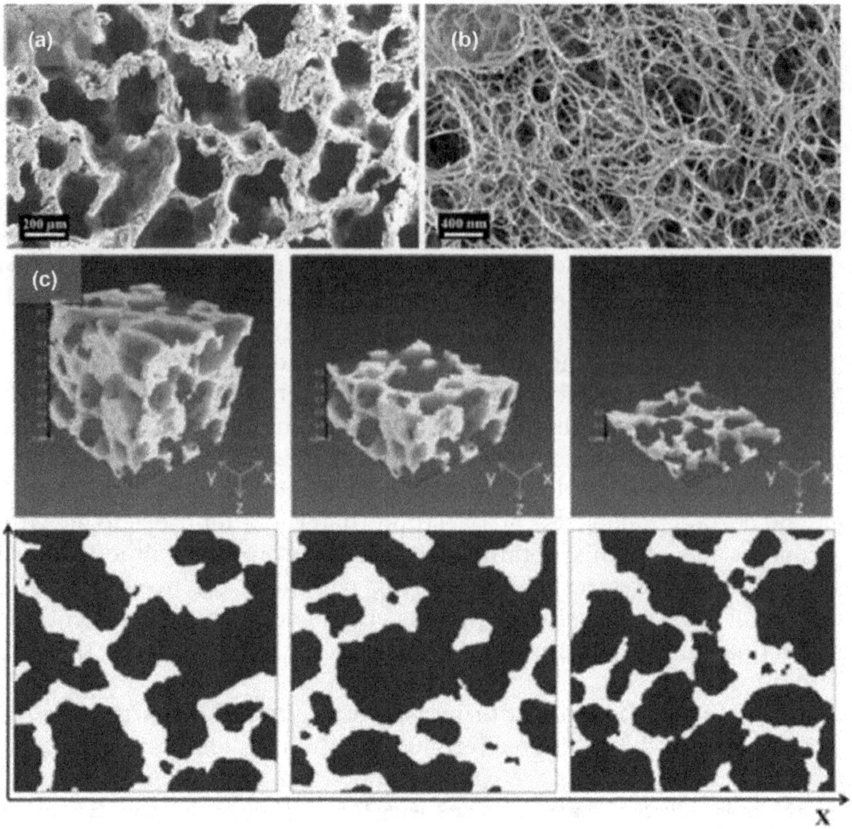

Figure 16.5 Macroporous cellulose aerogel scaffolds obtained by emulsion templating: SEM pictures showed that the material (a) had interconnected macroporosity as well as (b) mesoporosity; (c) micro-CT images confirmed a uniform macroporosity all along the unit cell measured. Reprinted from *Materials & Design*, **92**, K. Ganesan, A. Dennstedt, A. Barowski and L. Ratke, Design of aerogels, cryogels and xerogels of cellulose with hierarchical porous structures, 345–355, Copyright 2016, with permission from Elsevier.[82]

rise to higher macroporosity (up to 31%) and lower macropore sizes than that observed for faster depressurization. This technique still needs further research for implementation regarding an increase in macroporosity, a further control of the macropore size and a target of full interconnectivity.

Scaffolds for hard tissue repair can notably benefit from the use of aerogels. In this case, the mechanical properties of aerogel scaffolds should match the requirements of tissue strength loads, for instance large bones.[85] A compressive modulus of pure silk fibroin aerogels increased from 19.5 to 174 kPa as the fibroin content in the original aqueous solution changed from 2 to 6 wt.%.[71] For pure aerogel scaffolds, the effect on the mechanical properties of the presence of macropores in the aerogel structure needs to be compensated. In this case, a mechanical reinforcement of aerogel scaffolds

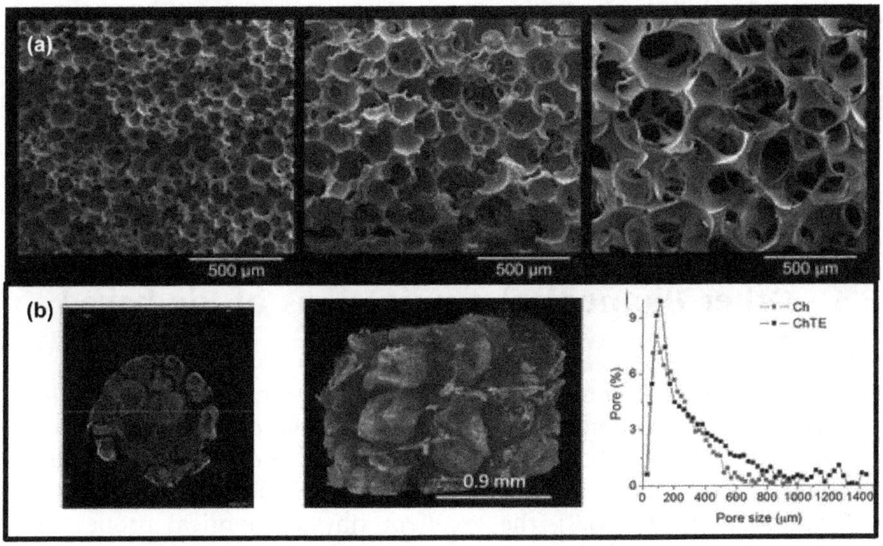

Figure 16.6 Strategies to obtain macroporous aerogel scaffolds: (a) Leaching of porogen particles (paraffin spheres) of increasing diameter (from left to right) led to larger macropores; and (b) agglomeration of chitin gel microspheres led to aerogel scaffolds (left and centre) of the controlled macropore size (right).
Part A adapted from ref. 83, http://dx.doi.org/10.1002/mame.201500048, published under the terms of the CC BY 4.0 licence, https://creativecommons.org/licenses/by/4.0/. Part B adapted from ref. 84 with permission from the Royal Society of Chemistry.

has been proposed.[68,86,87] In the first approach, bacterial cellulose aerogels were reinforced with synthetic biopolymers (PLA, cellulose acetate, and PMMA) through impregnation during the solvent exchange, followed by anti-solvent precipitation techniques with ethanol or $scCO_2$ and supercritical drying.[83,87] Densities of the reinforced cellulose aerogels linearly increased as the synthetic polymer content increased. Reinforced cellulose aerogels showed similar mechanical compression profiles as the original aerogel, but with increased stiffness and Young's moduli with higher synthetic polymer contents. In another approach, the reinforcement of several bio-based aerogel scaffolds was reported by a post-treatment consisting of the cross-linking of the fibres forming the aerogel backbone.[86] Unfortunately, no data on the improvement in the mechanical properties were provided.

Finally, the incorporation of growth factors in aerogel scaffolds is usually a cumbersome task, since growth factors are prone to lose activity during the solvent exchange steps and cannot be incorporated by supercritical impregnation postprocessing due to their low solubility in $scCO_2$. Supercritical foaming of synthetic polymers can surpass the abovementioned limitations to obtain growth factor-loaded scaffolds.[88,89] Supercritical foaming of scaffold formulations incorporating bio-based aerogel powders have thus been proposed.[90] Poly(ε-caprolactone) (PCL) scaffolds processed by supercritical foaming and containing increasing contents of starch aerogel microparticles had an impact on the increase of the scaffold porosity and a decrease in its bulk density. Notably, the presence of starch aerogel particles in the scaffolds significantly improved the size of the throat interconnection between the macropores and mesenchymal stem cells (MSCs) infiltration capacity, with a minor decrease in storage and loss moduli.

16.4 Other Biomedical Applications of Bio-based Aerogels

While aerogels are being currently applied in different technological fields, their use in biomedical sciences is still limited. However, interest in their use in health care sciences has been rising recently, and several applications are being studied.[27] In addition to drug delivery and tissue engineering (*cf.* Sections 16.2 and 16.3), the excellent physicochemical properties of aerogels make them suitable for other biomedical applications. Bio-based materials such as polysaccharides and proteins are the aerogel sources that demonstrate the best characteristics for application in the human body, since they are biocompatible and stable.[27,70]

16.4.1 Wound Care Applications

Wound healing is a complex and dynamic process in which devitalized and missing cellular structures and tissue layers are replaced. It is a normal, physiological response that happens when tissue integrity is compromised due to pathologies, surgical interventions and accidents. The final result of this process is the replacement of normal skin structures with scar tissue. Healing occurs in several steps that involve a great number of cells, extracellular elements, mediators such as cytokines and other molecules.[91] This process can be altered in chronic wounds such as diabetic foot ulcers, pressure ulcers and venous leg ulcers, and represents a worldwide problem. Chronic wounds may cause microbiological growth and infection spreading leading to fatal consequences such as amputation or even mortality.[92] Therefore, the development of materials able to act as wound dressings promoting the healing process is highly desirable. These dressings should stimulate tissue repair, facilitate the action of inflammatory cells, act as a barrier to microorganisms and provide a moist environment while removing the excess of exudate.

Aerogels represent a suitable option for wound healing as they are extremely porous materials able to absorb high amounts of aqueous fluids (Figure 16.7) such as exudates, which would prevent wounds from becoming infected. Specifically, bio-based aerogels have attracted the most interest in this field, due to their high stability, low toxicity, non-allergenic character and good biological performance.[43] During the exudate incorporation, the

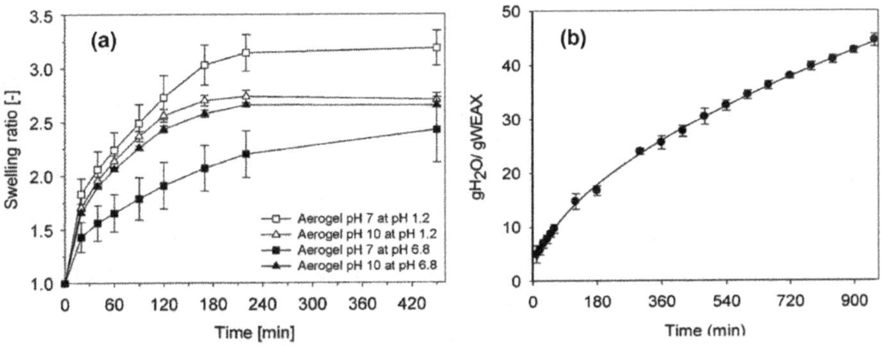

Figure 16.7 Water absorption capacity of bio-based aerogels: (a) whey protein, and (b) arabinoxylan aerogels.
Part A reprinted from *The Journal of Supercritical Fluids*, **72**, M. Betz, C. A. García-González, R. P. Subrahmanyam, I. Smirnova and U. Kulozik, Preparation of novel whey protein-based aerogels as drug carriers for life science applications, 111–119, Copyright 2012, with permission from Elsevier.[47] Part B reprinted from ref. 98, http://dx.doi.org/10.3390/molecules18055531, published under the terms of the CC BY 3.0 licence, https://creativecommons.org/licenses/by/3.0/.

solid matrix of the polysaccharide or protein aerogel swells, preventing the formation of water-filled pockets that act as a culture medium to bacteria. Moreover, bio-based aerogels usually permit the incorporation of substances that facilitate the process of healing, such as antimicrobial agents.

Several aerogel materials have been studied as wound dressings. For example, core-shell polysaccharide aerogels were designed to have an amidated pectin matrix and a shell of alginate with the ability to stimulate cytokine production by monocytes.[43] The core was loaded with doxycycline, an antibiotic drug with activity against Gram-positive and Gram-negative bacteria, in order to prevent infection in the wound area. Immersion in simulated wound fluid revealed that fluid uptake was directly related to the polysaccharide content. This fluid uptake was expressed as the weight ratio of the formed hydrogel and the dry aerogel, and reached the value of 8.1 w/w. The doxycycline loaded into the matrix provided a controlled drug release in a simulated wound fluid, with an initial burst effect followed by a slow release for 48 hours.

Chitosan, a polysaccharide derivative from chitin has been widely explored to prepare membranes, sponges and hydrogels for wound healing due to a certain antimicrobial activity.[73,93] The antimicrobial activity may rely on the crosslinking between the anions of the bacteria surface and the polycations of the chitosan, which results in the alteration of membrane permeability.[94] Chitosan aerogel particles obtained by chemical crosslinking with formaldehyde showed activity against *E. coli* in agar plates and liquid medium.[95] Aerogels prepared with chitosan permit the incorporation of drugs to its matrix and meet the requirements of biocompatibility and stability, being a suitable option to be applied in wound care. On the contrary, chitosan-silica hybrid aerogels were shown to be highly haemolytic.[96]

Collagen, a natural element of skin, is frequently used in wound care due to its biocompatibility and good adhesion properties. Composite aerogels prepared from dialdehyde nanocellulose and collagen by freeze-drying have been evaluated as wound dressings. The aerogels presented high porosity, bulk density and wettability (with a fluid uptake of nearly 40% w/w). Composite aerogels were also tested in terms of biocompatibility and cytotoxicity using a culture of L929 cells in Dulbecco's modified Eagle's Medium for this purpose. The resulting average cell activity was 96.8%, which means the aerogels were biocompatible and suitable for wound dressing applications.[97]

Other aerogel-based materials have not yet been specifically studied for wound healing applications, but some of their properties suggest that they could be useful for this purpose. This is the case of water extractable arabinoxylan, which presents a water uptake capability extended for 15 hours.[98]

16.4.2 Other Applications

In the last few decades, aerogels have gained a great deal of attention as implantable medical devices, ultra-sound contrast agents, and for biosensing, non-invasive imaging and cosmetics. However, there are few research studies that involve protein and polysaccharide-based aerogels with real application in these fields. Most research deals with their biocompatibility and biodegradability, and therefore refers only to their potential application.[70]

Starch aerogels have been proposed as an intermediate material for the production of chronic invasive electrodes. Aerogels are ideal materials for invasive electrodes as they present low impedance.[99] In these devices, the starch aerogel particles act as a sacrificial template for the synthesis of an aerogel from an electrical-conducting polymer (PEDOT). PEDOT aerogels did not cause cytotoxicity when tested with fibroblasts and can be useful as a component in invasive metallic electrodes for the electrical stimulation of nerve cells due to the observed significant reduction in the electrical impedance of the electrode with its incorporation.

For other purposes, pectin-based aerogels, where the polysaccharide-based matrix is combined with magnetic nanoparticles (maghemite), may find applications as contrast agents or for magnetic resonance imaging.[46]

16.5 Future Trends

The active research on the use of bio-based aerogels unveiled promising biomedical applications for these nanostructured materials. For pharmaceutical applications, the advantages of the use of bio-based aerogels as carriers have been already proved for several administration routes (oral, pulmonary and nasal). The next steps for this research are the studies of the handling of the material (flow properties, stability under storage, capsule filling, and integrity under tableting) and the implementation of the aerogel processing under good manufacturing practices (GMP)-conditions. For tissue engineering, important progress has been made in the last decade for bio-based aerogel scaffolds. Nevertheless, aerogel scaffolds should be regarded as an emerging alternative that are still under development. Aerogel scaffolds should be developed in interdisciplinary teams encompassing different technological and scientific expertise (engineering, pharmacy, chemistry, biology and biomechanics). Once the cytocompatibility and the tuning of the textural properties of aerogel scaffolds are confirmed, future work should be notably focused on obtaining a robust method to confer a controlled macroporosity to the aerogels and to improve the mechanical properties of the scaffolds to temporarily surrogate natural bone. The loading of growth factors into aerogel scaffolds with retained activity still remains a challenge. Finally, wound healing, chronic invasive electrodes, and non-invasive imaging represent other promising niche markets and the progress in these fields should be tracked in the coming years.

Abbreviations

API	active pharmaceutical ingredient
BCS	Biopharmaceutics Classification System
DPI	dry powder inhalation
EMA	European Medicines Agency
FDA	US Food and Drug Administration
GI	gastrointestinal
GMP	Good Manufacturing Practices
PBS	phosphate buffer solution
SAL	sterility assurance level
XRD	X-ray diffraction
MSCs	mesenchymal stem cells
PCL	poly(ε-caprolactone)
PLA	poly(lactic acid)
PMMA	poly(methyl methacrylate)
$scCO_2$	supercritical carbon dioxide
SEM	scanning electron microscopy

Acknowledgements

This work was funded by MINECO (SAF2017-83118-R), Xunta de Galicia (ED431F 2016/010) and (ED431C 2016/008), Agencia Estatal de Investigación (AEI) of Spain and FEDER. C. A. García-González acknowledges MINECO for a Ramón y Cajal Fellowship (RYC-2014-15239).

References

1. B. D. Ulery, L. S. Nair and C. T. Laurencin, *J. Polym. Sci., Part B: Polym. Phys.*, 2011, **49**, 832–864.

2. U. Shimanovich, G. J. L. Bernardes, T. P. J. Knowles and A. Cavaco-Paulo, *Chem. Soc. Rev.*, 2014, **43**, 1361–1371.
3. S. Tansaz and A. R. Boccaccini, *J. Biomed. Mater. Res., Part A*, 2016, **104**, 553–569.
4. S. Nivedita and V. Sivaprasad, *J. Inst. Eng. (India): Ser. E*, 2014, **95**, 57–61.
5. A. Basu, K. R. Kunduru, E. Abtew and A. J. Domb, *Bioconjugate Chem.*, 2015, **26**, 1396–1412.
6. J. Venkatesan, B. Lowe, S. Anil, P. Manivasagan, A. A. A. Kheraif, K.-H. Kang and S.-K. Kim, *Starch/Stärke*, 2015, **67**, 381–390.
7. J. Liu, S. Willför and C. Xu, *Bioact. Carbohydr. Diet. Fibre*, 2015, **5**, 31–61.
8. B. An, Y.-S. Lin and B. Brodsky, *Adv. Drug Delivery Rev.*, 2016, **97**, 69–84.
9. J. A. Burdick and G. D. Prestwich, *Adv. Mater.*, 2011, 23.
10. K. B. Chien and R. N. Shah, *Acta Biomater.*, 2012, **8**, 694–703.
11. S. Dumitriu and V. Popa, *Polymeric Biomaterials: 2 Volume Set*, Taylor & Francis, 3rd edn, 2013.
12. M. N. V. R. Kumar, R. A. A. Muzzarelli, C. Muzzarelli, H. Sashiwa and A. J. Domb, *Chem. Rev.*, 2004, **104**, 6017–6084.
13. M. T. Larsen, M. Kuhlmann, M. L. Hvam and K. A. Howard, *Mol. Cell. Ther.*, 2016, **4**, 3.
14. K. Y. Lee and D. J. Mooney, *Prog. Polym. Sci.*, 2012, **37**, 106–126.
15. G. Li, Y. Li, G. Chen, J. He, Y. Han, X. Wang and D. L. Kaplan, *Adv. Healthcare Mater.*, 2015, **4**, 1134–1151.
16. P. S. Li, I. Liang Lee, W. L. Yu, J. S. Sun, W. N. Jane and H. H. Shen, *Sci. Rep.*, 2014, 4.
17. Y. Luo and T. Wang, *Protein Byproducts*, Academic Press, 2016, pp. 147–160.
18. R. Mehvar, *J. Controlled Release*, 2000, **69**, 1–25.
19. F. Munarin, M. C. Tanzi and P. Petrini, *Int. J. Biol. Macromol.*, 2012, **51**, 681–689.
20. M. Nikkhah, M. Akbari, A. Paul, A. Memic, A. Dolatshahi-Pirouz and A. Khademhosseini, in *Biomaterials from Nature for Advanced Devices and Therapies*, John Wiley & Sons, Inc., 2016, pp. 37–62.
21. F. J. O'Brien, *Mater. Today*, 2011, **14**, 88–95.
22. R. Paliwal and S. Palakurthi, *J. Controlled Release*, 2014, **189**, 108–122.
23. G. F. Picheth, C. L. Pirich, M. R. Sierakowski, M. A. Woehl, C. N. Sakakibara, C. F. de Souza, A. A. Martin, R. da Silva and R. A. de Freitas, *Int. J. Biol. Macromol.*, 2017, **104**, 97–106.
24. A. Schiavi, R. Cuccaro and A. Troia, *J. Mech. Behav. Biomed. Mater.*, 2016, **53**, 119–130.
25. K. M. Zia, S. Tabasum, M. Nasif, N. Sultan, N. Aslam, A. Noreen and M. Zuber, *Int. J. Biol. Macromol.*, 2017, **96**, 282–301.
26. C. A. García-González, M. Alnaief and I. Smirnova, *Carbohydr. Polym.*, 2011, **86**, 1425–1438.
27. H. Maleki, L. Durães, C. A. García-González, P. del Gaudio, A. Portugal and M. Mahmoudi, *Adv. Colloid Interface Sci.*, 2016, **236**, 1–27.
28. Z. Ulker and C. Erkey, *J. Controlled Release*, 2014, **177**, 51–63.

29. C. A. García-González, M. C. Camino-Rey, M. Alnaief, C. Zetzl and I. Smirnova, *J. Supercrit. Fluids*, 2012, **66**, 297–306.
30. A. Ubeyitogullari and O. N. Ciftci, *Carbohydr. Polym.*, 2016, **147**, 125–132.
31. A. B. Paninho, C. Barbosa, I. D. Nogueira, V. Najdanovic-Visak and A. V. M. Nunes, *J. Supercrit. Fluids*, 2013, **83**, 35–40.
32. C. A. García-González, M. Jin, J. Gerth, C. Alvarez-Lorenzo and I. Smirnova, *Carbohydr. Polym.*, 2015, **117**, 797–806.
33. A. Veronovski, G. Tkalec, Z. Knez and Z. Novak, *Carbohydr. Polym.*, 2014, **113**, 272–278.
34. A. Muheem, F. Shakeel, M. A. Jahangir, M. Anwar, N. Mallick, G. K. Jain, M. H. Warsi and F. J. Ahmad, *Saudi Pharm. J.*, 2016, **24**, 413–428.
35. A. Singh, Z. A. Worku and G. Van den Mooter, *Expert Opin. Drug Delivery*, 2011, **8**, 1361–1378.
36. A. M. Hillery and K. Park, *Drug Delivery: Fundamentals & Applications*, CRC Press, Boca Raton, FL, USA, 2nd edn, 2017.
37. A. Veronovski, Ž. Knez and Z. Novak, *Int. J. Pharm.*, 2013, **454**, 58–66.
38. C. A. García-González and I. Smirnova, *J. Supercrit. Fluids*, 2013, **79**, 152–158.
39. T. Mehling, I. Smirnova, U. Guenther and R. H. H. Neubert, *J. Non-Cryst. Solids*, 2009, **355**, 2472–2479.
40. C. A. García-González, J. J. Uy, M. Alnaief and I. Smirnova, *Carbohydr. Polym.*, 2012, **88**, 1378–1386.
41. M. A. Marin, R. R. Mallepally and M. A. McHugh, *J. Supercrit. Fluids*, 2014, **91**, 84–89.
42. M. Pantić, P. Kotnik, Ž. Knez and Z. Novak, *J. Supercrit. Fluids*, 2016, **118**, 171–177.
43. F. De Cicco, P. Russo, E. Reverchon, C. A. García-González, R. P. Aquino and P. Del Gaudio, *Carbohydr. Polym.*, 2016, **147**, 482–489.
44. G. Tkalec, Z. Knez and Z. Novak, *J. Supercrit. Fluids*, 2015, **106**, 16–22.
45. S. Antonyuk, S. Heinrich, P. Gurikov, S. Raman and I. Smirnova, *Powder Technol.*, 2015, **285**, 34–43.
46. C. A. García-González, E. Carenza, M. Zeng, I. Smirnova and A. Roig, *RSC Adv.*, 2012, **2**, 9816–9823.
47. M. Betz, C. A. García-González, R. P. Subrahmanyam, I. Smirnova and U. Kulozik, *J. Supercrit. Fluids*, 2012, **72**, 111–119.
48. O. Veiseh, J. W. Gunn and M. Zhang, *Adv. Drug Delivery Rev.*, 2010, **62**, 284–304.
49. M. Vialpando, A. Aerts, J. Persoons, J. Martens and G. Van Den Mooter, *J. Pharm. Sci.*, 2011, **100**, 3411–3420.
50. C. M. Hentzschel, M. Alnaief, I. Smirnova, A. Sakmann and C. S. Leopold, *Eur. J. Pharm. Biopharm.*, 2012, **80**, 130–135.
51. A. Fernández Tena and P. Casan Clarà, *Arch. Bronconeumol.*, 2012, **48**, 240–246.
52. R. W. Dal Negro, *Multidisciplinary Respir. Med.*, 2015, **10**, 13.
53. K. P. Lee and G. L. Gould, Aerogel powder therapeutic agents, *U. S. Pat.* 6994842 B2, 2006.

54. R. M. Obaidat, B. M. Tashtoush, M. F. Bayan, T. Al Bustami and M. Alnaief, *AAPS PharmSciTech*, 2015, **16**, 1235–1244.
55. D. Mittal, A. Ali, S. Md, S. Baboota, J. K. Sahni and J. Ali, *Drug Delivery*, 2014, **21**, 75–86.
56. S. Grassin-Delyle, A. Buenestado, E. Naline, C. Faisy, S. Blouquit-Laye, L.-J. Couderc, M. Le Guen, M. Fischler and P. Devillier, *Pharmacol. Ther.*, 2012, **134**, 366–379.
57. L. Illum, *J. Controlled Release*, 2012, **161**, 254–263.
58. Y. W. Chien, *Nasal systemic drug delivery/Yie W. Chien, Kenneth S. E. Su, Shyi-Feu Chang*, Marcel Dekker, New York, 1989.
59. V. S. S. Gonçalves, P. Gurikov, J. Poejo, A. A. Matias, S. Heinrich, C. M. M. Duarte and I. Smirnova, *Eur. J. Pharm. Biopharm.*, 2016, **107**, 160–170.
60. M. Alnaief, M. A. Alzaitoun, C. A. García-González and I. Smirnova, *Carbohydr. Polym.*, 2011, **84**, 1011–1018.
61. M. Chaturvedi, M. Kumar and K. Pathak, *J. Adv. Pharm. Technol. Res.*, 2011, **2**, 215–222.
62. J. D. Smart, *Adv. Drug Delivery Rev.*, 2005, **57**, 1556–1568.
63. V. S. S. Gonçalves, A. A. Matias, J. Poejo, A. T. Serra and C. M. M. Duarte, *Int. J. Pharm.*, 2016, **515**, 1–10.
64. P. Tengamnuay, A. Sahamethapat, A. Sailasuta and A. K. Mitra, *Int. J. Pharm.*, 2000, **197**, 53–67.
65. G. K. B. Sándor, *Annu. Maxillofac. Surg.*, 2012, **2**, 8–11.
66. A. R. C. Duarte, V. E. Santo, A. Alves, S. S. Silva, J. Moreira-Silva, T. H. Silva, A. P. Marques, R. A. Sousa, M. E. Gomes, J. F. Mano and R. L. Reis, *J. Supercrit. Fluids*, 2013, **79**, 177–185.
67. C. A. García-González, A. Concheiro and C. Alvarez-Lorenzo, *Bioconjugate Chem.*, 2015, **26**, 1159–1171.
68. C. A. García-González, L. A. Díaz-Gómez, A. Concheiro and C. Alvarez-Lorenzo, *Recent Pat. Nanomed.*, 2015, **5**, 48–58.
69. E. Reverchon and S. Cardea, *J. Supercrit. Fluids*, 2012, **69**, 97–107.
70. J. Stergar and U. Maver, *J. Sol-Gel Sci. Technol.*, 2016, **77**, 738–752.
71. R. R. Mallepally, M. A. Marin, V. Surampudi, B. Subia, R. R. Rao, S. C. Kundu and M. A. McHugh, *Biomed. Mater.*, 2015, **10**, 035002.
72. E. Ozdemir, A. Sendemir-Urkmez and O. Yesil-Celiktas, *J. Supercrit. Fluids*, 2013, **75**, 120–127.
73. K. Rinki and P. K. Dutta, *J. Macromol. Sci., Part A: Pure Appl. Chem.*, 2010, **47**, 429–434.
74. H. Cai, S. Sharma, W. Liu, W. Mu, W. Liu, X. Zhang and Y. Deng, *Biomacromolecules*, 2014, **15**, 2540–2547.
75. L. Baldino, S. Concilio, S. Cardea and E. Reverchon, *Polymers*, 2016, 8.
76. L. Baldino, S. Concilio, S. Cardea, I. De Marco and E. Reverchon, *J. Supercrit. Fluids*, 2015, **103**, 70–76.
77. S. Quraishi, M. Martins, A. A. Barros, P. Gurikov, S. P. Raman, I. Smirnova, A. R. C. Duarte and R. L. Reis, *J. Supercrit. Fluids*, 2015, **105**, 1–8.
78. A. El Kadib and M. Bousmina, *Chem. – Eur. J.*, 2012, **18**, 8264–8277.

79. W. Czaja and D. D. Kyryliouk. Resorbable cellulose based biomaterial and implant, *WO Pat.* 2013126635A1, 2013.
80. F. Liebner, A. Potthast, P. Kosma, R. Wimmer and T. Rosenau. Porous solid body made of cellulose phosphate and carrying covalently bound phosphate groups on its entire surface and inner surface, useful as a cell growth substrate including bone implant- or cartilage- and bone regeneration-material, *DE Pat.* 102010011808, 2011.
81. S. S. Silva, A. R. C. Duarte, A. P. Carvalho, J. F. Mano and R. L. Reis, *Acta Biomater.*, 2011, 7, 1166–1172.
82. K. Ganesan, A. Dennstedt, A. Barowski and L. Ratke, *Mater. Des.*, 2016, **92**, 345–355.
83. N. Pircher, D. Fischhuber, L. Carbajal, C. Strauß, J. M. Nedelec, C. Kasper, T. Rosenau and F. Liebner, *Macromol. Mater. Eng.*, 2015, **300**, 911–924.
84. S. S. Silva, A. R. C. Duarte, J. F. Mano and R. L. Reis, *Green Chem.*, 2013, **15**, 3252–3258.
85. S. Cardea, P. Pisanti and E. Reverchon, *J. Supercrit. Fluids*, 2010, **54**, 290–295.
86. B. Ding, Y. Si, J. Ge, X. Tang, M. Huang, J. Zhu and J. Yu. Three-dimensional fiber-based airgel tissue engineering scaffold and production method thereof, *CN Pat.* 103285424A, 2013.
87. N. Pircher, S. Veigel, N. Aigner, J. M. Nedelec, T. Rosenau and F. Liebner, *Carbohydr. Polym.*, 2014, **111**, 505–513.
88. L. Diaz-Gomez, A. Concheiro, C. Alvarez-Lorenzo and C. A. García-González, *Carbohydr. Polym.*, 2016, **142**, 282–292.
89. L. Diaz-Gomez, F. Yang, J. A. Jansen, A. Concheiro, C. Alvarez-Lorenzo and C. A. García-González, *RSC Adv.*, 2016, **6**, 70510–70519.
90. L. Goimil, M. E. M. Braga, A. M. A. Dias, J. L. Gómez-Amoza, A. Concheiro, C. Alvarez-Lorenzo, H. C. de Sousa and C. A. García-González, *J. CO_2 Util.*, 2017, **18**, 237–249.
91. R. S. Kirsner and W. H. Eaglstein, *Dermatol. Clin.*, 1993, **11**, 629–640.
92. G. A. James, E. Swogger, R. Wolcott, E. d. Pulcini, P. Secor, J. Sestrich, J. W. Costerton and P. S. Stewart, *Wound Repair Regen.*, 2008, **16**, 37–44.
93. M. Kong, X. G. Chen, K. Xing and H. J. Park, *Int. J. Food Microbiol.*, 2010, **144**, 51–63.
94. A. K. Singla and M. Chawla, *J. Pharm. Pharmacol.*, 2001, **53**, 1047–1067.
95. K. Rinki, P. K. Dutta, A. J. Hunt, D. J. MacQuarrie and J. H. Clark, *Int. J. Polym. Mater. Polym. Biomater.*, 2011, **60**, 988–999.
96. M. R. Ayers and A. J. Hunt, *J. Non-Cryst. Solids*, 2001, **285**, 123–127.
97. T. Lu, Q. Li, W. Chen and H. Yu, *Compos. Sci. Technol.*, 2014, **94**, 132–138.
98. J. Marquez-Escalante, E. Carvajal-Millan, M. Miki-Yoshida, L. Alvarez-Contreras, A. R. Toledo-Guillén, J. Lizardi-Mendoza and A. Rascón-Chu, *Molecules*, 2013, **18**, 5531–5542.
99. R. Starbird, C. A. García-González, I. Smirnova, W. H. Krautschneider and W. Bauhofer, *Mater. Sci. Eng., C*, 2014, **37**, 177–183.

Subject Index